Recent Results in Cancer Research 107

Founding Editor
P. Rentchnick, Geneva

Managing Editors
Ch. Herfarth, Heidelberg · H. J. Senn, St. Gallen

Associate Editors
M. Baum, London · V. Diehl, Köln
C. von Essen, Villigen · E. Grundmann, Münster
W. Hitzig, Zürich · M. F. Rajewsky, Essen

Recent Results in Cancer Research

R. D. Issels   W. Wilmanns (Eds.)

Application of Hyperthermia in the Treatment of Cancer

With 118 Figures and 56 Tables

Springer-Verlag
Berlin Heidelberg New York
London Paris Tokyo

Dr. med. Dipl. biochem. Rolf D. Issels
Prof. Dr. med. Wolfgang Wilmanns

Medizinische Klinik III, Klinikum Großhadern
Ludwig-Maximilians-Universität
Marchioninistraße 15, 8000 München 70, FRG
und
Gesellschaft für Strahlen- und Umweltforschung (GSF)
Institut für Hämatologie
Landwehrstraße 61, 8000 München 2, FRG

ISBN 3-540-18486-4 Springer-Verlag Berlin Heidelberg New York
ISBN 0-387-18486-4 Springer-Verlag New York Berlin Heidelberg

Library of Congress Cataloging-in-Publication Data. Application of hyperthermia in the treatment of cancer. (Recent results in cancer research ; 107) Includes bibliographies and index. 1. Thermotherapy. 2. Cancer-Treatment. I. Issels, R. D. (Rolf D.), 1948–   .
II. Wilmanns, W. (Wolfgang), 1929–   . III. Series. [DNLM: 1. Hyperthermia, Induced.
2. Neoplasms-therapy. W1 RE106P v. 107 / QZ 266 A652]   RC261.R35 vol. 107
[RC271.T5]   616.99'4s   88-2035   [616.99'40632]

© Springer-Verlag Berlin Heidelberg 1988
Printed in Germany

Typesetting, printing, and binding: Appl, Wemding
2125/3140-543210

Preface

Hyperthermia combined with either radiation or systemic chemotherapy is rapidly becoming a clinical reality at many institutions as a form of the treatment of malignant disease. The book deals of the effects of temperature elevation on cells, tissues, and organisms, and with the problems of in vivo temperature measurement, dose calculation, and methods used to generate heat in superficial or deep-seated tumors. Especially in regard to deep-heating, the clinical results of several investigators using the annular phased array system are presented.

The clinical interest in hyperthermia is related to the ability of heat to destroy malignant cells situated in a physiologically deprived environment. Such cells are generally those most resistant to other forms of cancer therapy. Since heat has been demonstrated experimentally to sensitize cells to ionizing radiation and several chemotherapeutic agents, the clinical use of combined treatment modalities could be an attractive approach to enhance tumor control in the future.

This was the subject of the Mildred Scheel Symposium held at the University of Munich, 2–4 April 1987. The editors would like to thank all the authors for their contributions. We would also like to express our gratitude to the publishers, Springer-Verlag, for their assistance and rapid publication.

Munich, December 1987   Wolfgang Wilmanns
                                                        Rolf Issels

Contents

List of Contributors*

Arcangeli, G. 118[1]
Astrahan, M. 93, 136, 244
Baumhoer, W. 129
Berger, H. 263
Beuningen, van, D. 113
Bichler, K.-H. 222
Bleehen, N.M. 214
Brady, L.W. 147
Brinkmann, G. 267
Dahl, O. 157, 188
Dikomey, E. 104
Eichholtz-Wirth, H. 184
Eisler, K. 205
Emami, B. 244
Endrich, B. 44
Engelhardt, R. 27, 177, 200
Erdmann, W.D. 222
Feldmann, H.J. 129
Fiebig, H.H. 177
Field, S.B. 1
Fink, R.M. 17
Flüchter, S.H. 222
George, K.C. 113
Germann, R. 226
Hammersen, F. 44
Herrmann, D.J.B. 27
Howard, G.C.W. 217
Hürter, W. 99
Issels, R.D. 236, 263
Jung, H. 104

Kjellen, E. 152
Klaubert, W. 205
Klimpfinger, M. 226
Kneschaurek, P. 87
Knorr, H. 193
Koderhold, G. 209
Kolbabek, H. 209
Lam, K. 136
Landberg, T. 152
Lange, J. 205
Langholz, B. 136, 244
Lengfelder, E. 17
Lierse, W. 60
Liggett, P. 93
Lindholm, C.E. 152
Linsmeier, K.-D. 99
Lissner, J. 263
Lorenz, W.J. 99
Luxton, G. 93, 136
Markl, G. 263
Mehus, A. 188
Mella, O. 188
Messmer, K. 44
Mirtsch, S. 22
Molls, M. 129
Müller, M. 236
Müller, R.D. 129
Nespor, W. 209
Neumann, H.A. 27, 177
Nilsson, P. 152

---

* The address of the principal author is given on the first page of each
  contribution.
[1] Page on which contribution begins.

Permanetter, W. *193*
Persson, B. *152*
Petrovich, Z. *93, 136, 244*
Piroth, H.-D. *267*
Reinbold, F. *99*
Reinhold, H.S. *32*
Rossmeissl, G. *147*
Sack, H. *129*
Sauer, H. *170, 236*
Saxton, T. *249*
Schaefermeyer, T. *76, 249*
Schalhorn, A. *193*
Schlegel, W. *99*
Schnizer, W. *193*
Schöps, P. *193*
Seebass, M. *99*
Seegenschmiedt, M.H. *147*
Seichert, N. *193*

Steimann, J. *222*
Steindorfer, P. *226*
Streffer, C. *7, 22, 113*
Strohmaier, W.L. *222*
Strohmenger, U. *22*
Stupp-Poutot, G. *193*
Tiling, K. *236*
Tokita, H. *177*
Turner, P.F. *76, 249*
Valdagni, R. *123*
Vaupel, P. *65*
Voth, B. *170*
Wadepohl, M. *236*
Weisser, M. *87*
Wilmanns, W. *170, 205, 236*
Yamada, K. *177*
Yerushalmi, A. *141*
Zywietz, F. *60*

Biological, Biophysical and Technical Concepts of Hyperthermia

The Concept of Thermal Dose

S. B. Field

MRC Cyclotron Unit, Hammersmith Hospital, Du Cane Road,
London W12 OHS, Great Britain

Introduction

Dose is a difficult concept. The Oxford English Dictionary (1973) defines it as follows:
"A definite quantity of a medicine, or something regarded as analogous to a medicine,
given or prescribed to be given at one time". It is also "a definite amount of some ingre-
dient added to wine to give it a special character". For the present purpose, the first definition is perhaps the more important!

The purpose of a dose is to provide a number which relates to a biological response
and thus enable researchers, clinicians etc. to communicate regarding the likely effects of
a given amount of a medicine, or something analogous to a medicine. Principal requirements are that the biological response must be related to the dose in a relevant manner,
the dose should be a well-defined and measurable physical quantity and there should be
proper means of intercomparison.

Dose of Ionizing Radiation or Drugs

In considering the concept of thermal dose it is useful to examine the use of dose in other aspects of medicine, i.e. for ionizing radiation or drugs. The accepted dose unit for
ionizing radiation is based on the deposition of energy into tissue. Thus, the unit of
1 gray (Gy) is defined as 1 joule (J) of energy deposited in 1 kg of tissue. This applies, of
course, to a single treatment, but even with this restriction the biological response to exposure to ionizing radiation also depends on many factors other than the given dose.
Some of these are: the rate of application of the dose; the overall treatment time; the
size of the dose itself (large doses are more effective per Gy than small doses); various
modifying factors, e.g. oxygen; the previous history (tissue is more sensitive if pretreated); and individual susceptibility. If multiple doses are given, complex expressions must
be used in an attempt to describe the overall effect of fractionated treatment. These expressions are not the same for all tissues or cells and there is by no means universal
agreement on the best approach to solving this problem.

The giving of a medicine is normally described in terms of the mass of the drug and its
concentration. This definition also applies to a single application. As with ionizing radiation the response to a given dose of drug will depend on many factors other than the
drug dose. Some of these are: the period of administration of the drug; the route of ad-

Recent Results in Cancer Research, Vol. 107
© Springer-Verlag Berlin · Heidelberg 1988

ministration; the period of action; the breakdown products; the distribution of the drug in the body; previous history of the tissue; and individual susceptibility. The effects of multiple treatments are even more complex than with ionizing radiation and the means of computing the effects of such treatments are generally not available.

Nevertheless, despite these difficulties, there is universal acceptance of the use of the concept of dose in these two cases. The problems are noted but not solved, and we have become accustomed to using dose as defined above.

Concept of Dose in Hyperthermia

With hyperthermia, the principal determinants of biological effect are temperature and duration of heating. Deposited energy is not a primary determinant. For example, if a box containing cells were adiabatically shielded so that no energy could pass into or out of it, an increasing number of cells would die with increased time at hyperthermic temperature. Energy transfer is not required to cause cell death (Hahn 1982). On this basis, it is clear that for hyperthermia at a fixed and constant temperature, duration of heating is a reasonable way of expressing thermal dose. The unit of thermal dose would then be *time*. The use of this parameter adequately fulfills all the conditions stated above and is no less appropriate than energy deposition for ionizing radiation or mass and concentration for therapy by drugs.

The difficulty in using time at a given temperature for thermal dose is that a wide range of temperatures is used clinically and, in any case, temperature is not constant during treatment. The solution to this problem is not trivial. It is necessary to find some means of relating any treatment to an *equivalent time* at a chosen reference temperature. This requires a biological or biophysical transformation to give an equivalent time of heating or a thermal dose equivalent (TDE) or thermal isoeffect dose (TID).

Various attempts have been made to solve this problem. One of the earliest was by Atkinson (1977), who proposed relating heat dose to a standard heat survival curve. This is, however, not satisfactory, since heat survival curves are extremely variable, depending on conditions during heating, cell type, etc. Gerner (1985) proposed using a thermodynamic approach to this problem, based on the Arrhenius equation, i.e. the reaction rate constant $k = A e^{-H/RT}$, where H is the activation energy, T is temperature and R is the gas constant. The method is meaningful providing the relationship between $\log k$ and $\frac{1}{T}$ is linear, which does appear to be the case over the clinical range of temperatures. H is found to be approximately 600 kJ/mol, consistent with values for heat damage to proteins. However, below approximately 42.5° C the relationship exhibits a transition, so that either the steady-state conditions no longer apply or a different reaction becomes rate-limiting. The relationship also changes at very high temperatures.

Hahn (1982) derived a relationship between time of heating (t) and temperature based on the classical target theory approach, i.e. a cell is inactivated if its instantaneous kinetic energy exceeds a specific critical value at any time. The formula resulting from this approach is very similar to that deduced empirically by Dewey et al. (1977) based on in vitro observations:

$$\frac{t_2}{t_1} = R^{T_1 - T_2} \tag{1}$$

This equation provides a means of relating treatments to a given tissue with different temperatures and times of heating, but it takes no account of tissue sensitivity. Sapareto

Table 1. Mean value for the parameter R in the formula $\dfrac{t_2}{t_1} = R^{T_1 - T_2}$

	Above transition	Below transition
Summary of in vivo data	2.1 ± 0.07	6.4 ± 0.4
Summary of in vitro data	2.0 ± 0.08	5.9 ± 0.06

R is also the factor by which the time is changed equivalent to a change in temperature by $1°$ C. (Field and Morris, 1983)

and Dewey (1984) proposed that $43°$ C should be used as the reference temperature and that all treatments be described as equivalent minutes of heating at $43°$ C. By this means account can be taken of temperature variation, although there may be other complications (see below). Sapareto and Dewey (1984) tested the validity of Eq. 1 using cells in vitro. The formula did provide a reasonable fit, especially at temperatures above $43°$ C, although at lower temperatures there were some differences between the calculated and measured responses.

The validity of the approach was also considered by Field and Morris (1983), who reviewed the available literature giving isoeffect data in vivo or in vitro. It was found for almost all tissues that there was a transition in the time/temperature relationship, occurring between 42 and $43°$ C. The slopes of all curves were fairly similar above and below this transition, although it was clear that there are very large differences between tissues in absolute sensitivity. Table 1 gives the mean values for the constant R for the in vitro and in vivo results from this study. The mean temperature at which the transition occurred was $42.5°$ C.

These considerations, however, apply to hyperthermia given at constant temperature. In practice, it takes a finite time to reach a desired temperature, which is then difficult to keep constant. There is a finite time of cooling when the power is switched off. In practice, this is dealt with by integration of Eq. 1 over the whole treatment procedure. However, variations in temperature can also have a marked effect in determining the biological response. For example, one exposure may result in the development of a considerable resistance to further exposures, a phenomenon known as thermotolerance. Alternatively, if there is a drop in temperature from above the transition to below, the low temperature may become far more effective than usual, a phenomenon known as step-down sensitisation. Clearly, these effects have the potential to invalidate Eq. 1. The extent to which this is likely to happen in clinical practice has been tested experimentally.

The tails of young rats were heated by immersion in a temperature-controlled waterbath. The endpoint was necrosis at 6 weeks, assessed radiographically (Morris and Field 1985). Using this endpoint it was first shown that for heating at a constant temperature the relationship between heating time and temperature required to produce a given level of damage was consistent with the above equation. It was shown that the rat tail has a considerable ability to develop thermotolerance, the extent and time course of which are both related to the inducing (or priming) treatment. "Step-down sensitisation" was also clearly illustrated, the effect apparently being a loss of the transition in the time/temperature curve (Field and Morris 1984, 1985).

In order to simulate a temperature variation, animals' tails were immersed alternately in waterbaths of different temperatures. A total of 16 studies of this type were performed. Variables included immersion first at either the lower or the higher temperature,

Table 2. Differences between measured and predicted response of heated rat tails based on the formula $\frac{t_2}{t_1} = R^{T_1 - T_2}$ with $R = 2$ for $T \geqslant 42.5°$ C and $R = 6$ for $T < 42.5°$ C

Treatment mode	Matches for: equal heating times (T) equal effects (E)	Difference from prediction	
		Time factor	T (°C)
$(43 + 45) \times 1$	E	1.14	−0.2
$(43 + 45) \times 1$	T	1.13	−0.2
$(43 + 45) \times 2$	E	0.98	0
$(43 + 45) \times 2$	T	1.14	−0.2
$(43 + 45) \times 3$	E	1.00	0
$(43 + 45) \times 3$	T	1.00	0
$(42 + 45) \times 3$	E	1.21	−0.3
$(42 + 45) \times 3$	T	1.18	−0.2
$(43 + 45) \times 6$	E	1.02	0
$(43 + 45) \times 6$	T	1.06	−0.1
$(45 + 43) \times 1$	T	0.93	+0.1
$(45 + 43) \times 2$	E	0.92	+0.1
$(45 + 43) \times 2$	T	0.96	0
$(45 + 43) \times 3$	E	0.87	+0.2
$(45 + 43) \times 3$	T	0.93	+0.1

variation between 43° C and 45° C or between 42° C and 45° C. Times of treatment in the different waterbaths were either equal or adjusted to produce approximately equal effects at the different temperatures, and the number of cycles varied between one and six. The aim of the study was to match the effect to a single treatment according to Eq. 1.

The results of all 16 studies are summarised in Table 2. It is seen that when the high temperature was given first the effective temperature was slightly higher than that predicted by the formula. i. e. there was a net effect of step-down sensitisation. Conversely, when the lower temperature was the first treatment there was a net effect of thermotolerance. However, in both cases the effect was very small, despite the fact that both thermotolerance and step-down sensitisation play important roles in this experimental system. The equation, therefore, was shown to provide an acceptable prediction even in conditions of varying temperature. The *maximum* difference observed was approximately 20% in heating time, equivalent to an error of 0.3° C. Such a change is relatively small compared with other uncertainties in a typical hyperthermia treatment.

The validity of the approach has been further tested by computer simulation to allow for different values of the transition temperature and slope of the isoeffect curve in the low-temperature region. Based on a typical clinical treatment in which the tumour temperature varied within the range 44–45° C, normal tissue (not skin) 42–43° C and skin 40.5–42° C, it was found that the tumour isoeffect dose did not change significantly, whereas that for normal tissues with intermediate or lower temperatures was much more dependent on the choice of parameters. The degree of importance of these differences will depend on the particular treatment protocol.

It appears, therefore, that integration of the Dewey formula (Eq. 1) is reasonable providing the variations in temperature are not too extreme. Certainly, it will break down if there are violent temperature changes, e. g. giving the first half of the treatment at 44° C followed by the second half at 41° C. In cases such as this, step-down sensitisation would unquestionably invalidate the formula.

Thus, within reasonable limits, we have a useful formula for TDE or TID. But there are still numerous variables which will additionally effect the biological sensitivity (as is the case with ionizing radiation or therapy by drugs). Some of these are: rate of heating and cooling; overall time of treatment; distribution of temperature throughout the target volume (it is likely that the most important tumour temperature is the *minimum* and for normal tissue tolerance, the maximum); a variety of modifying factors; prior history and individual susceptibility. The problems of fractionated hyperthermia are outside the definition of dose, but are known to be of very great importance.

It cannot necessarily be assumed that the parameters for effects of heat alone are the same as those for the combination of hyperthermia with other modalities. For heat combined with X-rays far fewer data are available for the isoeffect relationships than for heat alone. From what is available, it is seen that the parameters appear to be fairly similar to those for heat alone, although the transition appears to occur at approximately 1° C lower for the combined treatment (Hume and Marigold 1985). We have no suitable method of estimating thermal dose for the combination of hyperthermia and chemotherapy.

Summary and Conclusions

1. In the ideal but theoretical situation, where hyperthermia is defined to be at a specific, constant temperature, duration of heating is a perfectly acceptable and satisfactory method of describing thermal dose.
2. In the real situation, the dose is far from constant. Time will be required to reach the desired temperature and to cool afterwards. During hyperthermia the temperature is variable. We are thus obliged to use a biologically equivalent thermal dose to account for these variations.
3. Various attempts have been made to solve this problem, but at present, the most feasible approach appears to be the integration of the biological isoeffect relationship (Dewey formula) over the whole treatment.
4. This method provides a practical and reasonable method of comparing hyperthermal treatments under conditions likely to be met in practice, i.e. moderate variation about a fairly steady temperature. However, the formula would certainly be inaccurate if there were a sudden reduction in temperature resulting in a significant effect of step-down sensitisation or a long treatment resulting in the development of thermotolerance. Also, it must be emphasised that the formula does not account for absolute differences in sensitivity among tissues. Nor does it address the problem of varying sensitivity throughout a course of fractionated heat treatments.
5. Less is known about the applicability of such a formula to the combination of heat and ionizing radiation when the two treatment modalities interact. The parameters are fairly similar to those for heat alone, although the transition appears to be lowered to 41.5° C. The formula cannot be applied to combinations of heat and chemotherapy.
6. Care should be taken not to allow this approach to become too firmly entrenched in our thinking. We may find a better method. Until then it is important to retain as many clinical hyperthermia data as practicable.

References

Atkinson (1977) Hyperthermia dose definition. J Bioeng 1: 487–492

Dewey WC, Hopwood LE, Sapareto SA, Gerweck LE (1977) Cellular responses to combinations of hyperthermia and radiation. Radiology 123: 463–474

Field SB, Morris CC (1983) The relationship between heating time and temperature: its relevance to clinical hyperthermia. Radiother Oncol 1: 179–183

Field SB, Morris CC (1984) Application of the relationship between heating time and temperature for use as a measure of thermal dose. In: Overgaard J (ed) Hyperthermic oncology 1984. Taylor and Francis, London, pp 183–186

Field SB, Morris CC (1985) Experimental studies of thermotolerance in vivo. I. The baby rat tail model. Int J Hyperthermia 1: 235–246

Gerner E (1985) Definition of thermal dose: biological isoeffect relationships and dose for temperature-induced cytotoxicity. In: Overgaard J (ed) Hyperthermic oncology. Taylor and Francis, London, pp 245–252

Hahn GM (1982) Hyperthermia and cancer. Plenum, New York

Hume SP, Marigold JCL (1985) Time-temperature relationships for hyperthermal radiosensitization in mouse intestine: influence of thermotolerance. Radiother Oncol 3: 165–171

Morris CC, Field SB (1985) The relationship between heating time and temperature for rat tail necrosis with and without occlusion of the blood supply. Int J Radiat Biol 47: 41–48

Oxford English Dictionary (1973) Oxford University Press, Oxford

Sapareto SA, Dewey WC (1984) Thermal dose determination in cancer therapy. Int J Radiat Oncol Biol Phys 10: 787–800

Aspects of Metabolic Change After Hyperthermia*

C. Streffer

Institut für Medizinische Strahlenbiologie, Universitätsklinikum Essen, Hufelandstraße 55, 4300 Essen 1, FRG

Introduction

Heat treatment of cells and tissues leads to instantaneous molecular and metabolic changes. These rapid changes are already observed at temperature ranges (40–45° C) which are used in tumour therapy. Such an immediate metabolic response is not seen after exposure to doses of ionizing radiation within the therapeutic range. Under these conditions metabolic changes are usually observed only several hours or days after exposure (Streffer 1969; Altman et al. 1970). On the other hand, it is very well understood today that DNA damage and effects on the chromosomal level are the most important molecular changes involved in the mechanism of reproductive cell death (Alper 1979; Streffer and van Beuningen 1985). The mechanism of heat-induced cell killing is less clear. However, it is evident that molecular processes in the cytoplasm as distinct from the cell nucleus are important for these mechanisms (Hahn 1982; Streffer 1982). Two principal effects can be observed during and after heat treatments (Streffer 1985):

1. Conformational changes, which lead to destabilization of macromolecules and multimolecular structures.
2. Elevated temperatures, which induce increased rates of metabolic reactions. As a consequence a rapid turnover of metabolites occurs which is followed by disregulation.

Conformational Changes of Molecular Structures

The conformation of biological multimolecular structures is stabilized by covalent bonds, by hydrogen bridges, by interaction of ionic groups and by hydrophobic interactions of molecular structures like side chains of amino acids in proteins. The three latter types of interactions and bonds are rather weak and depend very much on the microenvironment. They can be altered easily by an increase of temperature, changes of pH etc. Such conformational alterations can occur at the temperature range of 40–45° C (Privalov 1979; Lepock et al. 1983; Leeper 1985; Streffer and van Beuningen 1987).

In this connection changes of biological membranes after heat treatment are of great importance. Much attention has been paid to the lipids of membranes and their influ-

---

* The investigations were supported by the Bundesministerium für Forschung und Technologie, Bonn.

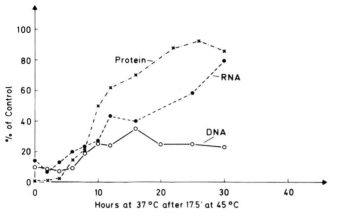

Fig. 1. Synthesis of *protein, RNA* and *DNA* in CHO cells in vitro after heating for 17.5 min at 45° C and recovery during incubation at 37° C. (Henle and Leeper 1979)

ence on membrane fluidity in relation to cell killing (Yatvin et al. 1982). Bowler et al. (1973) observed an increase in membrane permeability after hyperthermia and a loss of membrane-bound ATPase. These effects correlated with cell killing. Investigations on membrane-bound receptors have shown in several studies that they were inactivated or lost from the membranes. Concanavalin A induced capping, and cell survival responded in a similar way to hyperthermia (Stevenson et al. 1981). Calderwood and Hahn (1983) observed heat-induced inhibition of insulin binding to the plasma membrane of CHO cells. Cooperative effects between lipids and proteins are certainly of great importance in this connection, but conformational changes of proteins can influence these structures remarkably.

Microtubules of the cytoskeleton disaggregate during a hyperthermic treatment and reaggregate during a further incubation of the cells at 37° C (Lin et al. 1982). A correlation between cell killing and the disturbances of the cytoskeleton has been described; in dead cells no reassembly of the cytoskeleton occurs (Cress et al. 1982). Similar events occur with the spindle apparatus of mitotic cells. This explains the high thermosensitivity of cells in mitosis (Coss et al. 1982).

Several authors have described the heat sensitivity of DNA, RNA and protein synthesis (Streffer and van Beuningen 1987). Protein synthesis decreases even more rapidly than synthesis of nucleic acids. However, protein synthesis recovers faster than DNA synthesis at 37° C (Henle and Leeper 1979) (Fig. 1). All these processes are performed by complex multimolecular structures which are apparently disturbed at elevated temperatures and which can reaggregate at 37° C in cells in vitro.

Glucose and Energy Metabolism

Glucose metabolism plays a central role in the consideration of how to modify the thermosensitivity of cells. It is closely linked to the metabolism of lipids and some amino acids. Glucose is metabolized in all cells through the glycolytic pathway to pyruvate and under anoxic conditions to lactate (Fig. 2). Under oxic conditions pyruvate, the oxidized counterpart of lactate, is metabolized to acetyl-CoA, which is degraded in the citrate cy-

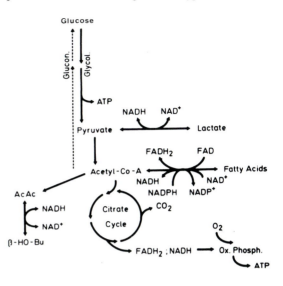

Fig. 2. Metabolic scheme of intermediary metabolism

cle. This pathway generates the reduced nucleotides which are needed as electron donors for oxidative phosphorylation. Thus these metabolic pathways are linked to various factors, such as pH and oxygen tension. In this connection it is interesting that in some tumours anaerobic glycolysis is the dominating pathway (Warburg et al. 1926), although this is not a general rule. It has been proposed that hyperthermia increases the glycolytic rate by which the lactate level should be increased in the tumour (von Ardenne 1980). Through this mechanism the intracellular pH should decrease which would enhance the cellular thermosensitivity.

Dickson and Calderwood (1979) heated Yoshida sarcomas in rats in situ to 42° C and measured the anaerobic glycolysis in vitro. Under these conditions a reduction of the glycolysis was observed. However, when the degradation of glucose was measured in mice under hyperthermic conditions (whole-body hyperthermia for 1 h at 40 and 41° C), an increased turnover was found (Schubert et al. 1982). In these experiments ^{14}C-glucose was injected before the hyperthermic treatment and the $^{14}CO_2$ output was measured (Fig. 3 A). On the other hand the glucose degradation was reduced when ^{14}C-glucose was injected after the hyperthermic treatment (Fig. 3 B). These data show clearly that during hyperthermia glucose metabolism, including glycolysis, is increased.

This is the case not only for glucose metabolism, but also for energy metabolism in general. Several authors have demonstrated that ATP levels decrease during heating of cells in vitro (Francesconi and Mayer 1979; Lunec and Cresswell 1983; Mirtsch et al. 1984). However, enhanced ATP synthesis is found in cells during hyperthermia when the incorporation of ^{3}H-adenosine into ATP is determined (Table 1). This increase of ATP synthesis is especially seen when under these conditions the specific activity is calculated (Table 1). Thus an increased energy supply is needed during the heating of cells and tissues which is provided by an enhanced ATP synthesis, resulting in enhanced ATP turnover. These observations are in agreement with the finding that the hepatic glycogen level decreases considerably during 1 h of whole-body hyperthermia at 40° C (Schubert et al. 1982).

Also comprehensive studies of glucose, glycolytic metabolites and lipids in a mammary adenocarcinoma on C57 Bl mice demonstrated this situation. A decrease of glucose as well as of glucose-6-phosphate was found after local hyperthermic treatment at

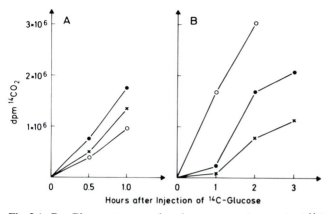

Fig. 3 A, B. Glucose turnover in mice, measured as expired $^{14}CO_2$ after intraperitoneal injection of ^{14}C-glucose. The animals received whole-body hyperthermia for 1 h at 37° C (O), 40° C (●) and 41° C (×). ^{14}C-glucose (6×10^{-2} μg/g body weight) was injected before (**A**) or after (**B**) heat treatment

Table 1. ATP content and incorporation of ^{3}H-adenosine *(^{3}H-AR)* into ATP in human melanoma cells in vitro. (Mirtsch et al. 1984)

	ATP content (fmol/cell)	^{3}H-AR incorporation (dpm/10^3 cells)	Specific activity in ATP (dpm/μmoles)
Controls	15	32	2.14×10^6
3 h at 42° C	9.5	40	4.30×10^6

Table 2. Metabolism in an adenocarcinoma (C57 Bl mouse) after hyperthermia at 43° C for 1 h (μmoles/g tissue)

	Control	Hyperthermia
Glucose	3.2	1.0
Glucose-6-phosphate	1.1	0.6
Glycerol-3-phosphate	0.39	0.52
Lactate	14.5	12.6
Lactate/pyruvate	39	66
β-Hydroxybutyric acid	0.09	0.25
Free fatty acids	8.1	5.4
Esterified fatty acids	23.5	14.5
Phospholipids	4.9	3.4

43° C for 1 h (Table 2). However, this treatment induced only a transient increase of the lactate level in the tumour and lactate was slightly decreased at the end of the hyperthermic treatment. Apparently the produced lactate was transported rapidly from the tumour via the blood flow to the liver (Fig. 4). The liver is the central organ from which glucose is provided for the peripheral organs, while the lactate which is produced in the periphery and not used in these tissues is metabolized through the hepatic gluconeogenesis or citrate cycle. On the other hand the lactate–pyruvate ratio is somewhat enhanced,

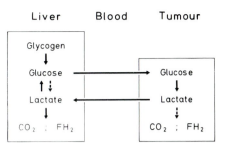

Fig. 4. Schematic diagram of flow of metabolites between liver and peripheral organs

Table 3. Lactate and pyruvate levels in human melanoma cells *(MeWo)* and their xenografts on nude mice after hyperthermia. (Mirtsch and Streffer, unpublished data)

	MeWo cells in vitro (μmoles/10^9 cells)		MeWo tumour (μmoles/g tissue)	
	Control	1 h after 44° C for 1 h	Control	2 h after 43° C for 30 min
Lactate	19.5	12.5	6.6	7.8
Pyruvate	1.6	1.8	0.25	0.20
Lactate/pyruvate	12.1	6.9	26.4	39.0

which indicates an increase of hypoxia within the tumour (Streffer 1985). The decrease of lipids in the tumour shows again that those metabolites which are needed for energy production are reduced after the hyperthermic treatment (Table 2). These data confirm the enormous need of energy during the heating of the tumour.

As the lactate levels are central in the discussion of glycolysis as well as for the possibility of modifying the cellular microenvironment by these processes in such a way that increased thermosensitivity results, the metabolic production of lactate was studied more carefully. The lactate and pyruvate levels were determined in human melanoma cells which were cultured in vitro. The data in Table 3 demonstrate that lactate decreases in the cells when the latter are heated for 1 h at 44° C; at the same time, pyruvate levels show only small changes. Therefore the redox ratio lactate/pyruvate decreases in the melanoma cells in vitro after the hyperthermic treatment. Apparently the reduced metabolites can transfer their electrons for oxidative phosphorylation, which is in agreement with the increased ATP synthesis as described before.

In contrast to this situation, a slight increase of the lactate level was observed when the same melanoma cells grew as a xenograft on nude mice. The tumour received localized heating at 43° C for 30 min with an ultrasound device. The pyruvate decreased somewhat in the xenograft after the hyperthermic treatment. The redox ratio increased after heating in the tumour (Table 3). A considerable difference was observed in cells in vitro and in the tumour in situ. Enough oxygen was available for the cells in vitro. Under the conditions in the tumour, apparently the oxygen supply was not sufficient. Thus the reduced metabolites accumulated in the tumour.

However, these metabolic changes vary very much in individual tumours and are apparently closely connected to the blood flow. Table 4 presents such data for three different human melanomas grown as xenografts on nude mice. In one tumour a doubling of the lactate level was observed, while in the others only a slight increase occurred. Inter-

Table 4. Metabolites (μmoles/g tissue) in xenografts of melanoma after hyperthermia *(Ht)* at 43°C, 1 h. (Mirtsch and Streffer, unpublished data)

	Bo		MeWo		Wi	
	Control	Ht	Control	Ht	Control	Ht
Glucose	1.8	1.1	1.9	1.9	1.6	–
Lactate	5.8	12.1	6.6	7.8	5.9	6.4
Pyruvate	0.19	0.17	0.25	0.20	0.15	0.14
Lactate/pyruvate	30.5	71.2	26.4	39.0	39.3	45.7

estingly, the glucose level decreased in the first tumour (Table 4). Apparently the blood flow was especially reduced in this tumour. Therefore glucose supply was reduced and lactate accumulated, as oxygen was not available and the reduced and acidic metabolite could not be transported from the tumour to the liver. However, in all three tumours an increase of the redox ratio lactate/pyruvate was found. These data strongly demonstrate the close connection and influences between physiological and metabolic processes which are not seen in cells in vitro. As these processes are very important for the modification of thermosensitivity, such studies in tumour systems are necessary.

Glycolysis After Glucose Loading

It has been observed that the extracellular pH decreases in transplantable animal tumours after glucose administration (von Ardenne 1980; Jähde and Rajewsky 1982). On the other hand, it has been demonstrated that the thermosensitivity of cells is considerably increased at pH 6.5–6.8 (Gerweck 1977). Therefore studies of glucose metabolism after glucose loading are of high interest, the rationale being that cellular thermosensitivity can be modified by these manipulations. It has been proposed that the decrease of pH is due to an increased lactate level after glucose application. Dickson and Calderwood (1979), however, reported that the heat-induced reduction of anaerobic glycolysis in a Yoshida sarcoma was even stronger after high glucose loading. Again the glycolysis was measured in vitro. On the other hand the degradation of glucose ($^{14}CO_2$ output after ^{14}C-glucose) was enhanced during whole-body hyperthermia when the mice received 1.5 or 6.0 mg glucose per g body weight (Schubert et al. 1982). Thus the situation is not clear at all.

Hengstebeck (1983) injected 6 mg glucose/g body weight into mice intraperitoneally and studied the glucose metabolism in the liver as well as in a transplanted mammary adenocarcinoma. Without hyperthermic treatment the glucose level rapidly increased in the tumour, more than threefold, and returned to normal values within 2 h. When the tumour was locally heated to 43°C for 60 min the peak level of glucose was smaller, but it took several hours before the normal value was reached again (Fig. 5). Under these conditions the lactate level also increased within the tumour and remained elevated for several hours, while such an increase was not seen after glucose injection or after heating of the tumour alone. As the glycolytic metabolites, especially fructose-1.6-diphosphate, also increased after the combined treatment (glucose and heat), it must be assumed that the increased lactate level originated from lactate production within the tumour.

The most dramatic changes were observed for the lactate/pyruvate ratio, which increased after the combined treatment and remained increased for several hours (Fig. 6).

GLUCOSE IN THE TUMOR

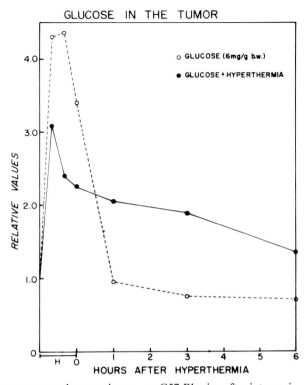

Fig. 5. Glucose levels in a transplanted mammary adenocarcinoma on C57 Bl mice after intraperitoneal injection of glucose (6 mg/g body weight) (O) and glucose injection followed directly by local hyperthermia for 1 h at 43° C (●)

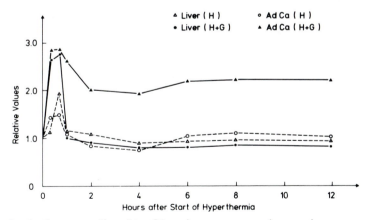

Fig. 6. Lactate/pyruvate ratios in the mouse liver (△, ●) and a mammary adenocarcinoma on C57 Bl mice (O, ▲) after local heat treatment of the tumour at 43° C for 1 h (△, O) or glucose injection (6 mg/g body weight) plus heat (●, ▲)

This ratio is linked to the intracellular redox equilibria and mirrors the situation of oxygenation in the cytoplasm through the following equilibrium constant:

$$K = \frac{[\text{Lact.}] \times [\text{NAD}^+]}{[\text{Pyr.}] \times [\text{NADH}] \times [\text{H}^+]}.$$

If oxygen supply is reduced, NADH cannot transfer its electrons to oxygen and thus increases, and as a consequence lactate is also enhanced. Furthermore, pH decreases in this situation. These effects might induce a reduction in blood flow which in turn reduces the transport of lactate from the tumour to the liver. Thus a cascade between these phenomena leads to an efficient interplay which apparently induces marked hypoxia in the tumour for several hours. The importance of this interplay between lactate, blood flow and pH can also be deduced from the finding that galactose, like glucose, reduces the blood flow in rodent tumours, but the decrease of pH was much more pronounced after glucose (Jain and Shah 1984, cited by Gerweck 1985). Furthermore, it has been demonstrated that heat damage to the microcirculation of a "sandwich" tumour is enhanced by glucose (Reinhold and van den Berg-Blok 1981).

These data demonstrate the close links between physiological and metabolic processes after hyperthermic treatment (Vaupel and Kallinowski 1987). By means of these processes a modification of thermosensitivity of tissues and tumours can be achieved by which the efficiency of a hyperthermic treatment might be increased. The data further demonstrate that in clinical therapy, radiation and probably also drugs should be given before a hyperthermic treatment, especially when it is combined with glucose, in order to avoid severe hypoxia during the irradiation.

Summary

Hyperthermia induces conformational changes of macromolecular structures. Such effects lead to a sudden inhibition of DNA, RNA and protein synthesis and a breakdown of membranes and of the cytoskeleton. These alterations can be very important for the mechanism of cell killing by hyperthermia.

Furthermore hyperthermia induces a number of immediate metabolic changes by increasing metabolic rates. These alterations have been studied especially in intermediary metabolism like glycolysis, citrate cycle, lipid metabolism and oxidative phosphorylation. An increased turnover of ATP has been observed in cells and tissues during heating. These changes lead to a depletion of energy reservoirs. Also, disregulations occur at certain metabolic key points. Thus, the pathway of pyruvate into the citrate cycle via acetyl-CoA is apparently reduced in heated melanoma cells in vitro. The redox ratios of lactate/pyruvate, NADH/NAD$^+$ and others are decreased.

When the same melanoma cells are grown as a xenograft on nude mice the metabolic rates are also enhanced; however, the lactate/pyruvate ratio increases during a localized heating of the tumour. The extent of this effect is very variable in individual tumours and is apparently correlated with the blood flow. These alterations can be enhanced by glucose loading and can be used as an indicator of hypoxia within the tumour. Thus, the micromilieu can be modified by these metabolic effects in such a way that the thermosensitivity is increased. The data show that metabolic processes are directly and indirectly involved in cell killing by hyperthermia.

References

Alper T (1979) Cellular radiobiology. Cambridge University Press, Cambridge
Altman D, Gerber GB, Okada S (1970) Radiation biochemistry. Academic, New York
Bowler K, Duncan CJ, Gladwell RT, Davison TF (1973) Cellular heat injury. Comp Biochem Physiol [A] 45: 953–963
Calderwood SK, Hahn GW (1983) Thermal sensitivity and resistance of insulin-receptor binding. Biochim Biophys Acta 756: 1–8
Coss RA, Dewey WC, Bamburg JR (1982) Effects of hyperthermia on dividing Chinese hamster ovary cells and on microtubules in vitro. Cancer Res 42: 1059–1071
Cress AE, Culver PS, Moon TE, Gerner EW (1982) Correlation between amounts of cellular membrane components and sensitivity to hyperthermia in a variety of mammalian cell lines in cultures. Cancer Res 42: 1716–1721
Dickson J, Calderwood SK (1979) Effect of hyperglycemia and hyperthermia on the pH, glycolysis and respiration of the Yoshida sarcoma in vivo. JNCI 63: 1371–1381
Francesconi R, Mayer M (1979) Heat- and exercise-induced hyperthermia: effects on high-energy phosphate. Aviat Space Environ Med 50: 799–802
Gerweck LE (1977) Modification of cell lethality at elevated temperatures: the pH effect. Radiat Res 70: 224–235
Gerweck LE (1985) Environmental and vascular effect. In: Overgaard J (ed) Hyperthermic oncology, vol 2. Taylor and Francis, London, pp 253–262
Hahn GM (1982) Hyperthermia and cancer. Plenum, New York
Hengstebeck S (1983) Untersuchungen zum Intermediärstoffwechsel in der Leber und in einem Adenocarcinom der Maus nach Hyperthermie. Dissertation, University of Essen
Henle KJ, Leeper DB (1979) Effects of hyperthermia (45° C) on macromolecular synthesis in Chinese hamster ovary cells. Cancer Res 39: 2665–2674
Jähde E, Rajewsky MF (1982) Sensitization of clonogenic malignant cells to hyperthermia by glucose-mediated, tumour-selective pH reduction. J Cancer Res Clin Oncol 104: 23–30
Leeper DB (1985) Molecular and cellular mechanisms of hyperthermia alone or combined with other modalities. In: Overgaard J (ed) Hyperthermic oncology 1984. Taylor and Francis, London, pp 9–40
Lepock JR, Cheng KH, Al-Qysi H, Kruuv J (1983) Thermotropic lipid and protein transitions in Chinese hamster lung cell membranes: relationship to hyperthermic cell killing. Can J Biochem Cell Biol 61: 421–427
Lin PS, Turi A, Kwock L, Lu RC (1982) Hyperthermia effect on microtubule organization. Natl Cancer Inst Monogr 61: 57–60
Lunec J, Cresswell SR (1983) Heat-induced thermotolerance expressed in the energy metabolism of mammalian cells. Radiat Res 93: 588–597
Mirtsch S, Streffer C, van Beuningen D, Rebmann A (1984) ATP metabolism in human melanoma cells after treatment with hyperthermia (42° C). In: Overgaard J (ed) Hyperthermic oncology 1984. Taylor and Francis, London, pp 19–22
Privalov PL (1979) Stability of proteins. Protein Chem 33: 167–241
Reinhold HS, van den Berg-Blok A (1981) Enhancement of thermal damage to the microcirculation of "sandwich" tumours by additional treatment. Eur J Cancer Clin Oncol 17: 781–795
Schubert B, Streffer C, Tamulevicius P (1982) Glucose metabolism in mice during and after whole-body hyperthermia. Natl Cancer Inst Monogr 61: 203–205
Stevenson MA, Minto KW, Hahn GM (1981) Survival and concanavalin-A-induced capping in CHO fibroblasts after exposure to hyperthermia, ethanol, and X-irradiation. Radiat Res 86: 467–478
Streffer C (1969) Strahlen-Biochemie. Springer, Berlin Heidelberg New York
Streffer C (1982) Aspects of biochemical effects by hyperthermia. Natl Cancer Inst Monogr 61: 11–16
Streffer C (1985) Metabolic changes during and after hyperthermia. Int J Hyperthermia 1: 305–319
Streffer C, van Beuningen D (1985) Zelluläre Strahlenbiologie und Strahlenpathologie (Ganz- und Teilkörperbestrahlung). In: Diethelm L, Heuck F, Olsson O, Strnad F, Vieten H, Zuppinger A (eds) Handbuch der medizinischen Radiologie, vol 20. Springer, Berlin Heidelberg New York, pp 1–39

Streffer C, van Beuningen D (1987) The biological basis for tumour therapy by hyperthermia and radiation. Recent Results Cancer Res 104: 24–70

Vaupel P, Kallinowski F (1987) Physiological effects of hyperthermia. Recent Results Cancer Res 104: 71–109

Von Ardenne M (1980) Hyperthermia and cancer therapy. Adv Pharmacol Chemother 10: 137–138

Warburg O, Wind F, Negelein E (1926) Über den Stoffwechsel von Tumoren im Körper. Klin Wochenschr 5: 829–834

Yatvin MB, Cree TC, Elso CE, Gipp JJ, Tegmo I-M, Vorpahl JW (1982) Probing the relationship of membrane "fluidity" to heat killing of cells. Radiat Res 89: 644–646

Hyaluronic Acid Degradation Mediated by Thiols and Effects of Hyperthermic Conditions

R. M. Fink and E. Lengfelder

Institut für Strahlenbiologie, Ludwig-Maximilians-Universität, Schillerstraße 42, 8000 München 2, FRG

Introduction

The glycosaminoglycan hyaluronic acid (HA) is an important structural component of connective tissue. It is composed of repeating disaccharide units of N-acetyl-β-D-glucosamine and β-D-glucuronate. HA differs from other glycosaminoglycans in three important respects: it has the highest molecular weight (4×10^3–8×10^6), it is not covalently bound to protein, and it lacks sulfate groups. Especially high concentrations of the ubiquitous HA are found in the vitreous humor, in the umbilical cord, in cartilage and synovial fluid of joints (Hukins 1984). Previous studies on the degradation of HA in vitro have demonstrated that activated oxygen species, especially the highly reactive hydroxyl radical, seem to represent the predominant degrading component. HA is not degraded by reducing radicals such as hydrated electrons, hydrogen atoms, or $\cdot COO^-$, nor by superoxide radicals or hydrogen peroxide (Kreisl and Lengfelder 1984; Betts et al. 1983). Activated oxygen species may also play an essential role in the mechanism of thermosensitization and cytotoxicity mediated by thiols. Some of the thiols used as radical scavengers or radioprotective agents in chemical and biological systems have been shown to be cytotoxic, especially at low concentrations (Issels et al. 1984, 1986; Biaglow et al. 1984). Increased temperatures of 40–44° C (hyperthermia), alone or combined with chemotherapy or radiation, have proved to be a useful therapeutic modality in the treatment of malignant disease (Hahn 1982).

The purpose of this study was to investigate the effect of various concentrations of thiols on HA degradation in the presence of chelated Fe-III and under the influence of hyperthermic conditions.

Materials and Methods

Degradation of HA was determined viscometrically using a Schott Micro Ubbelohde viscometer in a temperature-controlled waterbath ($\pm 0.05°$ C). All experiments were performed in 66 mM phosphate buffer at pH 7.4 purified by Chelex 100 treatment. All solutions, including the buffer, were prepared in double-distilled and additionally pyrolyzed water. The results are expressed as the percentage decrease in initial HA viscosity due to a particular substance or reaction. During the viscometric measurements reaction mixtures were continually gassed with oxygen. Hyaluronic acid (from human umbilical

nt Results in Cancer Research, Vol. 107
ringer-Verlag Berlin · Heidelberg 1988

cords) was from Serva, Heidelberg; T 77 was a gift from Inpharzam, Munich; and the other thiols were from Sigma, Munich.

Results

As shown by earlier investigations (Fink and Lengfelder 1986) thiols alone were not able to degrade HA. Only when chelated transition metals were present in the basic test system could the addition of thiols induce HA-degrading reactions. In the experiments reported here, each test system contained freshly prepared 0.1 mM FeCl$_3$/EDTA, which is also not able to degrade HA by itself (Fink and Lengfelder 1987).

The addition of thiols to reaction mixtures containing HA and FeCl$_3$/EDTA initiated a marked decrease in HA viscosity. As shown in Figs. 1–4 the extent of HA degra-

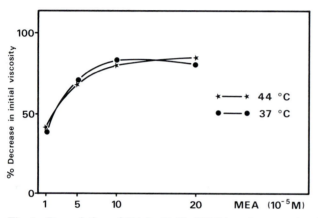

Fig. 1. Degradation of HA by FeCl$_3$/EDTA and cysteamine *(MEA)* at various concentrations and temperatures. The oxygen gassed reaction mixture contained, in 3 ml: 66 mM phosphate buffer pH 7.4, 2 mg HA, 0.1 mM FeCl$_3$/EDTA, 0.01–0.2 mM MEA. The temperatures investigated were 37° C (●) and 44° C (★). The results, expressed as percent decrease in initial HA viscosity, are values taken 3 min after thiol addition

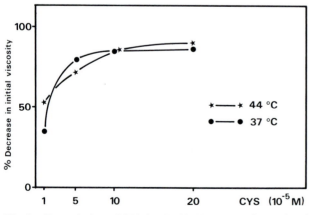

Fig. 2. Degradation of HA by FeCl$_3$/EDTA and cysteine free base *(CYS)* at various concentrations and temperatures. The reaction conditions were as described in Fig. 1

dation depended on the type of thiol as well as on the applied concentration of the investigated thiols. Raising the concentration of T 77, cysteamine (MEA), and cysteine (CYS) from 0.01 mM to 0.1 mM resulted in a marked increase in HA degradation. Higher concentrations of these thiols (up to 0.2 mM) merely caused a minor additional decrease of initial HA viscosity (Figs. 1–3). Therefore it seems that the enhancement of HA degradation by increasing the concentration of these thiols reaches a saturation level at relatively low concentrations. Of the applied thiols, T 77 and CYS increased the HA degradation rate to the greatest extent. In the case of MEA and CYS, increasing the temperature from 37° C to 44° C did not induce a significant enhancement of HA degradation under the experimental conditions (Figs. 1–2). Hyperthermia at 44° C led to a small increase of HA degradation mediated by Fe-III/EDTA and T 77 (Fig. 3).

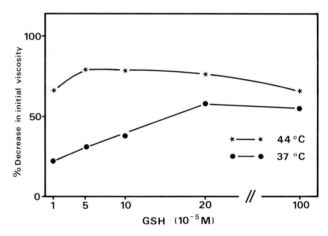

Fig. 3. Degradation of HA by FeCl$_3$/EDTA and T 77 at various concentrations and temperatures. The reaction conditions were as described in Fig. 1

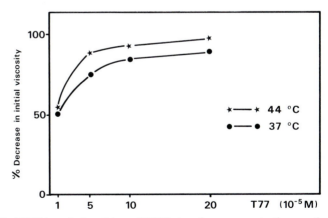

Fig. 4. Degradation of HA by FeCl$_3$/EDTA and glutathione *(GSH)* at various concentrations and temperatures. The oxygen gassed reaction mixture contained, in 3 ml: 66 mM phosphate buffer pH 7.4, 2 mg HA, 0.1 mM FeCl$_3$/EDTA and 0.01–1.0 mM GSH. The temperatures investigated were 37° C (●) and 44° C (★). The results, expressed as percent decrease in initial HA viscosity, are values taken 17 min after thiol addition

Glutathione (GSH), one of the most important cellular thiols, showed a completely different effect on HA degradation by interaction with EDTA-chelated Fe-III than the other thiols investigated. As shown in Fig. 4, increasing GSH concentrations induced considerable enhancement of HA degradation up to 0.2 mM GSH at 37° C. However, the relative extent of HA degradation caused by the various GSH concentrations was markedly reduced in comparison to the corresponding concentrations of MEA, CYS, and T 77. Additionally, significant differences in the extent of HA degradation by various GSH concentrations appeared only after reaction times much longer than those with MEA, CYS and T 77. Increasing the temperature from 37° C to 44° C resulted in a remarkable stimulation of HA degradation mediated by GSH, especially at the lower concentrations applied. The higher the concentration of GSH, the lower the relative extent of temperature-dependent HA degradation.

Discussion

As shown by previous studies (Kreisl and Lengfelder 1984; Betts et al. 1983; Fink and Lengfelder 1987) degradation of HA is an appropriate parameter to investigate the production and reactivity of OH˙-like species as well as the formation of ferryl ($Fe^{2+}O$) and/or perferryl ($Fe^{2+}O_2 \rightleftharpoons Fe^{3+}O_2^-$) species. Our present results demonstrate that the potential of various thiols in promoting the generation of highly oxidizing species in the presence of chelated ferric ions greatly depends on the type and effective concentration of thiol applied, as well as on the prevailing temperature.

As illustrated in Figs. 1–4, the decrease in the viscosity of HA is influenced by the available amount of thiol which reacts with Fe-III/EDTA resulting in the formation of Fe-II/EDTA and thiol radical RS˙. Autoxidation of the ferrous EDTA complex leads to the production of OH˙ radicals or similar highly oxidizing species which actually attack and degrade the HA macromolecules. In addition, we also assume a direct involvement of thiol radicals in the HA depolymerization process. The extent of thiol-radical-mediated HA degradation probably depends on the type of thiol administered. In the presence of constant amounts of EDTA-chelated Fe-III, the cellular nonprotein thiols CYS and MEA show similar effects on HA degradation after variation of their concentrations or temperature, respectively. T 77, however, differs from MEA and CYS in its temperature-stimulated effect on the HA depolymerization process. This may be due to an enhancement in the production and/or reaction rate of HA-degrading species in the T 77-containing in vitro system by increased temperature. Another explanation for the observed temperature effect of T 77-induced HA degradation may be provided by the hypothesis that the thiol-related hydrogen donation to radical sites in the HA macromolecules is not so effective at higher temperature. This results in an increase in net HA degradation caused by a reduced repair mechanism of damaged HA molecules. GSH, which occurs in substantial concentrations in mammalian cells, plays an important role in cellular protection. Therefore it is interesting that GSH affects in vitro HA depolymerization in a manner quite different from that of the other thiols investigated. At physiological temperature (37° C), GSH merely causes a relatively small degree of hyaluronic acid degradation. Hyperthermic conditions (e.g. 44° C), however, resulted in a remarkable increase of HA depolymerization, especially at the lower GSH concentrations applied. This stimulation of HA degradation caused by increased temperature was diminished with increasing GSH concentrations.

These findings support the concept that high cellular GSH levels may increase the cellular resistance toward certain toxic (including anti-tumor) agents, radiation, and oxidative effects, as proposed by Meister (1987). Thus the considerable damage to HA macromolecules by lowered GSH concentrations and higher temperatures may also result from insufficient rapair of attacked HA molecules, which is normally achieved by GSH-related hydrogen donation in order to protect the target molecule and detoxify the destructive agents simultaneously.

On the basis of our data we assume a concentration-dependent potential of thiol cytotoxicity and, in the case of GSH, a significant temperature effect. This temperature effect may be of special interest because some tumors show GSH levels close to the minimum required for survival (Meister 1987). Thus, therapy with hyperthermia probably promotes the capacity of low-level GSH-related cytotoxic effects. This phenomenon may play an important role in the multiple modifications of metabolism, transport and cellular protection caused by hyperthermic treatment.

References

Betts WH, Cleland LG, Gee DJ, Whitehouse MW (1983) Iron-associated hydroxyl radical production: influence of metal chelators, copper (II) and thiols. In: Cohen G, Greenwald RA (eds) Oxy radicals and their scavengers systems, vol 1. Elsevier, New York, pp 95–100

Biaglow JE, Issels RD, Gerweck LE, Varnes ME, Jacobsen B, Mitchell JB, Russo A (1984) Factors influencing the oxidation of cysteamine and other thiols: implications for hyperthermic sensitization and radiation protection. Radiat Res 100: 298–312

Fink RM, Lengfelder E (1986) Hyaluronic acid degrading reactions under hyperthermic conditions and effects of thiols. In: Rotilio G (ed) Superoxide and superoxide dismutase in chemistry, biology and medicine. Elsevier, Amsterdam, pp 60–63

Fink RM, Lengfelder E (1987) Hyaluronic acid degradation by ascorbic acid and influence of iron. Free Radiat Res Commun, (in press)

Hahn GM (1982) Hyperthermia and cancer. Plenum, New York, pp 232–242

Hukins DWL (1984) Connective tissue matrix. Verlag Chemie, Weinheim

Issels RD, Biaglow JE, Epstein L, Gerweck LE (1984) Enhancement of cysteamine cytotoxicity by hyperthermia and its modification by catalase and superoxide dismutase in Chinese hamster cells. Cancer Res 44: 3911–3915

Issels RD, Fink RM, Lengfelder E (1986) Effects of hyperthermic conditions on the reactivity of oxygen radicals. Free Radiat Res Commun 2: 7–18

Kreisl C, Lengfelder E (1984) Hyaluronic acid degradation by reactions producing activated oxygen species. In: Rotilio G, Bannister JV (eds) Oxidative damage and related enzymes. Harwood, London, pp 81–86 (Life chemistry reports)

Meister A (1987) Effects of modulation of glutathione levels and metabolism (Abstr). 2nd International Conference of Anticarcinogenesis and Radiation Protection, March 8–12, Gaithersburg

Glutathione Level in Melanoma Cells and Tissue*

S. Mirtsch, U. Strohmenger, and C. Streffer

Institut für Medizinische Strahlenbiologie, Universitätsklinikum Essen, Hufelandstraße 55, 4300 Essen 1, FRG

Introduction

Glutathione (L-γ-glutamyl-L-cysteinyl-glycine) is found in animal cells and also in most plants and bacteria. Its functions in various fields, such as radiation biology and cancer therapy, are presently the subject of extensive studies by several groups of researchers (Meister and Anderson 1983).

Most cellular glutathione exists in the reduced thiol form (GSH), although various mixed disulfides, thioester and to a lesser extent glutathione disulfide (GSSG) contribute to the total cellular pool of glutathione. GSH is known to protect against oxygen-induced free radicals and peroxides that are normally produced in small amounts as a result of oxygen metabolism (Chance et al. 1979).

The depletion of cellular GSH by inhibition of γ-glutamylcysteine synthetase might make tumour cells more susceptible to radiation and chemotherapeutic agents (Meister and Griffith 1979; Dethmers and Meister 1981).

Mitchell et al. (1983) have reported the effects of elevated temperatures on the GSH levels in cells. Continuous heating or acute heat exposure resulted in a rapid elevation of cellular GSH content up to about twice that of the control values. However, the GSH content returned to normal values subsequently. We have studied the effect of hyperthermia on the GSH and GSSG levels of two melanomas (MeWo and Be 11) in vitro and in vivo. Since it has been shown that Be 11 cells are less radiosensitive than MeWo cells, we also examined the combined effects of irradiation and hyperthermia on the Be 11 cell lines.

Material and Methods

GSH, GSSG, 5,5'-dithiobis(2-nitrobenzoic) acid (DTNB), NADPH and yeast glutathione reductase were obtained from Sigma (Munich, FRG). 2-Vinylpyridine was purchased from Aldrich (Steinheim, FRG). The assay for glutathione was carried out under conditions similar to those described by Tietze (1969) and Griffith (1980).

* This work was supported by the Bundesministerium für Forschung und Technologie, Bonn.

Preparation of Cell Culture and Tumour

Melanoma cells (MeWo and Be 11) were cultured in vitro as described by Streffer et al. (1979). For studies on xenografts the human melanoma cells were inoculated into nu^+/nu^+ mice. When the tumours reached the volume of 500–700 mm^3 they were excised and small pieces of tumour tissue were transplanted into nu^+/nu^+ mice for further experiments.

Hyperthermia and Radiation Conditions

Local hyperthermia of the tumours was induced by ultrasound at 43° C for 30 min. Cells in vitro were incubated for 3 h at 42° C, beginning 21 h after start of the culture.

Cells were X-irradiated with 3.7 Gy 21 h after start of the culture at a dose rate of about 0.94 Gy min^{-1} (100 R min^{-1}) using a Siemens Stabilipan X-ray machine operated at 220 kVp and 15 mA with a 0.5 mm Cu filter and from a distance of 50 cm. For combined treatment, 21-h-old cells were irradiated with 3.7 Gy immediately followed by hyperthermia at 42° C for 3 h.

The glutathione levels of the cells were measured at 24-h intervals from 24 to 96 h after start of the culture. Immediately and 2 h after heat treatment the glutathione content was studied in both melanoma xenografts. Tumours and cells without hyperthermia served as controls.

Results and Discussion

The GSH levels were highest during the early exponential growth phase in MeWo cells and decreased continuously with increasing time of culture. At 96 h it was lowered to about 35% of the level at 24 h (data not shown). After heat treatment for 3 h at 42° C higher glutathione levels were observed. The GSH levels were about 10%–25% higher than in untreated MeWo cells. After hyperthermia the GSSG contents in the MeWo cells were increased about 30%–50% in comparison to the controls. The ratios of GSH to GSSG were about 25% lower after heat treatment than in untreated MeWo cells (Table 1).

Table 1 also shows glutathione levels in xenografts of these cells in nude mice before and after hyperthermia. In the tumour the GSH levels also increased slightly after heat treatment. However, GSSG contents decreased in this case after heat treatment. Therefore, the ratio of GSH to GSSG was found to be increased in MeWo tumours.

Table 1. Glutathione levels in human melanoma cells *(MeWo)* (μmoles/10^9 cells) and their xenografts on nude mice (μmoles/g) after hyperthermia

	MeWo-tumours Time after 43° C, 30 min			MeWo-cells	
	Untreated	0 h	2 h	Untreated	Directly after 3 h, 42° C
GSH	1.2 ±0.13	1.4 ±0.12	1.57 ±0.13	8.7 ±0.25	9.7 ±0.20
GSSG	0.025±0.006	0.019±0.007	0.017±0.004	1.05±0.09	1.65±0.07
GSH/GSSG	48.0	73.7	92.3	8.3	5.9

Table 2. Glutathione levels in human melanoma cells *(Be 11)* (μmoles/10^9 cells) and their xeno-grafts on nude mice (μmoles/g) after hyperthermia

	Be 11-tumours Time after 43° C, 30 min			Be 11-cells	
	Untreated	0 h	2 h	Untreated	Directly after 3 h, 42° C
GSH	1.81 ±0.23	1.68±0.19	1.71 ±0.2	25.8 ±0.31	25.1±0.4
GSSG	0.036±0.009	0.02±0.003	0.016±0.005	1.65±0.06	2.2±0.9
GSH/GSSG	50.2	84.0	106.8	15.6	11.4

The GSH contents in Be 11 cells are three times higher than in MeWo cells. Directly after hyperthermia the amounts of GSH were the same in both heat-treated and untreated Be 11 cells, but it increased during the period 48–96 h in heat-treated cells (data not shown). An increase of about 20%–40% in the level of GSSG was observed after hyperthermia compared to untreated cells (Table 2).

Table 2 shows the GSH and GSSG levels in Be 11 tumours and Be 11 cells after heating. The GSH levels in both xenografts and Be 11 cells were not significantly affected directly after hyperthermia. However, the GSSG contents were reduced in the xenografts and increased in Be 11 cells in vitro. Further, it can be seen that the ratios of GSH to GSSG again increased in the tumour and decreased in the cells after heating.

Figure 1 shows the levels of GSH and GSSG in Be 11 cells as a function of the culture time after X-ray treatment with 3.7 Gy and after combined treatment with 3.7 Gy plus 42° C for 3 h. X-irradiation does not appear to markedly affect the GSH levels in Be 11 cell lines. However, after combined treatment an increase in the amount of GSSG to about 50% over the control level was observed. Directly after combined treatment the ratio of GSH to GSSG was about 50% and at later intervals about 30% lower than in untreated Be 11 cells.

The data presented here show that hyperthermia treatment produces an increase in glutathione levels in both melanoma cell lines. This effect has also been reported previously for V 79 cells by Mitchell and Russo (1983).

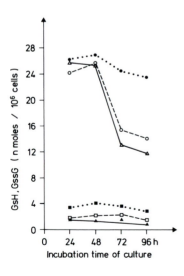

Fig. 1. GSH and GSSG levels in Be 11 cells as a function of culture times *(h)*, △ GSH, ▲ GSSG controls; ○ GSH, □ GSSG after radiation alone (3.7 Gy); ● GSH, ■ GSSG with hyperthermia at 42° C for 3 h given immediately after radiation (3.7 Gy). Each point represents the mean of five to eight experiments

Hyperthermia leads to different effects in the human melanoma cells in vitro and in their xenografts on nude mice. In both cell lines the ratios of GSH to GSSG decrease after heat treatment. This effect is in agreement with a decrease of the lactate/pyruvate ratio under the same conditions. Apparently a general decrease of the redox ratios occurs, including NADH/NAD$^+$ and NADPH/NADP$^+$, and this has consequences for energy metabolism.

In contrast, the redox ratios increased in the tumour in vivo after hyperthermia. Under these conditions the availability of oxygen is apparently insufficient, so that the reduced metabolites cannot transfer their electrons to the oxygen. Blood flow seriously interferes with these phenomena. There is a strong interrelation between blood flow and the described metabolic conditions. Further, these results demonstrate that data obtained with cells in vitro can be extrapolated to tumours in vivo only with great caution. In some instances it is apparently even impossible, as the complex interplay between physiological and biochemical factors does not exist in vitro.

Summary

We studied the effects of hyperthermia by measuring the content of reduced (GSH) and oxidized (GSSG) glutathione separately in two human melanoma cell lines (MeWo, Be 11) and the xenografts of the same melanomas on nu$^+$/nu$^+$ mice.

The Be 11 cell lines are less radiosensitive but more thermosensitive than MeWo cells. Therefore the levels of glutathione were also studied in Be 11 cells after combined treatment with 3.7 Gy plus 42° C for 3 h.

The levels of GSH were lower in both untreated MeWo cells and MeWo tumour than in Be 11 cells and Be 11 xenograft. After heating the cells in vitro at 42° C for 3 h and the tumour for 30 min at 43° C the levels of GSH and of GSSG increased in both melanoma cell lines. In the MeWo and Be 11 tumours hyperthermia did not markedly influence GSH levels but the GSSG levels decreased. From these data it followed that the ratios of GSH to GSSG were decreased in both cell lines, whereas the ratios increased in both tumours.

X-irradiation had no significant effects on GSH level in Be 11 cell lines, but the content of GSSG increased markedly after combined treatment.

References

Chance B, Sies H, Boveris S (1979) Hydroperoxide metabolism in mammalian organs. Physiol Rev 59: 527–605

Dethmers JK, Meister A (1981) Glutathione export by human lymphoid cells: depletion of glutathione by inhibition of its synthesis decreases export and increases sensitivity to irradiation. Proc Natl Acad Sci USA 78: 7492–7496

Griffith OW (1980) Determination of glutathione and glutathione disulfide using glutathione reductase and 2-vinylpyridine. Anal Biochem 106: 207–212

Meister A, Anderson ME (1983) Glutathione. Annu Rev Biochem 52: 711–760

Meister A, Griffith OW (1979) Effects of methionine sulfoximine analogs on the synthesis of glutathamine and glutathione: possible chemotherapeutic implications. Cancer Treat Rep 63: 1115–1121

Mitchell JB, Russo A (1983) Thiols, thiol depletion, and thermosensitivity. Radiat Res 95: 471–485

Mitchell JB, Russo A, Kinsella TJ, Glatstein E (1983) Glutathione elevation during thermotolerance induction and thermosensitization by glutathione depletion. Cancer Res 43: 987–991

Streffer C, van Beuningen D, Zamboglou N (1979) Cell killing by hyperthermia and radiation in cancer therapy. In: Abe M, Sakamoto K, Philips TL (eds) Treatment of radioresistant cancers. Elsevier/North Holland, Amsterdam, pp 55–70

Tietze F (1969) Enzymic method for quantitative determination of nanogram amounts of total and oxidized glutathione: applications to mammalian blood and other tissues. Anal Biochem 27: 502–522

Changes of Plasma Acetate Levels in Whole-Body Hyperthermia

H. A. Neumann[1], D. J. B. Hermann[1], und R. Engelhardt[2,*]

[1] St. Josef Krankenhaus, Ruhr-Universität, Gudrunstraße 56, 4630 Bochum, FRG
[2] Medizinische Klinik, Albert-Ludwigs-Universität, Hugstetter Straße 55, 7800 Freiburg, FRG

Introduction

High acetate production and low acetate utilization have been described as metabolic characteristics of malignant tumor metabolism (Bush 1953; Elliot and Greig 1937; Fredman and Graff 1958; Hepp et al. 1966; Katz et al. 1976; Pardee et al. 1959). These observations were made in the murine system. Hepp et al. (1966) described a disequilibrium between the acetate-activating and acetate-inactivating enzymes (acetate thiokinase and acetyl-coenzyme A deacylase respectively) and demonstrated that the inactivation was more pronounced.

A high acetate level in patients was therefore expected. However, Skutches et al. (1979) did not find an elevated concentration of acetate in patients with several malignant diseases compared to healthy persons.

Little information is available about the acetate concentrations under hyperthermia. We therefore investigated acetate concentrations in normal volunteers and in tumor patients under hyperthermic treatment.

Methods

Patients and Healthy Volunteers

Venous plasma acetate concentrations were determined in 15 healthy volunteers, in 18 tumor patients with small cell carcinoma of the lung and in four patients with Hodgkin's disease stage IV. All patients and volunteers gave their informed consent.

Hyperthermic Treatment

Hyperthermia was induced using the Siemens cabinet (Pomp, 1978). This consists of a plexiglass cabin preheated with hot air to 50–60° C. At the bottom of the cabin a coil field electrode is placed connected to a 27-MHz radiofrequency generator. The core tem-

* The authors thank Mrs. Marlies Braun for excellent technical assistance.

perature was monitored with a rectal thermometer. Temperature, pulse rate and blood pressure were monitored at 10-min intervals. A core temperature was reached after 40–60 min. The hyperthermic treatment was maintained for 1 h.

Blood samples were drawn before treatment (t_0), when a core temperature of $40°$ C was reached (t_1), after 1 h at a temperature of 40–$40.5°$ C (t_2), 1 h after completion of hyperthermia (t_{2a}) and 24 h after hyperthermic treatment (t_3).

The patients with small cell carcinoma of the lung received chemotherapy with adriamycin (60 mg/ml) vincristine (2 mg) intravenously when a core temperature of $40°$ C was reached. Seven patients received three cycles of thermochemotherapy.

The patients with Hodgkin's disease were treated with mustargen and vincristine. Acetate was monitored in a group of tumor patients that received adriamycin and vincristine under normal temperatures.

Determination of Plasma Acetate

Plasma acetate levels were determined as described by Bergmeyer and Möllering (1966). Briefly, blood was drawn in a heparinized syringe. Plasma was obtained by centrifugation at 3500 rpm, deproteinized with perchloric acid and then neutralized with $KHCO_3$. Afterwards the samples were distilled. Acetate was determined enzymatically using acetate kinase.

Statistical analysis was performed with the use of Student's test.

Results

Healthy volunteers and tumor patients with small cell carcinoma of the lung were subjected to whole-body hyperthermia of $40.5°$ C for 2 h. Hyperthermic treatment was tolerated well, and no cardiopulmonary complications or liver damage were observed.

The plasma acetate levels in healthy persons were lower than in the tumor patients. Acetate levels increased during whole-body hyperthermia in both groups and returned to lower values after hyperthermic treatment. In tumor patients the initial plasma acetate level was increased compared to the normal persons (Fig. 1, Table 1). A group of seven patients with small cell carcinoma of the lung was monitored over three chemotherapy cycles. The plasma acetate levels in six of these patients showed a decrease of the plasma acetate level from the first to the third cycle. These patients responded to the therapy. One patient did not show an increased level at the initiation of treatment and did not respond to the therapy (Fig. 2). An individual example of acetate concentrations during the therapy is shown in Fig. 3.

In six patients receiving chemotherapy without hyperthermia, no changes in plasma acetate levels were observed when blood samples were examined at the same time intervals as in the hyperthermia group.

Table 1. Acetate levels before, during and after hyperthermia (μmol/ml)

	t_0	t_1	t_2	t_3
Healthy volunteers ($n=15$)	90 ± 21	126 ± 34	125 ± 23	89 ± 19
Tumor patients ($n=22$)	129 ± 40	183 ± 51	190 ± 20	140 ± 12

Results expressed as mean $\pm$ SEM.

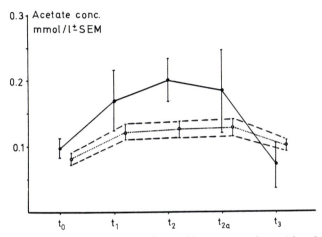

Fig. 1. Course of plasma acetate in normal volunteers *(open circles)* and in tumor patients *(closed circles)* during whole-body hyperthermia treatment. t, before hyperthermia; t_1, core temperature of 40.5° C reached; t_2, after 1 h of hyperthermic treatment; t_{2a}, 1 h after completion of hyperthermic treatment; t_3, 24 h after hyperthermia

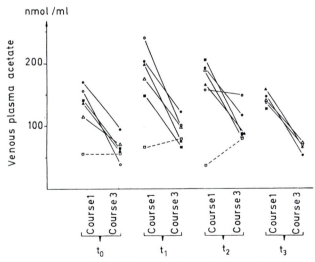

Fig. 2. Plasma acetate levels in seven tumor patients with small cell carcinoma of the lung at the first and the third course of thermochemotherapy. In six patients acetate levels decreased from the first to the third course. These patients showed a partial remission. In one patient acetate levels increased. This patient did not respond to therapy

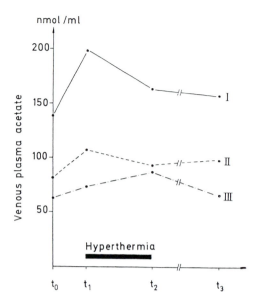

Fig. 3. Acetate levels from one patient with small cell carcinoma of the lung after thermo-chemotherapy (adriamycin, vincristine). *I* first, *II* second, *III* third course of chemotherapy under whole-body hyperthermia

Discussion

Earlier experiments had shown that tumor cells produce more acetate and utilize less acetate than normal cells. This led to the postulation that acetyl-coenzyme A is in part hydrolyzed and does not enter the citric cycle. Hereby, due to lack of aerobic energy, higher activity of the glycolytic pathway is induced. A higher level of acetate in tumor patients might be expected. Skutches et al. (1979), investigating a number of patients with different malignomas, could not confirm this hypothesis. In contrast, the pretreatment plasma acetate level in our patients was 129 ± 4 µmol/l ($n = 22$), whereas in a group of 15 healthy volunteers the level was 90 ± 21 µmol/l.

Under hyperthermic treatment, tumor patients and healthy persons showed an increase of plasma acetate level that was significant within the groups ($p < 0.02$). In six patients with small cell carcinoma of the lung treated with three cycles of chemotherapy and hyperthermia, the acetate levels before the third cycle of thermochemotherapy were lower than before treatment. Obviously, the release of acetate is reduced by reducing the tumor mass. In one patient with small cell carcinoma of the lung who did not respond to treatment, the acetate level increased in the course of the thermochemotherapy. It is well known that low pH has an important role in the thermal enhancement of some drugs (Hahn and Shiu 1983; Overgaard 1976), and the increase of acetate during hyperthermia might play an important part in this enhancement.

References

Bergmeyer HU, Möllering H (1966) Enzymatische Bestimmung von Azetat. Biochem Z 344: 167
Bush H (1953) Studies on the metabolism of acetate $-1-C^{14}$ in tissues of tumor bearing rats. Cancer Res 13: 789
Elliot KAC, Greig ME (1937) The metabolism of lactic and pyruvatic acids in normal and tumor tissues. Biochemistry 31: 1021

Fredman AD, Graff S (1958) The metabolism of pyruvate in the aricarboxylic acid cycle. J Biol Chem 233: 292

Hahn GM, Shin EC (1983) Effect of pH and elevated temperature on the cytotoxicity of some chemotherapeutic agents on Chinese hamster cells in vitro. Cancer Res 43: 5789

Hepp D, Prüsse E, Weiss H, Wieland O (1966) Aerobic formation of acetate in tumor cells: mechanism and possible significance. Adv Enzyme Regul 4: 89

Katz J, Brand K, Golden S, Rubinstein D (1974) Lactate and pyruvate metabolism and reducing equivalent transfer in Ehrlich ascites tumor. Cancer Res 34: 872

Overgaard J (1976) Influence of extracellular pH on the viability and morphology of tumor cells exposed to hyperthermia. INCI 56: 1234

Pardee AB, Heidelberger C, Potter R (1959) The oxidation rate of acetate $1-C^{14}$ by time in vitro. J Biol Chem 186: 625

Pomp H (1978) Clinical application of hyperthermia in gynecological malignant tumors. In: Streffer C, et al. (eds) Cancer therapy by hyperthermia and radiation. Urban and Schwarzenberg, Munich, p 326

Skutches CL, Holroyde CP, Myers RN, Parl P, Reichard GA (1979) Plasma acetate turnover and oxidation. J Clin Invest 64 (3): 708

Warburg O (1926) Über den Stoffwechsel von Tumoren. Springer, Berlin

Physiological Effects of Hyperthermia

H. S. Reinhold

Radiobiological Institute TNO, P. O. Box 5815, 2288 W Rijswijk, The Netherlands

For an efficient application of hyperthermia to tumours, the intensity of the tumour circulation is of utmost importance. Generally, it is assumed that the tumour microcirculation is inferior to that in healthy tissues. This certainly holds if the criterion for comparison is the morphological vascular architecture. There is overwelming evidence, derived using morphological as well as angiographic methods, that the tumour circulation is plainly chaotic. This holds for human tumours as well as for experimental tumours, regardless of the method of investigation used. However, a chaotic vascular pattern in tumours does not per se imply that the blood circulation in the tumours is inferior to, or lower than, the normal tissues in which the tumour grows. Obviously the differences in blood flow between tumour and normal tissue depend upon the tumour and the normal tissue concerned. With regard to the tumours, not only the type of tumour is of importance, but also its size and the site of growth. On the other hand, normal tissue blood flow depends not only on the tissue type, but, for many tissues, also very much on the physiological conditions. In contrast to tumours, many normal tissues can adapt easily to changing demands. In many tissues, such as muscle, fat, skin and intestine, there may be 10- to 30-fold differences, depending upon the conditions. Thus, if one compares the blood flow rate in a tumour, e.g. mammary carcinoma, with that in certain surrounding tissues under normal conditions, e.g. cool skin at $34°$ C, the circulation in the tumour may appear to be higher than in the normal tissue. However, under conditions of heating this relationship may change considerably. This will be discussed below.

Figure 1 gives an idea of this relationship. The values used in this figure are derived from many different sources, including textbooks. The normal tissues are listed in increasing order of blood flow. In the lower part some tumour values are shown with the values of the tissue of origin, while in the upper part some published values on a variety of tumours is depicted. Many of the data were compiled by Jain and Ward-Hartley (1984) and Patterson and Strang (1979). Other values were derived for breast carcinoma from Beaney et al. (1984), who found an average value of about 19 ml/100 g/min, and for brain and cerebral tumours from Ito et al. (1982) and Lammertsma et al. (1985). Data were taken from Mäntylä (1979) for lymphomas, anaplastic carcinomas and differentiated malignomas, and from Touloukian et al. (1971) for lymphangiomas. Finally, data were derived from Plengvanit et al. (1972) and Taylor et al. (1979) for liver tumours and liver metastases respectively.

What happens if one heats a tumour? Depending upon the temperature and the exposure time, the tumour physiology changes essentially during the hyperthermic treat-

Recent Results in Cancer Research, Vol. 107
© Springer-Verlag Berlin · Heidelberg 1988

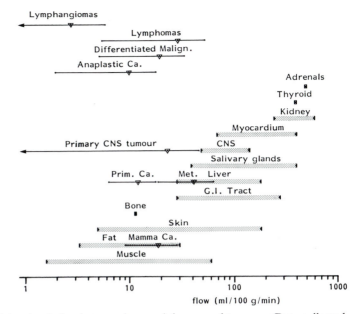

Fig. 1. Composite diagram of the circulation in several normal tissues and tumours. Data collected from various sources which are discussed in the text

ment. There is a certain similarity in this change in tumour physiology between normal tissues and tumour, but it has frequently been observed that the tumour physiology, in this instance the tumour circulation, adapts to a much lesser extent to the heat stimulus than do tissues like muscle and skin. Figure 2 gives an example of what can be observed if a tumour is treated with hyperthermia, when the temperature is gradually increased. This experiment deals with the rhabdomyosarcoma BA 1112 growing in observation chambers. In this instance the circulation was evaluated by means of a laser-doppler fluxmeter, giving an indication of the total erythrocyte flux in the tumour. As can be seen (in Fig. 2, when the temperature is increased, the tumour blood flow first increases, then maintains a plateau for some time but, under the conditions of the continuous temperature increase, the flow index again starts to rise. Then, at a given heat dose, the circulation collapses, and even when the temperature is not increased further a complete vascular stasis develops. This type of vascular stasis has been observed in the majority of the experimental tumour systems that have been investigated for this purpose. A visual representation of this occurrence is given in Fig. 3. Here again, the rhabdomyosarcoma BA 1112 is transplanted into a thin observation chamber (tumour thickness 200–300 μm; tumour diameter 2–3 mm). If such a tumour is heated at 42.5° C for prolonged periods of time, after 1–3 h the microcirculation in the tumour may show stasis. This is inevitably followed by the development of necrosis. It should be realized that this vascular stasis, with its subsequent necrosis, confines itself mainly to the central parts of the tumour, and that the periphery of the tumour seems largely unaffected. This is in keeping with the idea that the microenvironment of the centre of tumours is unfavourable, with a low supply of nutrients and a low pH and therefore great sensitivity to hyperthermia. Not all tumour types react in the same way, but even with the same tumour type growing in similar conditions, sizeable differences in reactions among the individual tumours can

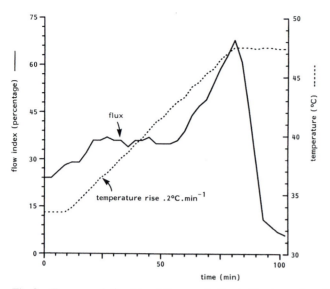

Fig. 2. Tumour relative blood flow, as indicated by laser-doppler flux of a rat rhabdomyosarcoma growing in a "sandwich" observation chamber. The diagram depicts the temperature, which increased during the experiment, as well as the flux. Note that after a given temperature dose (in this instance consisting of a gradual increase to 47° C over 65 min) the flux sharply declines. This indicates a sudden, complete vascular collapse

be observed. Nonetheless, it has been shown possible to derive the Arrhenius slope for this effect (Fig. 4). The data points in Fig. 4 were calculated with statistical methods based on estimation of the probability of events happening, in this case the failure rate of the microcirculation. This is the general way of handling such data in cellular biology, radiobiology and tumour biology. It is frequently not recognized that the methods used by physiologists to derive their stimulus-response data are of an entirely different nature. An example of such was given in Fig. 2. The great variety of methods of determining dose-time relationships for circulation parameters in hyperthermic treatment may explain the wide scatter of points that can be discerned in Fig. 5. In this figure the effects are divided into those, seen in observation chambers, indicated by inverted triangles, physiological measurements, indicated by triangles, and observations in humans, indicated by squares. If an effect was observed, i. e. a decrease of flow or stasis of the tumour microcirculation, the symbol is dark. Open symbols indicate that no effect was observed. The lines indicate published isoeffect lines. As can be seen from this figure, most experimental tumours treated with a dose of 43° C for 1 h, show a decrease in microcircula-

Fig. 3a–d. Sequence of events with a hyperthermic treatment (42.5° C) on the microcirculation of ▷ the experimental rhabdomyosarcoma BA 1112 in a "sandwich" observation chamber (diameter 3 mm, thickness 200 μm). **a** After 1 h at 42.5° C, there is good, visible circulation in all blood vessels. **b** After 2 h at 42.5° C, the circulation in some vessels starts to decrease. **c** After 3 h at 42.5° C, there is vascular collapse in most of the centrally located vessels and development of some areas of decreased viability (cloudiness). End of treatment, return to 33.5° C. **d** Two days later, those (central) areas showing vascular stasis in the former picture now show extensive necrosis. Note that the periphery is largely unaffected. (From Reinhold et al. 1978)

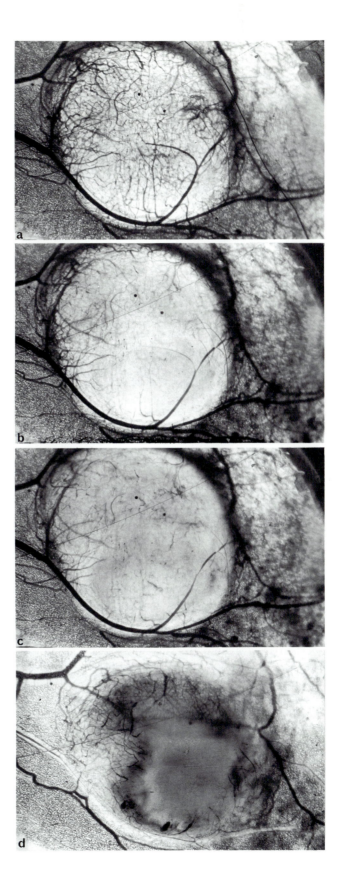

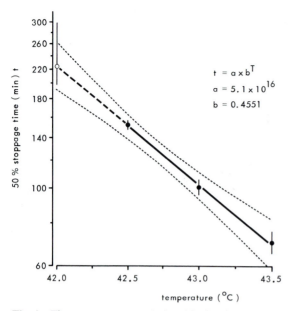

$$t = a \times b^T$$
$$a = 5.1 \times 10^{16}$$
$$b = 0.4551$$

Fig. 4. Time-temperature relationship for the development of vascular stasis in the rat rhabdomyosarcoma BA 1112 in "sandwich" observation chambers. The slope of this curve is very similar to the slopes determined for many other endpoints such as cell killing (cell survival) and normal tissue response. t, time; a, intercept; b, slope. (From van den Berg-Blok and Reinhold 1984)

tion. A notable exception is the Walker carcinoma, which requires 45° C for 1 h. Also very interesting are the human data derived by Waterman et al. (1987) (open squares), who, in a small series of human tumours, found no indication of flow reduction even if the temperatures were as high as 45° C. On the other hand, in whole-body hyperthermia in humans at 41.5° C for 2 h (black squares), Karino et al. (1984) found a decrease in circulation in some tumours. Also apparent from Fig. 5 is that some tumours are much more sensitive then others and may respond to exposure times of a few minutes at temperatures ranging from 41–43° C.

With regard to the response of healthy tissue, as mentioned, the general reaction is to increase flow as a response to temperature increase, and if the temperature dose (consisting of a given temperature for a given time) is high enough, then the circulation in the normal tissues may decrease. Not all normal tissues have been investigated in this way, but data are available on skin, subcutis and granulation tissue, muscle, brain and intestine. The skin of the rat has been investigated by Lokshina et al. (1985), Song et al. (1980), Dickson and Calderwood (1983), and Ham and Hurley (1968), and the mouse skin by Stewart and Begg (1983). The subcutis and granulation tissue have been investigated by Dewhirst et al. (1984) and Dudar and Jain (1984). Investigations of muscle are available on rodents from Lokshina et al. (1985), Bicher et al. (1980), Badylak et al. (1985), Rappaport and Song (1983), Song et al. (1984) and Ham and Hurley (1968), while Milligan and Panjehpour (1983) and Voorhees and Babbs (1982) investigated the effect of heat on canine muscle flow. The effects of heat on the microcirculation efficiency, i.e. the oxygenation, of the rabbit brain was investigated by Bicher et al. (1980). Finally, the effect on the intestine was investigated by Falk (1983) with morphological methods, Hume et al. (1981) with autoradiographic measurements and Peck and Gibbs

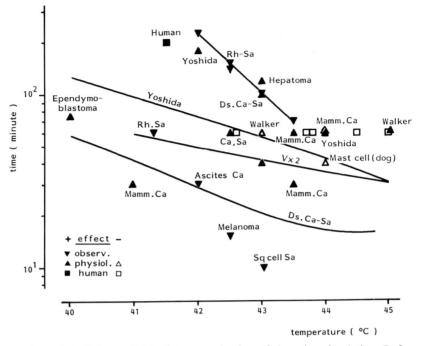

Fig. 5. Time-temperature plot of the available data on reduction of the microcirculation. References for the various tumour types indicated in the diagram are as follows: Ascites carcinoma (Scheid 1961); Carcinoma NT and MT (Stewart and Begg (1983); DS carcinoma (von Ardenne and Krüger 1980; Vaupel et al. 1980); Ependymoblastoma (Sutton 1976); Hepatoma (Karino et al. 1984); Human tumours (various), whole-body hyperthermia (Karino et al. 1984); Human tumours (various), local hyperthermia (Waterman et al. 1987); Mammary carcinoma *(Mamm. Ca)* (Song et al. 1980; Robinson et al. 1982; Rappaport and Song 1983; Shrivastav et al. 1983; Dewhirst et al. 1984; Kallinowski et al. 1984); Mast cell tumours (dog) (Milligan et al. 1983); Rhabdomyosarcoma *(Rh. Sa)* (Endrich et al. 1979; Reinhold et al. 1978; Emami et al. 1981; van den Berg-Blok and Reinhold 1984); Sarcoma F and Fa. (Stewart and Begg 1983); Sarcoma S-180 (Tanaka et al. 1984); Squamous cell carcinoma *(Sq cell Sa)* (Eddy 1980); VX$_2$ carcinoma (Dudar and Jain 1984); Walker carcinoma (Song 1978; Gullino 1980); Yoshida sarcoma (Dickson and Calderwood 1980; Vaupel et al. 1983; Kallinowski et al. 1984). (From Field and Franconi 1987)

(1983) with ^{133}Xe washout. Surprisingly enough, no data are available on the effects of heat on the microcirculation of fat. This is all the more surprising in view of the fact that fat is one of the dose-limiting tissues in many hyperthermic applications.

With regard to the results, Dudar and Jain (1984) concluded from their observations on granulation tissue that the magnitude and time of vascular stasis are a bimodal function of temperature and time of hyperthermia, while Stewart and Begg (1983) found that the relative perfusion of normal skin was progressively increased during heating with moderate dosages. The same observation was made by Song et al. (1980). On the other hand, Dickson and Calderwood (1980) found that the flow rate, as determined by ^{86}Rb declined after 1 h of heating at 42° C. Dewhirst et al. (1984) made the observation that vascular stasis in normal tissue can occur at temperatures as low as 42.6° C, provided the heating rate is 1° C/min. Milligan et al. (1983) found, using a thermal clearance method to measure blood flow rate, that the peak blood flow rate increased to 40, 59 and

183 ml/100 g/min at 43, 45 and 47° C respectively. The effects of heat on most normal tissues can be envisaged as follows. With increasing temperature a rise in blood flow and vascular volume develops, depending on both the treatment temperature and the exposure time. With moderate temperatures, up to 42–43° C, an increase in blood flow develops which gradually reaches a maximum that can be 4–6 times the normal value.

In the case of higher temperatures, especially over 43° C, the blood flow increases faster, but after a certain period of time this is followed by a decrease to preheating levels or even lower. When temperatures of 45–47° C are applied, the flow values at these temperatures could very well break down to blood flow levels that are lower than under normal conditions (Song et al. 1984; Milligan et al. 1983).

With regard to the investigations on mouse intestine, some seemingly controversial observations have been made. Falk (1983), with his morphological method using the length of the visible venous tree as an endpoint, found mouse intestine very sensitive, while Hume et al. (1981), using autoradiographic labelling of the arteriolar endothelium, found little response.

What measurable effects other than those on blood flow have been found in hyperthermic tumour physiology? A large number of parameters have been investigated, and virtually all showed an effect. The oxygenation of tumours was used by Bicher and Mitagvaria (1981) as an indicator for tumour circulation. They encountered, while investigating the effect of various temperatures, a "breaking point" at 42° C, below which the oxygenation increased and above which it decreased. These investigations were performed using micro-oxygen electrodes. Using pH electrodes, the pH of tumours was investigated by Bicher et al. (1980), Song et al. (1980) and Vaupel (1982). The experiments by Song et al. and Vaupel are replotted in Fig. 6. Here it can be seen that when an experimental tumour is heated to 43° C, the pH gradually drops. This implies that the sensitivity of the tumour tissue to hyperthermia must increase during the therapy. Vaupel (1982) suggested that a minimum temperature of 42° C was required to induce this effect. The mechanisms behind the decrease in pH are generally explained as a combination of increase in metabolic rate and decrease of egress of anaerobic metabolites. Indeed, increases in lactic acid and β-hydroxybutyric acid have been demonstrated by Streffer and colleagues (Streffer et al. 1981; Steffer 1985), Ryu et al. (1982) and Lee et al. (1986). It is therefore likely that during hyperthermic treatment the pH of the tumour tissue decreases, although human data on this subject are lacking. There is, however, evidence that human tumour pH is lower than normal tissue pH. For reviews, see Thistlethwaite et al. (1985) and Wike-Hooley et al. (1984). The only data on human tumour pH during hyperthermia were obtained during whole-body hyperthermia (Van der Zee et al. 1983) and did not show a change in pH. Another point which is not yet explained is that during hyperthermic treatment there is a slight increase in tumour volume. An example of this is shown in Fig. 7. No explanation for this phenomenon has yet been offered, although several authors, including Vaupel et al. (1983) report the development of oedema in the tumour interstitium during hyperthermic treatment. One would, however, in that case expect some correlation between the circulation parameters and the volume increase, but such correlation is presently lacking (Reinhold 1987, unpublished material). Another factor that may be very important is that the arterial venous pressure gradient over the tumour tends to decrease with hyperthermia. This was observed by Endrich (1984) with micropipette intravascular pressure measurements. During hyperthermia of an amelanotic hamster melanoma, growing in observation chambers, the blood pressure on the arterial side decreased, while on the venous side the intravascular pressure increased. This resulted in a decrease of the arterial-venous pressure gradient, and doubt-

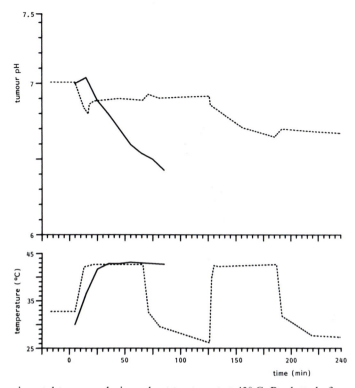

Fig. 6. Decrease in pH in experimental tumours during a heat treatment at 43° C. Replotted after data from Song et al. (1980) and Vaupel (1982). *Upper part*, interstitial tumour pH; *lower part*, temperature manipulations; *solid lines*, C3H mammary adenocarcinoma (Vaupel 1982); *dashed lines*, Walker carcinoma (Song et al. 1980)

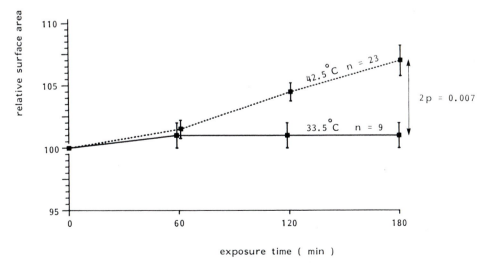

Fig. 7. Relative increase in size in tumours during hyperthermia. The experimental rhabdomyosarcoma BA 1112 was grown in "sandwich" observation chambers, and during heating at 42.5° C or sham-heating the tumour size was very accurately recorded with optical markers. It appears that the tumour size increases slightly, but significantly, during the 3-h heating period

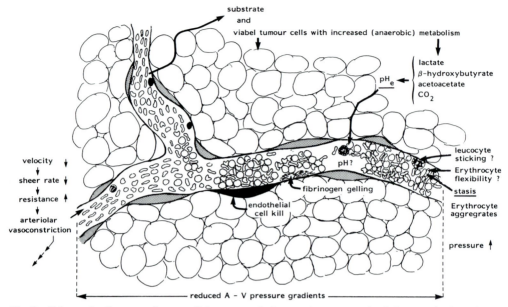

Fig. 8. Schematic diagram of some of the reported changes in tumour physiology and microcirculation during a hyperthermic treatment. (From Field and Franconi 1987)

less therefore in a decrease in erythrocyte velocity in the capillaries. In these conditions the sheer rate of the blood flow decreases, and it has been shown by Schmidt-Schönbein et al. (1984) that under these circumstances, in combination with a low pH, the flexibility of the erythrocytes in the capillaries decreases. This might very well enhance any ongoing vascular stasis mechanism. The same holds for the arterial venous shunting that has been found by Dewhirst et al. (1984). Recently, another possible mechanism for the development of vascular stasis in tumours has been formulated by Lee et al. (1986). They found that under hyperthermic conditions fibrin accumulation in the tumours took place, and they referred to experiments by Copley (1980), who showed that under conditions of low pH, as may occur close to the wall of capillaries in an acidified tumour, the fibrin may convert directly into fibrinogenin, a gel-like material. This mechanism should also be considered as one of the factors contributing to the development of vascular stasis during hyperthermia. In addition, aggregation of erythrocytes has been observed by Endrich et al. (1984), leukocyte sticking by Endrich et al. (1979), thrombosis by Storm (1979) and, finally, arteriolar vasoconstriction by Endrich (1984). This last phenomenon, however, occurred in sequence after the decrease of capillary flow. It is therefore very likely that the arteriolar vasoconstruction respresents a secondary phenomenon in the complex of factors that determine the blood flow in tumours during hyperthermic treatment. It should be realized that virtually all phenomena that have been observed in the response of tumour microcirculation to hyperthermia could equally be explained by endothelial cell destruction. This, however, is still an open question. With regard to the heat sensitivity of endothelial cells in the sense of cell survival, determined using in vitro systems, somewhat differing conclusions were drawn by Rhee and Song (1984) and Fajardo et al. (1985). Moreover, the many studies with direct observation of the microcirculation of tumours in "observation chambers" never reveal the endothelial cell as being particularly sensitive to heat. The same holds for electron-microscopic studies (Hammersen, quoted by Reinhold and Endrich 1986). The various interactions that may take place in a tumour during hyperthermia are schematically depicted in Fig. 8. Here, it should be realized again that, while in the majority of experimental tumours extensive damage to the tumour's microvascular bed has been demonstrated, in human tumours the situation may be different. This may be due to the problems inherent in adequate heating of human tumours for prolonged periods of time, but one could also think in terms of specific species differences or of differences between spontaneous and transplanted tumours. More information on this subject will be required.

References

Badylak SF, Babbs CF, Skojac TM, Voorhees WD, Richardson RC (1985) Hyperthermia-induced vascular injury in normal and neoplastic tissue. Cancer 56: 991–1000

Beaney RP, Lammertsma AA, Jones T, McKenzie CS, Halnan KE (1984) Positron emission tomography for in vivo measurement of regional blood flow, oxygen utilisation, and blood volume in patients with breast cancinoma. Lancet 21: 131–134

Bicher HI, Mitagvaria N (1981) Circulatory responses of malignant tumors during hyperthermia. Microvasc Res 21: 19–26

Bicher HI, Hetzel PW, Sandhu TS, Frinak S, Vaupel P, O'Hara MD, O'Brian T (1980) Effects of hyperthermia on normal and tumour microenvironment. Radiology 137: 523–530

Copley AL (1980) Fibrinogen gel clotting, pH and cancer therapy. Thromb Res 18: 1–6

Dewhirst MW, Sim DA, Grochowski KJ (1984) Thermal influence on radiation induced complications vs. tumor response in a phase III randomized trial. In: Overgaard J (ed) Hyperthermic oncology 1984. Taylor and Francis, London, pp 313–316

Dickson JA, Calderwood SK (1980) Temperature range and selective sensitivity of tumors to hyperthermia: a critical review. Ann NY Acad Sci 335: 180–205

Dickson JA, Calderwood SK (1983) Thermosensitivity of neoplastic tissues in vivo. In: Storm FK (ed) Hyperthermia in cancer therapy. Hall, Boston, pp 63–140

Dudar TE, Jain RK (1984) Differential response of normal and tumor microcirculation to hyperthermia. Cancer Res 44: 605–612

Eddy HA (1980) Alterations in tumor microvasculature during hyperthermia. Radiology 137: 515–521

Emami B, Nussbaum GH, Hahn N, Piro AJ, Dritschilo A, Quimby F (1981) Histopathological study on the effects of hyperthermia on microvasculature. Int J Radiat Oncol Biol Phys 7: 343–348

Endrich B (1984) Mikrozirkulation maligner Tumoren. Ph. D. thesis, University of Heidelberg

Endrich B, Hammersen F (1986) Morphologic and hemodynamic alterations in capillaries during hyperthermia. In: Anghileri W, Robert J (eds) Hyperthermia in cancer treatment. CRC, Boca Raton, pp 17–47

Endrich B, Zweifach BW, Reinhold HS, Intaglietta M (1979) Quantitative studies on microcirculatory function in malignant tissue: influence of temperature on microvascular hemodynamics during the early growth of the BA 1112 rat sarcoma. Int J Radiat Oncol Biol Phys 5: 2021–2030

Endrich B, Zweifach BW, Intaglietta M, Reinhold HS (1984) Influence of temperature on microvascular hemodynamics during the early growth of the BA 1112 rat sarcoma (Abstr). Microvasc Res 17(3): S 120

Fajardo LF, Egbert B, Marmor J, Hahn GM (1980) Effects of hyperthermia in a malignant tumor. Cancer 45: 613–623

Fajardo LF, Schreiber AB, Kelly NI, Hahn GM (1985) Thermal sensitivity of endothelial cells. Radiat Res 103: 276–285

Falk P (1983) The effect of elevated temperature on the vasculature of mouse jejunum. Br J Radiol 56: 41–49

Field SB, Franconi C (eds) (1987) Physics and technology of hyperthermia. Nijhoff, Dordrecht

Gullino PM (1980) Influence of blood supply on thermal properties and metabolism of mammary carcinoma. NY Acad Sci 335: 1–18

Ham KH, Hurley JV (1968) An electron-microscope study of the vascular response to mild thermal injury in the rat. J Pathol Bacteriol 151: 175–183

Hume SP, Marigold JCL, Hirst DG (1981) The effect of hyperthermia on one aspects of the response of mesenteric blood vessels to radiation. Int J Radiat Biol 39: 321–327

Ito M, Lammertsma AA, Wise RJS, Bernardini S, Frackowiak RSJ, Heather JD, McKenzie CG, Thomas DGT, Jones T (1982) Measurement of regional cerebral tumours using ^{15}O and positron emission tomography: analytical techniques and preliminary results. Neuroradiology 23: 63–74

Jain RK, Ward-Hartley K (1984) Tumor blood flow – characterization, modifications and role in hyperthermia. IEEE Trans Sonics Ultrason 31: 504–526

Kallinowski F, Vaupel P, Schaefer C, Benzing H, Mueller-Schauen LW, Fortmeyer HP (1984) Hyperthermia-induced blood flow changes in human mammary carcinomas transplanted into nude (rnu/rnu) rats. In: Overgaard J (ed) Hyperthermic oncology 1984, vol 1. Taylor and Francis, London, pp 133–136

Karino M, Koga S, Maeta M, Hamazoe R, Kumane T, Oda M (1984) Experimental and clinical studies on effect of hyperthermia on tumor blood flow. In: Overgaard J (ed) Hyperthermic oncology 1984, vol 1. Taylor and Francis, London, pp 173–176

Lammertsma AA, Wise RJS, Cos TCS, Thomas DGT, Jones T (1985) Measurement of blood flow, oxygen utilisation, oxygen extraction ratio, and fractional blood volume in human brain tumours and surrounding oedematous tissue. Br J Radiol 58: 725–734

Lee SY, Song CW, Levitt SH (1985) Change in fibrinogen uptake in tumors by hyperthermia. Eur J Cancer Clin Oncol 12: 1507–1513

Lee SY, Ryu K, Kang MS, Song CW (1986) Effect of hyperthermia on the lactic acid and β-hydroxybutyric acid contect in tumor. Int J Hyperthermia 2: 213–222

Lokshina A, Song CW, Rhee JG, Levitt SH (1985) Effect of fractionated heating on the blood flow in normal tissues. Int J Hyperthermia 1: 117–129

Mäntylä MJ (1979) Regional blood flow in human tumors. Cancer Res 39: 2304–2306

Milligan AJ, Panjehpour M (1983) Canine normal and tumor tissue blood flow during fractionated hyperthermia. In: Broerse JJ, Barendsen GW, Kal HB, van der Kogel AJ (eds) Proceedings of the 7th International Congress on Radiation Research. Nijhoff, The Hague

Milligan AJ, Conran PB, Ropar MA, McCulloch HA, Ahuja RK, Dobelbower RR (1983) Predictions of blood flow from thermal clearance during regional hyperthermia. Int J Radiat Oncol Biol Phys 9: 1335–1343

Patterson J, Strang R (1979) The role of blood flow in hyperthermia. Int J Radiat Oncol Biol Phys 5: 235–241

Peck JW, Gibbs FA (1983) Capillary blood flow in murine tumors, feet and intestines during localized hyperthermia. Radiat Res 96: 65–81

Plengvanit U, Suwanik R, Chearani O, Intrasupt S, Sutayavanich S, Kalayasiri C, Viranvatti V (1972) Regional hepatic blood flow studied by intrahepatic injection of ^{133}Xenon in normals and in patients with primary carcinoma of the liver, with particular reference to the effect of hepatic artery ligation. Aust NZ J Med 1: 44–48

Rappaport DS, Song CW (1983) Blood flow and intravascular volume of mammary adenocarcinoma 13726 A and normal tissues of rat during and following hyperthermia. Int J Radiat Oncol Biol Phys 9: 539–547

Reinhold HS, van den Berg-Blok AE (1983) Hyperthermia-induced alteration in erythrocyte velocity in tumors. Int J Microcirc Clin Exp 2: 285–295

Reinhold HS, Endrich B (1986) Tumour microcirculation as a target for hyperthermia: a review. Int J Hyperthermia 3: 111–137

Reinhold HS, Blachiewicz B, van den Berg-Blok AE (1978) Decrease in tumor microcirculation during hyperthermia. In: Streffer C (ed) Cancer therapy by hyperthermia and radiation. Urban and Schwarzenberg, Munich, pp 231–232

Rhee JG, Song CW (1984) Thermosensitivity of bovine aortic cells in cultures: in vitro clonogenicity study. In: Overgaard J (ed) Hyperthermic oncology 1984. Taylor and Francis, London, pp 157–160

Rhee JG, Kim TH, Levitt SH, Song CW (1984) Changes in acidity of mouse tumor by hyperthermia. Int J Radiat Oncol Biol Phys 10: 393–399

Robinson JE, McCulloch D, McCready WA (1982) Blood perfusion of murine tumors at normal and hyperthermal temperatures. Natl Cancer Inst Monogr 61: 211–215

Ryu KHL, Song CW, Kang MS, Levitt SH (1982) Changes in lactic acid content in tumors by hyperthermia. Radiat Res 91: 319–320

Scheid B (1961) Funktionelle Besonderheiten der Mikrozirkulation im Karzinom. Bibl Anat 1: 327–335

Schmid-Schönbein H, Singh M, Malotta H, Leschke D, Teitel P, Driessen G, Scheidt-Bleichert H (1984) Subpopulations of rigid red cells in hyperthermia and acidosis: effect on filtrability in vitro and on nutritive capillary perfusion in the mesenteric microcirculation. Int J Microcirc Clin Exp 3: 497

Shrivastav S, Kaelin WG Jr, Joines WT, Jirtle RL (1983) Microwave hyperthermia and its effect on tumor blood flow in rats. Cancer Res 43: 4665–4669

Song CW (1978) Effect of hyperthermia on vascular functions of normal tissues and experimental tumors: brief communication. JNCI 60: 711–713

Song CW, Kang MS, Rhee JG, Levitt SH (1980) The effect of hyperthermia on vascular function, pH, and cell survival. Radiology 137: 795–803

Song CW, Lokshina A, Rhee JG, Patten M, Levitt SH (1984) Implication of blood flow in hyperthermic treatment of tumours. IEEE Trans Biomed Eng 31: 9–16

Stewart F, Begg A (1983) Blood flow changes in transplanted mouse tumours and skin after mild hyperthermia. Br J Radiol 56: 477–482

Storm FK, Harrison WH, Elliott RS, Morton RL (1979) Normal tissue and solid tumor effects of hyperthermia in animal models and clinical trials. Cancer Res 39: 2245–2251

Streffer C (1985) Metabolic changes during and after hyperthermia. Int J Hyperthermia 1: 305–319

Streffer C, Hengstebeck S, Tamulevicius P (1981) Metabolic studies in an experimental tumor and liver of mice after hyperthermia and glucose. Strahlentherapie 157: 623

Sutton CH (1976) Necrosis and altered blood flow produced by microwave-induced tumor hyperthermia in a murine glioma. Proc Annu Meet Am Soc Clin Oncol 17: 63

Tanaka Y, Hasegawa T, Murata T (1984) Effect of irradiation and hyperthermia on vascular function in normal and tumor tissue. In: Overgaard J (ed) Hyperthermic oncology 1984. Taylor and Francis, London, pp 145–148

Taylor I, Bennet R, Sheriff S (1979) The blood supply of colorectal liver metastases. Br J Cancer 39: 749–756

Thistlethwaite AJ, Leeper DB, Moylan DJ III, Nerlinger RE (1985) pH distribution in human tumors. Int J Radiat Oncol Biol Phys 11: 1647–1652

Touloukian RJ, Rickert RR, Lange RC, Spencer RP (1971) The microvascular circulation of lymphangiomas: a study of Xe^{133} clearance and pathology. Pediatrics 48: 36–40

Van den Berg-Blok AE, Reinhold HS (1984) Time-temperature relationship for hyperthermia-induced stoppage of the microcirculation in tumors. Int J Radiat Oncol Biol Phys 10: 737–740

Van der Zee J, van Rhoon GC, Wike-Hooley JL, Faithfull NS, Reinhold HS (1983) Whole-body hyperthermia in cancer therapy: a report of a phase I-II study. Eur J Cancer Clin Oncol 19: 1189–1200

Vaupel P (1982) Einfluß einer lokalisierten Mikrowellen-Hyperthermie auf die pH-Verteilung in bösartigen Tumoren. Strahlentherapie 158: 168–173

Vaupel P, Ostheimer K, Müller-Klieser W (1980) Circulatory and metabolic responses of malignant tumors during localized hyperthermia. J Cancer Res Clin Oncol 98: 15–29

Vaupel P, Müller-Klieser W, Otte J, Manz R, Kallinowski F (1983) Blood flow, tissue oxygenation, and pH distribution in malignant tumors upon localized hyperthermia. Basic pathophysiological aspects and the role of various thermal doses. Strahlentherapie 159: 73–81

Von Ardenne M, Krüger W (1980) The use of hyperthermia within the frame of cancer multistep therapy. Ann NY Acad Sci 335: 356–361

Voorhees WD III, Babbs CF (1982) Hydralazine-enhanced selective heating of transmissible veneral tumor implants in dogs. Eur J Cancer Clin Oncol 18: 1027–1033

Waterman FM, Nerlinger RE, Moylan DJ III, Leeper D (1987) Response of human tumor blood flow to local hyperthermia. Int J Radiat Oncol Biol Phys 13: 75–82

Wike-Hooley JL, Haveman J, Reinhold HS (1984) The relevance of tumour pH to the treatment of malignant disease. Radiother Oncol 2: 343–366

Hyperthermia-Induced Changes in Tumor Microcirculation

B. Endrich[1,3], F. Hammersen[2], and K. Messmer[3]

[1] Abteilung für Chirurgie, Kreiskrankenhaus, Alte Waibstadter Straße 2, 6920 Sinsheim, FRG
[2] Institut für Anatomie, Technische Universität, 8000 München, FRG
[3] Abteilung für Experimentelle Chirurgie, Universität Heidelberg, 6900 Heidelberg, FRG

Introduction

A biological rationale for treating malignant tumors with hyperthermia has been provided by a great number of studies in recent years. It became obvious that particularly nutritionally deprived, hypoxic and acidic tumor cells were very sensitive to heat (for a review see Dickson and Calderwood 1980; Overgaard and Bichel 1977; Overgaard 1981; Rhee et al. 1984; Song et al. 1980a; Streffer 1985; Vaupel et al. 1983). This effect was not limited to tumor cells alone, but the deleterious effect of hyperthermia on microscopic blood channels suggested an inverse relationship between blood flow and thermal sensitivity (Dewhirst et al. 1984; Emami and Song 1984; Gullino et al. 1982; Pence and Song 1986; Reinhold and Endrich 1986; Song 1980b). Up to the present, however, only a few studies have evaluated directly the influence of heat on tumor capillaries.

Such a study seems to be needed because the hastily arranged tumor microvasculature, with capillaries consisting only of a single-layered lining of endothelial cells without a basal lamina, is likely to react differently to hyperthermia (Dudar and Jain 1984). Moreover, since even tumor cells are found to complete the vascular lining (for a review see Hammersen et al. 1983, 1985), the dissipation of heat via the microvascular system associated with possible alterations of microhemodynamic function will significantly affect any therapeutic response. When such small vascular channels are severely stressed by hyperthermia, a sudden leakage of plasma at damaged sites might result in a relative increase of the hematocrit within tumor capillaries and thus render the blood flow sluggish. Rupture of the vessel wall would entail petechial hemorrhage and further aggravate stasis in microscopic tumor vessels.

A thorough understanding of this sequence of events, as well as the underlying mechanisms, can be achieved only by means of direct microscopic analysis, because only this technique will allow *measurements at the level of single tumor capillaries*. Consequently, intravital microscopy and quantitative video image analysis were utilized in conjunction with a transparent chamber technique to carry out detailed measurements of microvascular flow in the amelanotic melanoma A-Mel-3 of the hamster. These results were combined with data of local tissue oxygen tension to provide a detailed survey of peripheral vascular hemodynamics associated with local hyperthermia.

Recent Results in Cancer Research, Vol. 107
© Springer-Verlag Berlin · Heidelberg 1988

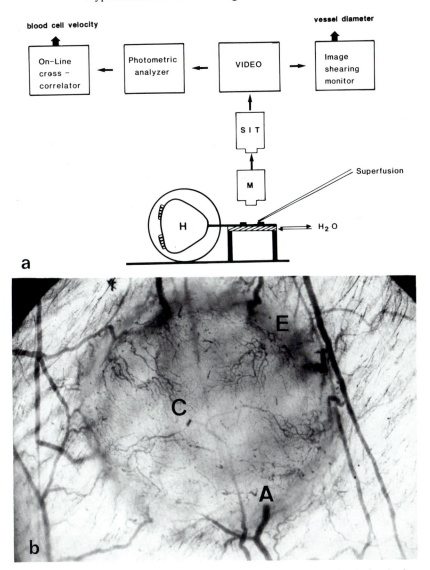

Fig. 1. a Graphical illustration of the experimental model to study a tumor microcirculation by intravital microscopy. The conscious animal *(H)* is kept in a plastic tube with the chamber on a microscopic stage that is perfused by H_2O. At the same time, the preparation is irrigated *(Superfusion)*. Selected regions of the microcirculation within the chamber are observed through a microscope *(M)* and recorded by a video camera *(SIT)*. This information is stored on video tape *(VIDEO)*. From the video tape, the vessel diameter can be measured using an image shearing monitor (Intaglietta and Tompkins 1973). The blood cell velocity can be analyzed by a photometric analyzer and on-line cross-correlation (Intaglietta et al. 1975).   **b** A close-up picture of the amelanotic melanoma 5 days after implantation into a transparent skin fold chamber. This picture reflects the typical chaotic tumor microvasculature. Note the abrupt diameter changes in tumor capillaries *in the center* of the photograph *(C)*; *in the lower part* of the picture, an artery seems to be occluded *(A)*. In the zone of tumor advance, there is quite considerable tissue edema *(E)*. Magnification approximately × 40

Methods and Material

The techniques for intravital microscopy and microhemodynamic analysis in tumors have been reported previously (Fig. 1) (Endrich et al. 1980, 1982 a). A transparent double-frame-type aluminum chamber was implanted in the dorsal skin fold of 28 Syrian golden hamsters during pentobarbital anesthesia (50–70 mg/kg i. p.). Permanent indwelling catheters were inserted into the right superior vena cava and the ascending aorta via the right jugular vein and carotid artery respectively. After a recovery period of

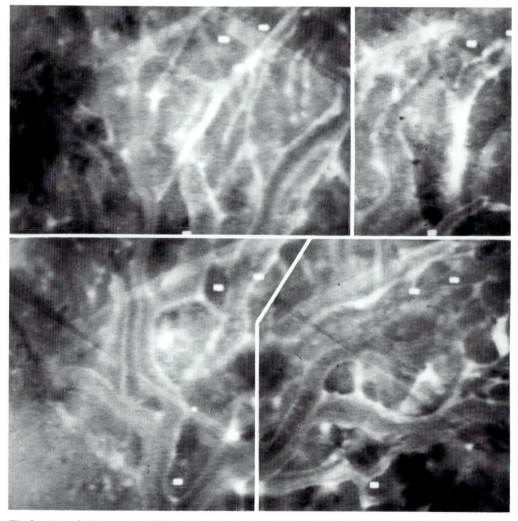

Fig. 2. A typical example of the microcirculation of the amelanotic melanoma A-Mel-3 of the hamster (from the animal H 12-9-80 G₂) after visualizing the small vessels with FITC dextran. A small region at the tumor's edge is shown in a photomontage; such an arrangement of pictures was necessary for the determination of capillary density. With fluorescence microscopy, it was seen that the dye leaked into the perivascular tissue after it had brightened the vascular wall for quite some time. The *white dots* represent the photometric windows that were purposely placed on each picture to facilitate orientation. Magnification approximately × 200

Table 1. The microcirculation of the amelanotic melanoma A-Mel-3 during hyperthermia: experimental techniques of the study

Parameter	Technique	Reference
Local PO_2	Platinum multiwire electrode	Kessler et al. (1976)
Blood cell velocity	Video densitometry	Intaglietta et al. (1975)
Vessel diameter	Video densitometry	Baez (1966), Intaglietta and Tompkins (1973)
Microvascular pressure	Direct puncture and servo-nulling method	Wiederhielm et al. (1964), Intaglietta et al. (1970)
Capillary density	Morphometry	Schmid-Schönbein et al. (1977)

48 h, 4×10^4 cells of the amelanotic melanoma A-Mel-3 were implanted into the subcutaneous tissue within the chamber. Any measurement was done in the conscious but immobilized animal as described earlier (Endrich et al. 1980, 1982a, b).

Representative areas of the microvasculature were examined during the whole course of each experiment. We identified a region to be studied within the chamber on the basis of tissue transparency and visibility of large blood vessels (20–40 μm) and visualized the capillary network utilizing a fluorescent marker. The injection of FITC (fluorescein isothiocyanate) dextran improved the contrast between the wall of tumor capillaries, the adjoining tissue, and the blood cells, which were delineated as dark structures against a bright yellow background of plasma (Fig. 2). The intraluminal vessel diameter served as criterion for the classification of blood vessel segments. The terminal vascular bed was categorized according to Zweifach (1974), using tumor capillaries as reference structure and dividing the arterial and venous segments in two sections each.

Table 1 summarizes the measurements performed during this study; it also provides appropriate references for a detailed description of the techniques used to quantify changes in the tumor microcirculation.

Experimental Protocol

After a control period of 30 min at 30° C, the tumor temperature was increased within 10 min to 35° C and kept constant for 15 min. Heating was accomplished by perfusing a heat exchanger beneath the chamber with distilled water and isopropyl alcohol (70%). To obtain an accurate reading of the tumor temperature, the cover glass was removed prior to each experiment. Thereafter, the entire tissue preparation was irrigated with the temperature of the superfusate precisely adjusted to the level given by the protocol. Further temperature increments included measurements 15 min after tumor temperatures of 38, 40, and 42.5° C were established (for further technical information see Endrich and Hammersen 1986).

Preparation of the Melanoma for Electron Microscopy

To evaluate the structural effects of hyperthermia on tumor vessels, skin fold chambers of the hamsters bearing the amelanotic melanoma A-Mel-3 were exposed to hyperthermia (42.5° C) for 10 and 40 min and fixed immediately thereafter by intraarterial perfusion. For this purpose, a 2.5% glutaraldehyde solution in 0.1 M sodium cacodylate buffer (pH 7.4) was used. Ultrathin sections of selected regions within the tumor and in tissue surrounding the melanoma were cut with diamond knives on an LKB III ultra-tome and viewed with a Zeiss EM-10 A electron microscope operated at 60 kV (Ham-mersen et al. 1985).

Results

Hyperthermia-Induced Changes in Tissue Oxygenation

Figure 3 summarizes the data of local tissue PO_2 in chamber preparations void of tumor tissue (control) and on the surface of the amelanotic melanoma A-Mel-3. Already at a local temperature of 30° C, significant (Kolmogorow-Smirnow test) differences between control preparations and the melanoma were found. While the sum histogram of the control revealed a homogeneous distribution of local PO_2 values (mean value 19.0 mmHg), severe tissue hypoxia was evident on the surface of the A-Mel-3. The mean value was only 8.8 mmHg and 32% of all values were found within the hypoxic range between 0 and 5 mmHg.

As the local temperature was increased to 35° C, tissue oxygenation improved in both the tumor and in chambers void of melanoma tissue. This improvement was statis-tically significant; in the hamster A-Mel-3, however, 18% of all values were still within the hypoxic range.

Tissue hypoxia or even anoxia became quite evident 15 min after a temperature of 42.5° C was reached in the tumor. At that time, 52% of all values were found between 0 and 5 mmHg, with the mean value down to only 6.6 mmHg.

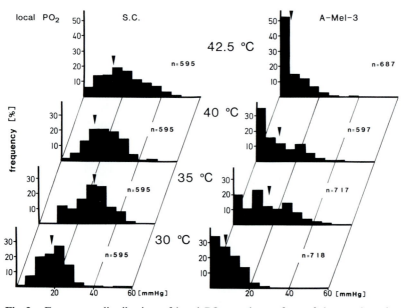

Fig. 3. Frequency distribution of local PO_2 on the surface of the amelanotic melanoma A-Mel-3 and in chamber preparations without tumor tissue *(S. C.)* 15 min after reaching different tempera-ture values. *n*, number of single determinations from five animals in each group; ▼, arithmetic mean

There appeared to be a difference in "thermal sensitivity." As an addition to the sum histograms obtained, original recordings with a stepwise increase in temperature were compared for subcutaneous tissue and the melanoma. In the normal microcirculation, a decrease in local PO_2 associated with a deterioration of microvascular function was observed at a temperature of approximately 44° C. By contrast, a sudden decrease was seen in the melanoma at a tumor temperature of only 40.6° C.

Microvascular Flow upon Local Hyperthermia

Figure 4 illustrates the changes of blood cell velocity in tumor capillaries. It should be noted that we did not observe significant changes of the mean values as long as the temperature did not exceed 40° C. However, as the temperature was slowly increased in the melanoma, there was a remarkable increase of blood cell velocity values in the range usually referred to as "prestasis" (0.05–0.2 mm/s) (Fig. 4). This portion of tumor capillaries was approximately 50% at 38° C and reached 60% at a temperature of 40° C.

When temperature was kept constant at 42.5° C for at least 15 min in the melanoma, there was a pronounced reduction of blood flow. The mean value was as low as

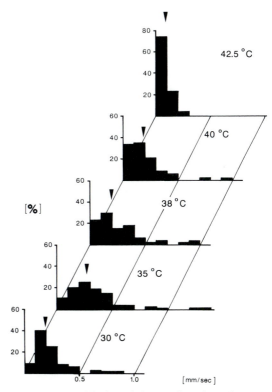

Fig. 4. Frequency distribution of blood cell velocities in capillaries of the amelanotic melanoma A-Mel-3 of the hamster. The tumor was exposed to various temperatures for 15 min each; the measurements were obtained off line from a video tape recording. The histograms are based on 112 single measurements from 12 animals in which the tumor was exposed to graduate hyperthermia. ▼, mean value

Table 2. Pressure measurements (mmHg) in the microcirculation of the amelanotic melanoma A-Mel-3 as implanted in a hamster dorsal skin fold chamber and exposed to local hyperthermia

Vessel segment	30° C	35° C	42.5° C
Arterioles (n = 8)	37.9 ± 1.5	37.5 ± 1.5	34.8 ± 2.0
Precapillaries (n = 8)	36.8 ± 1.5	33.1 ± 1.5	30.3 ± 2.2
Capillaries (n = 20)	24.7 ± 1.4	21.1 ± 1.5	26.9 ± 1.2
Postcapillaries (n = 14)	12.8 ± 1.2	10.8 ± 1.1	20.6 ± 1.9
Collecting venules (n = 11)	11.9 ± 1.1	7.4 ± 0.6	16.9 ± 2.0

Mean value ± SEM; data derived from 12 animals; n, number of single determinations. The statistical evaluation was done using analysis of variance and the H-test of Kruskal and Wallis. The increase of postcapillary and venular pressure at 42.5° C is statistically significant ($p < 0.05$).

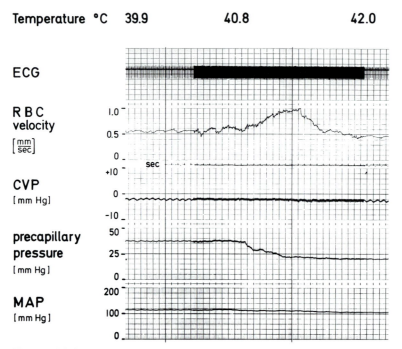

Fig. 5. Original recording of ECG, red blood cell *(RBC)* velocity in a tumor capillary (mm/sec) (fed by the precapillary punctured), central venous pressure *(CVP)*, mean arterial pressure *(MAP)*, and the pressure in a 14-μm precapillary during a slow increase of tumor temperature. The vessel punctured was located just at the edge of the amelanotic melanoma A-Mel-3. The body temperature of the animal was enhanced at the end of the experiment but the difference was not statistically significant. The drop in precapillary pressure observed at a tumor temperature of 41° C resulted from a slight arteriolar constriction observed further upstream

0.08 mm/s and more than 50% of the previously perfused tumor capillaries were at a complete standstill. As a result, the capillary density was reduced by 50%.

Table 2 summarizes our data on microvascular pressure derived from 12 animals. Moreover, figure 5 illustrates an original recording of precapillary pressure during hyperthermia. At a tumor temperature of 30° C, the mean arteriolar pressure was 37.9 mmHg. Microvascular pressure in inflow vessels did not change at 35° C, but was remarkably lower 15 min after a temperature of 42.5° C was reached in the melanoma. By contrast, in outflow vessels we noted a slight decrease at moderate hyperthermia but a significant increase at 42.5° C. Consequently, the arteriolar-venular pressure gradient was much lower under extreme hyperthermia.

Based on measurements of the mean arterial pressure and local microvascular flow and pressure, the vascular resistance throughout the tumor microcirculation could be calculated. Figure 6 demonstrates a pronounced increase of precapillary and to a lower extent of postcapillary resistance of the melanoma.

Most notable was the almost complete arteriolar constriction which occurred 10–30 min after a temperature of 42.5° C was reached in the melanoma. At the same time, venules and small veins dilated at the edge of the tumor. The constriction of all vessels feeding the tumor was seen despite pronounced tumor hypoxia (Fig. 7); it persisted for at least 72 h.

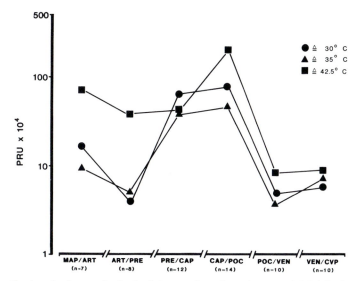

Fig. 6. Segmental resistance in the terminal vascular bed of the amelanotic melanoma A-Mel-3 of the hamster. This parameter was calculated from the systemic pressure as well as the local microvascular pressure and blood flow. *MAP,* mean arterial pressure; *ART,* arterioles; *PRE,* precapillaries; *CAP,* capillaries; *POC,* postcapillaries; *VEN,* collecting venules; *CVP,* central venous pressure; *n,* number of single determinations for each microvascular segment. Note that the change in resistance upon hyperthermia is most pronounced in arterioles/precapillaries and in postcapillaries

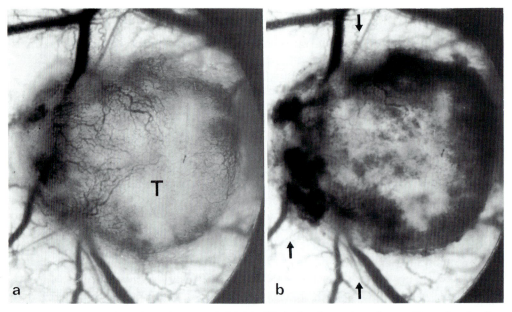

Fig. 7 a, b. The amelanotic melanoma A-Mel-3 *(T)* during hyperthermia. **a** Control; **b** after 30 min hyperthermia at 42.5° C. There is petechial bleeding at the edge of the tumor and capillaries are barely visible at hyperthermic temperatures. All arterioles feeding the tumor seem to have decreased their diameter *(arrows)* while the adjacent draining veins showed a tendency to dilate. Magnification approximately × 30

Ultrastructural Changes During Hyperthermia

Upon application of heat for 10 min, tumor capillaries did not show discontinuities to a significantly greater extent than seen in untreated controls. There was, however, considerable red cell aggregation with the endothelium having lost already most of its usual organelles. The mitochondria revealed focal lightenings; on occasion they were completely translucent, residing as irregularly outlined vacuoles within the cytoplasm. A great number of vacuoles were also seen in adjacent tumor cells; these changes were due to rarefaction of mitochondrial cristae and swelling of the organelle (Fig. 8 a).

Following 40 min of hyperthermia (42.5° C), destructive changes of the striated fibers of the cutaneous muscle coat were observed (Fig. 8 b). The mitochondria showed considerable swellings together with rarefaction of their cristae; myofibrils appeared splintered and broken into small fragments. Capillaries close to or between these muscle fibers were mostly stuffed with red blood cells. Their endothelium, however, maintained its continuity despite some damage.

Fig. 8. **a** Cross-sectioned tumor blood vessel (amelanotic melanoma) after 10 min exposure to hy- ▷ perthermia (42.5° C). The endothelium is still continuous; it lacks, however, the normal electron opacity of its cytomembranes. Note swollen and translucent mitochondria (➤), × 6000. **b** Small vessel adjacent to a muscle fiber *(MF)* of the cutaneous muscle layer from a tumor-bearing skin fold chamber exposed to hyperthermia (42.5° C) for 40 min. Note disrupted fibrils together with severely damaged mitochondria in myofiber. The endothelium has retained its continuity despite some structural deterioration. × 3800

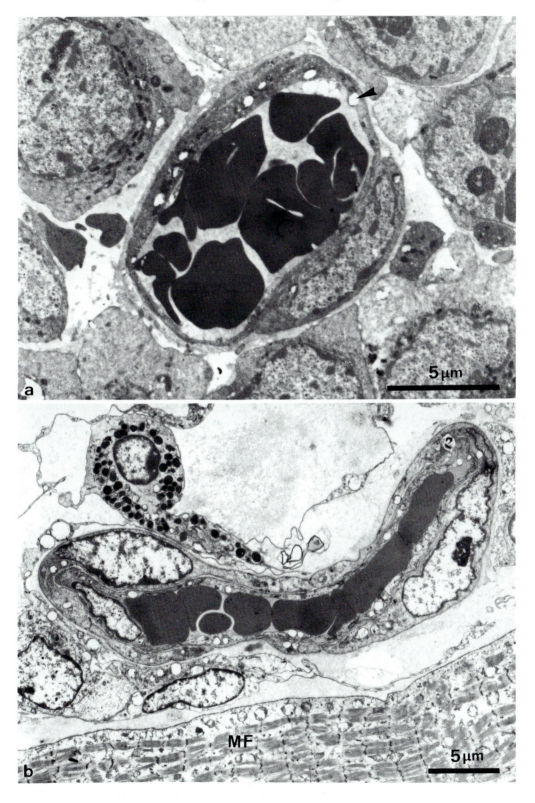

MF

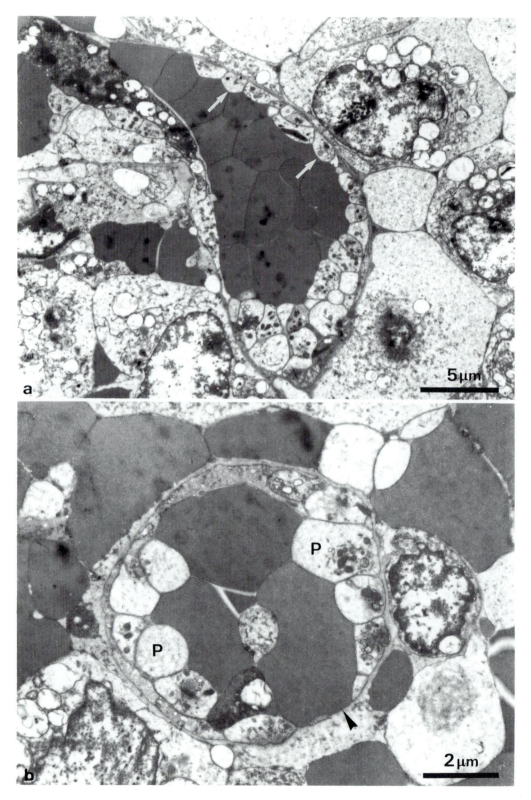

a   5 μm

b   2 μm

Table 3. Time-dependent alterations of the ultrastructure of tumor capillaries after a 42.5° C hyperthermia

Structural features	42.5° C	
	10 min	40 min
Continuity of the endothelium	+	−
Swelling of the endothelium	−	−
Extravasation of erythrocytes	−	+
Damage within the endothelium	+	+ +
Aggregation of blood cells	+	+ + +
Tumor cells within the endothelium	+	+
Formation of a pseudoendothelium	−	+

By contrast, the tumor capillaries looked entirely different. They were also packed with erythrocytes whose aggregates were often surrounded by orderly aligned platelets (Fig. 9a). Obviously this served to substitute the endothelial lining, because we could not identify vascular endothelium in a true sense. The cytoplasmic profiles bordering the red cell aggregates may represent remnants of virtual endothelial cells. However, there is no reliable proof for this assumption, since they displayed a remarkable similarity to adjoining tumor cells. Finally, the vessels disintegrate completely, leaving their contents as an indicator for their former location (Fig. 9b). It should be noted that pronounced ultrastructural alterations on tumor vessels were seen only *after hyperthermia-induced microhemodynamic changes* became evident. We have summarized the time-dependent differences of morphologic changes in tumor capillaries in table 3.

Discussion

Tissue oxygenation is one of the determinants of heat sensitivity; changes of tumor oxygenation appear to be primarily caused through alterations in tumor blood flow, because (1) identical variations of tumor blood flow and oxygenation were observed in all tumors studied, and (2) a similar correlation was obtained when only mean tissue PO_2 was measured during temperature elevation.

The improvement of tumor oxygenation under moderate hyperthermia can be explained (a) by a change in the amount of O_2 made available to tumor cells and (b) by alteration in the oxygen consumption of tumor cells.

The enhanced O_2 availability is due to the increase in nutritive blood flow and, possibly to a much smaller extent, to the improved oxygen release because of a shift to the right of the O_2 dissociation curve after moderate elevation of tumor temperature. It is now well established that the enhanced oxygen availability is predominant, while meta-

◁ **Fig. 9.** **a** Small vessel completely stuffed with red blood cells within the center of a tumor exposed to hyperthermia (42.5° C for 40 min). Over longer distances, the endothelium is replaced by a string of closely adjacent platelets (⟶) which in general are not seen in untreated tumors. × 4000. **b** Vessel similar to that shown in **a** in cross-section, displaying an endothelial defect (➤) with wedged-in parts of a red blood cell. Platelets *(P)* are either closely attached to an intact endothelium *(left)* or replace the latter *(right)*. Note numerous extravasated erythrocytes. × 10000

bolic effects, such as the ability of cells to consume oxygen, have only a minor influence on improved tissue oxygenation (Gullino et al. 1982; Vaupel et al. 1980; Vaupel 1982).

As a consequence, the function of the tumor microcirculation is of paramount importance not only for the nourishment of malignant cells, but in particular for homogeneous heat dissipation within the entire tumor. However, even in untreated experimental tumors, microvascular flow varies considerably from one region to another (Endrich et al. 1979a), resulting in nonuniform fluid and heat transfer (Gullino et al. 1982; Müller-Klieser and Vaupel 1984). Moreover, it was shown in great detail that the ultrastructure of tumor microvessels is different compared to vessels in nonmalignant tissue. Therefore, it is likely that malignant lesions have a greater sensitivity to heat particularly because of specific features of the microvascular network.

In fact, a difference in heat sensitivity has first been demonstrated by Scheid, who reported in 1961 no change of blood flow in the mesentery of mice at a temperature of 42° C. By contrast, in an ascites carcinoma implanted in the mesentery, vascular stasis was seen after approximately 35 min. Subsequently, a great amount of data confirmed that the microvascular system of tumors is more sensitive to heat than is a normal microcirculation. In a series of experiments, Reinhold et al. (1978) demonstrated in rats a slowly progressing vascular insufficiency in the BA 1112 sarcoma (for further reviews see Berg-Blok and Reinhold 1984; Reinhold and Endrich 1986; Wike-Hooley et al. 1984). Using the same tumor, we reported in 1979 the first quantitative study of capillary blood flow after local hyperthermia. As observed by means of intravital microscopy, leukocytes were adhering to the endothelium of "tumorous" blood channels in increasing numbers, so that the capillary circulation became languid. Later, stasis was seen (Eddy 1980; Emami et al. 1980; Emami and Song 1984; Endrich et al. 1979b), occurring in the melanoma primarily at microvascular bifurcations (Endrich and Hammersen 1986).

The flow inhibition associated with a tumor temperature above 42° C seems to be a consistent finding. Moreover, there is strong evidence that the tumor microenvironment will affect the heat sensitivity (Overgaard and Bichel 1977; Overgaard and Nielsen 1980; Vaupel 1982; Vaupel et al. 1980, 1983). Tissue pH and PO_2 will correlate with the nutritive blood flow in tumor capillaries, thus changes detected at the capillary level serve as an very early indicator of the effect(s) of heat in tumors.

Similar to other tumors, the microcirculation of the amelanotic melanoma was found to react differently than that of a "normal tissue" preparation. With pronounced tissue hypoxia prior to heating, the following changes were consistently seen in the amelanotic melanoma A-Mel-3 of the hamster:

1. Moderate hyperthermia resulted in a significant improvement of tumor oxygenation only at a tumor temperature of 35° C, but we did not observe a significant overall improvement of nutritive flow in tumor capillaries.
2. A rise in temperature to 42.5° C was followed by a quick reduction of blood flow in capillaries, postcapillaries and collecting venules. Moreover, we have measured a considerable increase of intravascular pressure in capillaries and outflow channels and have observed the aggravation of tissue/tumor hypoxia. Finally, despite tissue hypoxia, arteriolar constriction was seen to persist for 72 h.

This consistent microscopic observation seems to disprove the hypothesis given earlier on the mechanism of microvascular obstruction (for a review see von Ardenne 1986). Arteriolar constriction could occur if vascular smooth muscle is stimulated during heat application and contracts as a response to changes in intravascular pressure, in much as the same manner as any stretching of other smooth muscle cells leads to their contraction (Johnson 1980). It is important to note that feeding vessels of a tumor originate mostly from "host vessels"; thus, they will be encircled by one or more layers of smooth muscle cells. These vessels, however, do not show a spontaneous rhythmic contraction like it occurs in normal arterioles, but they are instead maximally dilated (Endrich et al. 1979b, 1982b). A permanent maximum dilation of arterioles causes the microvascular resistance to be substantially higher (Funk et al. 1983). Furthermore, the network originating from such arterioles has to be considered as an arrangement of rigid tubes in which flow is primarily a function of the arteriolar-venular pressure gradient and changes in blood viscosity. These factors will be the only limiting parameters of tissue perfusion during hyperthermia; since it is not possible to measure microvascular pressure or local viscosity in a clinical setup, the knowledge of the *actual temperature in the tumor and of the effect this temperature has is of great importance.* Moreover, it should be noted that even under only moderate hyperthermia, there is an increase of tumor capillaries with prestatic flow in the amelanotic melanoma A-Mel-3.

Conclusions

From these new data, we draw the following conclusions:

1. If hyperthermia is used as an adjuvant measure to increase tumor perfusion, the temperature should not exceed 40° C. Furthermore, one should take into account that the number of tumor capillaries with prestatic flow could already be enhanced at a temperature of 38° C; thus, specific methods might be added to lower blood viscosity in order to prevent the development of prestatic flow conditions observed under moderate hyperthermia alone. We found systemic and isovolemic hemodilution to induce a considerable reduction of the number of tumor capillaries with prestatic flow (Fig. 10) (Oda et al. 1984).

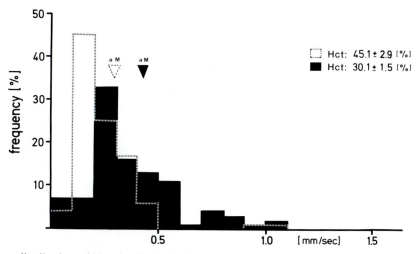

Fig. 10. Frequency distribution of blood cell velocity in tumor capillaries as measured in a preselected field of microscopic observation before and after isovolemic hemodilution (dextran 60). These histograms are derived from 89 measurements in 14 animals in a separate set of experiments. *Hct,* systemic hematocrit, mean value ± SEM; *aM,* mean value

2. If hyperthermia is used as a "cytotoxic measure" and the temperature is increased to approximately 42° C, there will be a quick, pronounced and longer-lasting deterioration of microvascular flow. This dysfunction could contribute significantly to tumor hypoxia and tissue acidosis. Since hyperthermia is used in conjunction with other modalities, such as radiation, the almost total obstruction of nutritive tumor capillaries will not permit any access to the tumor via the vascular system until these channels are completely rebuilt. Neovascularization was observed to begin in our model approximately 3 days after local hyperthermia.

References

Baez S (1966) Recording of microvascular dimensions with an image-splitter television microscope. J Appl Physiol 21: 299–301

Berg-Blok AE, Reinhold HS (1984) Time-temperature relationship for hyperthermia induced stoppage of the microcirculation in tumors. Int J Radiat Oncol Biol Phys 10: 737–740

Dewhirst M, Gross JF, Sim D, Arnold P, Boyer D (1984) The effect of rate of heating or cooling prior to heating on tumor and normal tissue microcirculatory flow. Biorheology 21: 539–558

Dickson JA, Calderwood SK (1980) Temperature range and selective sensitivity of tumors to hyperthermia: a critical review. Ann NY Acad Sci 335: 180–205

Dudar TE, Jain RK (1984) Differential response of normal and tumor microcirculation to hyperthermia. Cancer Res 44: 605–612

Eddy HA (1980) Alterations in tumor microvasculature during hyperthermia. Radiology 137: 515–521

Emami B, Song CW (1984) Physiological mechanisms in hyperthermia: a review. Int J Radiat Oncol Biol Phys 10: 289–295

Emami B, Nussbaum GH, TenHaken RK, Hughes WL (1980) Physiological effects of hyperthermia: response of capillary blood flow and structure to local tumor heating. Radiology 137: 805–809

Endrich B, Hammersen F (1986) Morphologic and hemodynamic alterations in capillaries during hyperthermia. In: Anghileri LJ, Robert J (eds) Hyperthermia in cancer treatment, vol 2. CRC, Boca Raton, pp 17–47

Endrich B, Reinhold HS, Gross JF, Intaglietta M (1979a) Tissue perfusion inhomogeneity during early tumor growth in rats. JNCI 62: 387–395

Endrich B, Zweifach BW, Reinhold HS, Intaglietta M (1979b) Quantitative studies of microcirculatory function in malignant tissue: influence of temperature on microvascular hemodynamics during the early growth of the BA 1112 rat sarcoma. Int J Radiat Oncol Biol Phys 5: 2021–2030

Endrich B, Asaishi K, Götz A, Meßmer K (1980) Technical report. A new chamber technique for microvascular studies in unanesthetized hamsters. Res Exp Med 177: 125–134

Endrich B, Götz A, Meßmer K (1982a) Distribution of microflow and oxygen tension in hamster melanoma. Int J Microcirc Clin Exp 1: 81–99

Endrich B, Hammersen F, Götz A, Meßmer K (1982b) Microcirculatory blood flow, capillary morphology and local oxygen pressure of the hamster amelanotic melanoma A-Mel-3. JNCI 68: 475–485

Funk W, Endrich B, Meßmer K, Intaglietta M (1983) Spontaneous arteriolar vasomotion as a determinant of peripheral vascular resistance. Int J Microcirc Clin Exp 2: 11–25

Gullino PM, Jain RK, Grantham FH (1982) Relationship between temperature and blood supply or consumption of oxygen and glucose by rat mammary carcinomas. JNCI 60: 519–533

Hammersen F, Osterkamp-Baust U, Endrich B (1983) Ein Beitrag zum Feinbau terminaler Strombahnen und ihrer Entstehung in bösartigen Tumoren. Prog Appl Microcirc 2: 15–51

Hammersen F, Endrich B, Meßmer K (1985) The fine structure of tumor blood vessels. I. Participation of non-endothelial cells in tumor angiogenesis. Int J Microcirc Clin Exp 4: 31–43

Intaglietta M, Tompkins WR (1973) Microvascular measurements by video image shearing and splitting. Microvasc Res 5: 309–312

Intaglietta M, Pawula RF, Tompkins WR (1970) Pressure measurements in the mammalian micro-vasculature. Microvasc Res 2: 212–220

Intaglietta M, Silverman NR, Tompkins WR (1975) Capillary flow velocity measurements in vivo and in situ by television method. Microvasc Res 10: 165–179

Johnson P (1980) The myogenic response. In: Bohr DF, Somlyo AP, Sparks HV (eds) Handbook of physiology, vol 2, sect 2. American Physiological Society, Bethesda, pp 409–442

Kessler M, Höper J, Krumme BA (1976) Monitoring of tissue perfusion and cellular function. Anesthesiology 45: 184–197

Müller-Klieser W, Vaupel P (1984) Effect of hyperthermia on tumor blood flow. Biorheology 21: 529–538

Oda T, Lehmann A, Endrich B (1984) Capillary blood flow in the amelanotic melanoma of the hamster after isovolemic hemodilution. Biorheology 21: 509–520

Overgaard J (1981) Effect of hyperthermia on the hypoxic fraction in an experimental mammary carcinoma in vivo. Br J Radiol 54: 245–249

Overgaard J, Bichel P (1977) The influence of hypoxia and acidity on the hyperthermic response of malignant cells in vitro. Radiology 123: 511–514

Overgaard J, Nielsen OS (1980) The role of tissue environmental factors on the kinetics and morphology of tumor cells exposed to hyperthermia. Ann NY Acad Sci 335: 254–278

Pence DW, Song CW (1986) Effects of heat on blood flow. In: Anghileri LJ, Robert J (eds) Hyperthermia in cancer treatment, vol 2. CRC, Boca Raton, pp 1–16

Reinhold HS, Endrich B (1986) Tumour microcirculation as a target for hyperthermia. Int J Hyperthermia 2: 111–137

Reinhold HS, Blachiewicz B, Berg-Blok A (1978) Decrease in tumor microcirculation during hyperthermia. In: Streffer C (ed) Cancer therapy by hyperthermia and radiation. Urban and Schwarzenberg, Munich, pp 231–232

Rhee JG, Kim TH, Levitt SH, Song CW (1984) Changes in acidity of mouse tumor by hyperthermia. Int J Radiat Oncol Biol Phys 10: 393–399

Scheid P (1961) Funktionelle Besonderheiten der Mikrozirkulation im Karzinom. Bibl Anat 1: 327–335

Schmid-Schönbein GW, Zweifach BW, Kovalchek S (1977) The application of stereological principles to morphometry of the microcirculation in different tissues. Microvasc Res 14: 303–317

Song CW, Kang MS, Rhee JG, Levitt SH (1980a) Effect of hyperthermia on vascular function in normal and neoplastic tissues. Ann NY Acad Sci 335: 35–47

Song CW, Kang MS, Rhee JG, Levitt SH (1980b) The effect of hyperthermia on vascular function, pH and cell survival. Radiology 137: 795–803

Streffer C (1985) Metabolic changes during and after hyperthermia. Int J Hyperthermia 1: 305–319

Vaupel P (1982) Einfluß einer lokalisierten Mikrowellenhyperthermie auf die pH-Verteilung in bösartigen Tumoren. Strahlentherapie 158: 168–173

Vaupel P, Ostheimer K, Müller-Klieser W (1980) Circulatory and metabolic responses of malignant tumors during normothermia and hyperthermia. J Cancer Res Clin Oncol 98: 15–29

Vaupel P, Müller-Klieser W, Otte J, Manz R, Kallinowski F (1983) Blood flow, tissue oxygenation, and pH distribution in malignant tumors upon localized hyperthermia. Basic pathophysiological aspects and the role of various thermal doses. Strahlentherapie 159: 73–81

Von Ardenne M (1986) The present developmental state of cancer multistep therapy (CMT): Selective occlusion of cancer tissue capillaries by combining hyperglycemia with two stage regional or local hyperthermia using the CMT Selectotherm technique. In: Anghileri LJ, Robert J (eds) Hyperthermia in cancer treatment, vol 3. CRC, Boca Raton, pp 1–24

Wiederhielm CA, Woodbury JW, Kirk ES, Rushmer RF (1964) Pulsatile pressure in the microcirculation of the frog's mesentery. Am J Physiol 207: 173–176

Wike-Hooley JL, Zee J, Rhoon GC, Berg AP, Reinhold HS (1984) Human tumor pH changes following hyperthermia and radiation therapy. Eur J Cancer Clin Oncol 20: 619–623

Zweifach BW (1974) Quantitative analysis of microcirculatory structure and function. I. Analysis of pressure distribution in the terminal vascular bed in cat mesentery. Circ Res 34: 843–857

Changes in Tumor Vasculature Under Fractionated Radiation – Hyperthermia Treatment

F. Zywietz[1] and W. Lierse[2]

[1] Institut für Biophysik und Strahlenbiologie, Universitäts-Krankenhaus Eppendorf, Martinistraße 52, 2000 Hamburg 20, FRG
[2] Institut für Neuroanatomie, Universitäts-Krankenhaus Eppendorf, Martinistraße 52, 2000 Hamburg 20, FRG

Introduction

The advantages of combining radiation and hyperthermia in the treatment of cancer are well documented for experimental and human tumors (Dewhirst et al. 1983; Arcangeli et al. 1983; Lindholm et al. 1985; Overgaard 1985). The role of blood flow in tumors and normal tissues in hyperthermia has been described by Song (1978, 1983), Vaupel et al. (1983), Bicher et al. (1983). Histopathological studies on the effects of hyperthermia on the microvasculature by Eddy 1980 and Emami et al. (1981) have shown that the vascular network changes with temperature and heating time. Aspects of tumor microcirculation as a target for hyperthermia have been described in a review by Reinhold and Endrich (1986). To our knowledge only two papers have reported on the effects of radiation and hyperthermia on vascular function in tumor tissue after single doses of radiation and heat (Reinhold and van den Berg-Blok 1984; Tanaka et al. 1984).

Following our investigation into the response of a murine tumor to fractionated radiation–hyperthermia treatment (Zywietz and Jung 1986), the present study was undertaken to ascertain the histopathological changes of the vascular network during treatment for 5 weeks with fractionated radiation, either alone or in combination with local hyperthermia.

Material and Methods

The studies were performed upon the rhabdomyosarcoma R1H of the rat superficially growing in the right flank of the animals. R1H ist a radioresistant tumor which is isogeneic in origin in the rat strain WAG/Rij. Details of this tumor system have been described elsewhere (Jung et al. 1981). Tumors 22 days after implantation with a volume of 1.8 ± 0.3 cm^3 and a diameter of about 1.5 cm were selected throughout this study. For treatment the rats were anesthetized with 6 mg/kg b.w. Rompun in combination with 50 mg/kg b.w. Ketavet.

Irradiation of the tumors was carried out using a Co-γ-unit (Gammatron, Siemens, FRG) with a dose rate of 1.00 Gy/min and a field size of 3×3 cm^2. Local hyperthermia in the tumors was induced with a temperature-controlled, 2450-MHz microwave heating system (Zywietz et al. 1986). Tumors were heated at 43° C for 60 min using a contact applicator with skin surface cooling.

Recent Results in Cancer Research, Vol. 107
© Springer-Verlag Berlin · Heidelberg 1988

The studies were carried out in trials of two arms. Tumors were randomized to one of the following schedules:

1. Radiotherapy alone: A total 75-Gy tumor dose was given in 25 fractions of 3 Gy 5 times a week for 5 weeks.

2. Radiotherapy and hyperthermia: Radiation treatment was administered as above. Hyperthermia – 43° C, 60 min – was additionally given 2 times each week (Monday, Friday) for 5 weeks. Hyperthermia was started 10 min after the γ-irradiation, based on previous studies of the decay of thermoradiosensitization in this tumor (Jung et al. 1986). The time interval of the two hyperthermia fractions each week is based on dielectric measurements of tumor and skin (Zywietz and Knöchel 1986).

The vascular network of the tumors was prepared after each week of treatment by injecting intra vitam an indian ink solution into the thoracic aorta. The caudal vena cava was dissected so that the animal with the tumor in situ was perfused. Thereafter these vessels were clamped and the whole animal was fixed in a solution of glutaraldehyde. After histological procedures the vascular pattern was determined by transilluminating central slices of the tumor of about 2 mm. The same tissue samples were further processed for histological examination using routine paraffin and Masson-Goldner staining. It must be noted that the figures presented here are not a full representation of either the three-dimensional observations or the histological studies.

Results

Figure 1 shows the changes in R1H tumor vasculature and in tumor tissue during fractionated radiation with ^{60}Co-γ-rays for 5 weeks. The untreated rhabdomyosarcoma R1H (Fig. 1 a) is a solid spherical tumor with circularly arranged blood vessels in the periphery and a radial vascular system in the center. The radially arranged vascular network is feature of the viable intermediate area, while in areas where degeneration and necrosis are developing vessels are reduced in number or are absent. The histological picture of the tumor center shows dividing pleomorphic and widely anaplastic cells. Cell membranes were relatively indistinct. Tumor cells with a high nuclear to cytoplasmic ratio varied in shape from round/oval to fusiform. There were occasional foci with cells in various stages of necrosis. Blood vessels were relatively uniformly distributed and were more concentrated in the periphery of the tumor than in its center.

Two weeks after the beginning of the irradiation (Fig. 1 b), when the tumor regained its starting volume, a large central necrosis was found. Dilatated vessels with some petechiae within and at the tumor margin were observed. Cell edema, pyknotic cells and increased vacuolation were seen more frequently. The fraction of mitotic cells is reduced. Blood vessels appeared to be intact but dilatated.

After the 5th week, at the end of irradiation (Fig. 1 c) – the tumor had shrunk to one-fifth of the pretreatment volume – the tumor showed marked dilatation of blood vessels at the hilus and reduced caliber of vessels in the periphery zone. In the histological picture the vessels seemed morphologically to be intact. The tumor tissue included high fractions of pleomorphic cells and of cells with vacuolation or with more than one nucleus. Among them a few isolated mitotic cells were present. These proliferative cells and the remaining vascular supply could be responsible for tumor regrowth, as after a radiation dose of 75 Gy no tumor control was observed.

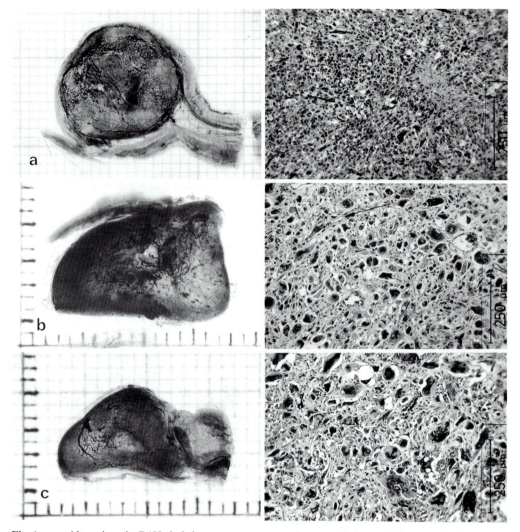

Fig. 1a–c. Alterations in R1H rhabdomyosarcoma vasculature and tissue during a radiation treatment with ^{60}Co-γ-rays for 5 weeks: macroscopic and microscopic results. **a** Vascular pattern before treatment; **b** after 2 weeks (30 Gy/10 fractions); **c** after 5 weeks (75 Gy/25 fractions)

Figure 2 shows the changes in tumor R1H vasculature and in tumor tissue during the combined treatment with ^{60}Co-γ-rays and microwave hyperthermia for 5 weeks. The untreated tumor (Fig. 2a) is as described above.

Two weeks after the combined treatment (Fig. 2b) macroscopically a massive necrosis was observed in the part of the tumor which was close to the microwave applicator. No identifiable microvaculature existed in this area which is encircled by dilatated and ruptured blood vessels. The part of the tumor proximal to the body of the animal showed an intensive radial vasculature. The histologic section showed prominent hyperemia and extensive hemorrhaging. The majority of tumor cells were undergoing pyknosis and karyorrhexis with some areas of marked necrosis. Only a few mitotic cells were visible in the tumor margin.

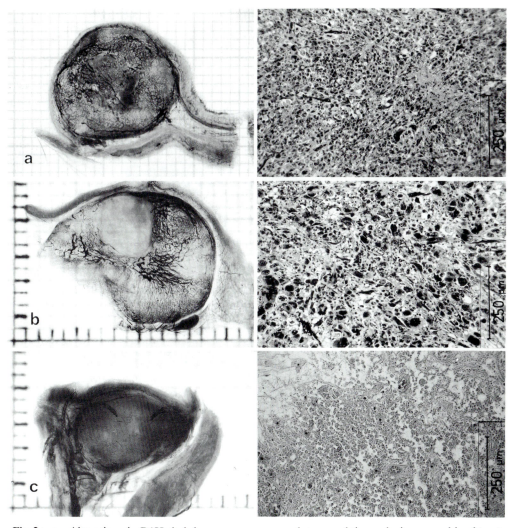

Fig. 2a–c. Alterations in R1H rhabdomyosarcoma vasculature and tissue during a combined treatment with ^{60}Co-γ-rays and microwave hyperthermia for 5 weeks: macroscopic and microscopic results. **a** Vascular pattern before treatment; **b** after 2 weeks (30 Gy/10 fractions + 4 × 43° C; 60 min); **c** after 5 weeks (75 Gy/25 fractions + 10 × 43° C, 60 min)

After the 5th week, at the end of combined treatment (Fig. 1 c), the tumor was reduced to about one-third of its starting volume. The vascular network was drastically reduced in caliber, particularly in the center of the tumor. There were massive hemorrhagic necrosis that involved most of the central areas of the tumor mass. Remaining small blood vessels showed only modest dilatation and occasional rupture. In the tumor periphery some dilatated and less severely damaged blood vessels existed. The histological picture showed small pyknotic cells, edema and lakes of hemorrhagic necrotic cell debris. Mitotic cells were no longer present. The whole tumor had become a necrotic mass with a massively damaged network. The result after such combined treatment was either partial or complete response of the tumor.

In general, the effects observed after the combined treatment with radiation and hyperthermia were always more intense than those after radiation alone. The enhanced cellular and vascular damage can be explained by the raised local temperatures in the R1H tumors during the microwave heat treatments (Zywietz et al. 1986).

References

Arcangeli G, Cividalli D, Nervi C, Creton G, Lovisolo G, Mauro F (1983) Tumor control and therapeutic gain with different schedules of combined radiotherapy and local external hyperthermia in human cancer. Int J Radiat Oncol Biol Phys 9: 1125–1134

Bicher HI, Hetzel FW, Sandhu TS (1983) Physiology and morphology of tumor microcirculation in hyperthermia. In: Storm FK (ed) Hyperthermia in cancer therapy. Hall, Boston, pp 207–222

Dewhirst MW, Sim DA, Wilson S, de Young D, Parsells JL (1983) Correlation between initial and long-term responses of spontaneous pet animal tumors to heat and radiation or radiation alone. Cancer Res 43: 5735–5741

Eddy HA (1980) Alterations in tumor microvasculature during hyperthermia. Radiology 137: 515–521

Emami B, Nussbaum GH, Hahn N, Piro AJ, Dritschilo A, Quimby F (1981) Histopathological study on the effects of hyperthermia on microvasculature. Int J Radiat Oncol Biol Phys 7: 343–348

Jung H, Beck HP, Brammer I, Zywietz F (1981) Depopulation and repopulation of the R1H rhabdomyosarcoma of the rat after X-irradiation. Eur J Cancer 17: 375–386

Jung H, Dikomey E, Zywietz F (1986) Ausmaß und zeitliche Entwicklung der Thermoresistenz und deren Einfluß auf die Strahlenempfindlichkeit von soliden Transplantationstumoren. In: Streffer C, et al. (eds) Lokale Hyperthermie. Deutscher Ärzte-Verlag, Cologne, pp 23–38

Lindholm C, Kjellen E, Landberg T, Nilsson P, Hertzman S, Persson B (1985) Microwave-induced hyperthermia and radiotherapy. Clinical results. In: Overgaard J (ed) Hyperthermic oncology 1984, vol 1. Taylor and Francis, London, pp 341–344

Overgaard J (1985) Rationale and problems in the design of clinical studies. In: Overgaard J (ed) Hyperthermic oncology 1984, vol 2. Taylor and Francis, London, pp 325–338

Reinhold HS, Endrich B (1986) Tumour microcirculation as a target for hyperthermia. Int J Hyperthermia 2: 111–137

Reinhold HS, van den Berg-Blok AE (1984) Heat-induced microcirculatory stoppage strongly enhances tumor control by radiation. In: Overgaard J (ed) Hyperterhmic oncology 1984, vol 1. Taylor and Francis, London, pp 149–152

Song CW (1978) Effect of hyperthermia on vascular functions of normal tissues and experimental tumors. JNCI 60: 711–713

Song CW (1983) Blood flow in tumors and normal tissues in hyperthermia. In: Storm FK (ed) Hyperthermia in cancer therapy. Hall, Boston, pp 187–206

Tanaka Y, Hasegawa T, Murata T (1984) Effect of irradiation and hyperthermia on vascular function in normal and tumor tissue. In: Overgaard J (ed) Hyperthermic oncology 1984, vol 1. Taylor and Francis, London, pp 145–148

Vaupel P, Müller-Klieser W, Otte J, Manz R, Kallinowski F (1983) Blood flow, tissue oxygenation, and pH distribution in malignant tumors upon localized hyperthermia. Basic pathophysiological aspects and the role of various thermal doses. Strahlentherapie 159: 73–81

Zywietz F, Jung H (1986) Response of a murine tumour to different fractionation schedules of ^{60}Co-γ-rays in combination with microwave hyperthermia. Int J Hyperthermia 2: 417

Zywietz F, Knöchel R (1986) Dielectric properties of Co-γ-irradiated and microwave-heated rat tumour and skin measured in vivo between 0.2 and 2.4 GHz. Phys Med Biol 31: 1021–1029

Zywietz F, Knöchel R, Kordts J (1986) Heating of a rhabdomyosarcoma of the rat by 2450 MHz microwaves: technical aspects and temperature distributions. Recent Results Cancer Res 101: 36–46

Pathophysiology of Tumors in Hyperthermia*

P. Vaupel, F. Kallinowski, and M. Kluge**

Abteilung für Angewandte Physiologie, Universität Mainz, Saarstraße 21, 6500 Mainz, FRG

Introduction

The response of tumor cells to hyperthermia is critically influenced by a number of pathophysiological factors both in vitro and in vivo. The most relevant factors in this context are tumor blood flow, tissue oxygenation, the energy status, and the pH distribution, which in turn define the cellular microenvironment.

In many solid tumors, this microenvironment is characterized by hypoxia (and often anoxia), acidosis, and energy deprivation, which are known to enhance the effect of heat even on single cells in vitro. These characteristic features of the cellular microenvironment are mostly determined by the tumor microcirculation, which is heterogeneously distributed and is generally insufficient for larger tumors. This flow pattern is typical of solid tumors and has two relevant consequences: (1) it induces the hostile cellular microenvironment described above, thus rendering tumor cells in vivo more heat-sensitive than normal tissue, and (2) it limits heat dissipation from the tumor tissue, thus also limiting the energy input required to reach a therapeutic tissue temperature. The latter fact often implies the possibility of heating the tumor tissue relatively selectively.

Because of the relative susceptibility of tumors to heat due to their hostile micromilieu, changes of the microenvironment upon heating are of interest since an additional deterioration of the nutritive blood flow would further sensitize the cells to heat (for reviews, see Mueller-Klieser and Vaupel 1984; Jain and Ward-Hartley 1984; Song 1984; Reinhold and Endrich 1986; Vaupel and Kallinowski 1987; Vaupel et al. 1987a) making a feedback mechanism possible. Because of the practical relevance of these pathophysiological aspects during heat treatment, hyperthermia-induced effects on tumor blood flow as well as microcirculation-related changes of the tumor microenvironment are discussed here.

---

*   Supported by the Bundesministerium für Forschung und Technologie (grant 01 VF 034).
** We gratefully acknowledge the expert laboratory assistance of Gabriele Berg and the help of Anne Deutschmann-Fleck in preparation of this manuscript and the artwork.

Tumor Blood Flow in Hyperthermia

The first quantitative data on tumor blood flow considering both tissue temperature level and exposure time were derived from rodent tumor systems (Scheid 1961; Vaupel et al. 1977). Thereafter many investigations have been performed on different tumor types, utilizing various heating techniques and protocols and devices for thermometry. Whereas data on hyperthermia-induced changes in blood flow derived from animal tumor systems are abundant, blood flow investigations in human tumors during hyperthermic treatment are scarce. Most studies have shown that the tumor perfusion rate decreases as a result of hyperthermic treatment. This flow decline is occasionally preceded by an initial transient flow increase when mild levels of hyperthermia and/or short exposure times are applied (for a recent review, see Vaupel and Kallinowski 1987).

The data obtained during the past decade usually show that heat-induced changes in tumor blood flow are considerably different from those obtained in most normal tissues. Despite a certain uniformity in the heat-induced tumor blood flow changes, there are great differences in the exposure times and hyperthermia levels required to achieve this effect. Furthermore, it is obvious that the blood supply to some tumors fails *during* heating, whereas in other tumors the breakdown of perfusion appears only *after* heating. These differences are of great relevance for the order and timing of multifraction heat therapy, and during combined treatment with radiation or anticancer drugs, since tumor blood flow is of paramount importance for radiosensitivity, for intratumor pharmacokinetics, and for pharmacodynamics. The observed shutdown of tumor blood flow is frequently irreversible (particularly after high thermal doses), but, depending on the type of tumor, blood flow may be restored 1–3 days after heat treatment.

In the relevant literature on murine tumors, it is striking that there seem to be only two significant exceptions to the usual blood flow pattern found upon tumor hyperthermia: (1) in tissue-isolated tumor preparations connected to only one artery and one vein, no significant flow changes occurred at 40°–43° C for 30–60 min (Gullino 1980; Gullino et al. 1978), and an initial flow increase of approximately 39% was followed by only a minimal decline of about 12% at 44° C for 20–30 min (Vaupel et al. 1982a); and (2) in Walker 256 carcinosarcomas, no significant alterations in blood flow could be observed unless tissue temperatures of 45° C were maintained for over 1 h. The latter finding implies that the circulation in Walker tumors cannot be shut down significantly without simultaneously disturbing the perfusion of the normal tissue surrounding the tumor.

The scarce clinical data available today are derived from sporadic observations rather than from systematic studies. They do not allow conclusive statements to be made concerning changes of tumor blood flow upon hyperthermia in patients or comparisons to be drawn between animal and human tumors. Although a few human tumors are apparently adequately perfused and so cannot be treated efficiently, other investigations have indicated a blood flow decrease in patient tumors comparable to that obtained in animal neoplasms (for a recent review, see Vaupel and Kallinowski 1987).

Since far-reaching conclusions cannot be drawn from these pilot studies, we have tried to close the gap between isotransplanted rodent tumors and human tumors in situ by investigating the heat-induced blood flow changes of xenotransplanted human breast carcinomas in the subcutis of T-cell-deficient rnu/rnu rats (for details concerning the implantation technique, see Vaupel et al. 1987b).

Local hyperthermia was induced using an ultrasound feedback control system (1.7 MHz). Tissue temperature in the center of the subcutaneous tumors (mean wet weight approximately 3 g) was monitored with a miniaturized thermocouple. The mean arterial blood pressure (MABP) was measured with a pressure transducer through a catheter in the thoracic aorta. Tumors were heated to either 40°, 42°, or 44° C for 60 min. During this period, the power to the ultrasound transducer was shut off at intervals and the subsequent thermal washout was recorded. In the control series, the tumor temperature was elevated to 38.5° C for 60 s and the thermal washout curves were then recorded. This short-term temperature increase had no significant effect on tumor blood flow, as shown by comparison with the [85]Kr-clearance technique.

The blood flow patterns of squamous cell carcinoma xenografts after 1, 20, 40, and 60 min hyperthermia at 40°, 42°, and 44° C are presented in Fig. 1, together with the local flow under control conditions (tissue temperature 38.5° C for 60 s). During treatment, a time- and temperature-dependent decrease in flow was observed. Compared with the control values, after 60 min-hyperthermia significant reductions of tumor blood flow were obtained at all hyperthermia levels, the reduction being most pronounced at the highest tissue temperature. A total stoppage of flow occurred in approximately 10% of the tumors heated to 40° C, in 25% of the tumors with 42° C hyperthermia, and in 50% of the tumors at 44° C.

Blood flow changes upon hyperthermia in medullary breast cancer xenografts are shown in Fig. 2. Here, an initial increase of tumor blood flow was obvious after 1 min at 44° C. Subsequently, a time- and temperature-dependent decrease in flow again occurred during treatment. However, the flow reduction was somewhat less pronounced than that observed in the squamous cell carcinomas. After 60 min of treatment, significant reductions were seen at 42° and 44° C. At that time, a total flow stoppage was obvious in 10% of the tumors heated to 40° C, in 20% of the tumors with a tissue temperature of 44° C, and in 30% of the xenografts heated to 42° C.

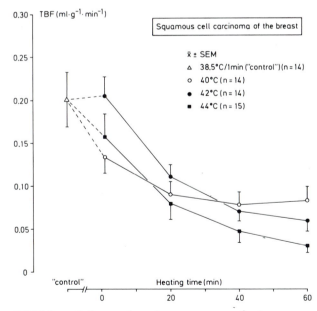

Fig. 1. Changes in tumor blood flow *(TBF)* through human breast cancer xenografts (squamous cell carcinomas) in rnu/rnu rats as a function of exposure time and tissue hyperthermia (40° C, *circles;* 42° C, *dots;* 44° C, *squares;* ultrasound heating). "Control" flow *(triangle)* is measured using thermal washout curves after a 60-s heating period up to 38.5° C. Values are given as mean ± standard error of the mean ($\bar{x} \pm$ SEM). *n*, number of tumors investigated

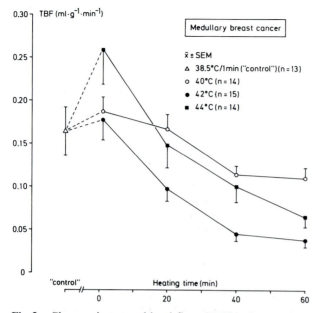

Fig. 2. Changes in tumor blood flow *(TBF)* in human breast cancer xenografts (medullary breast cancers) in rnu/rnu rats as a function of exposure time and tissue temperature level. For further details see legend to Fig. 1

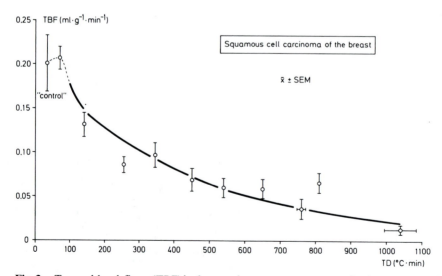

Fig. 3. Tumor blood flow *(TBF)* in human breast cancer xenografts (squamous cell carcinomas) as a function of thermal dose *(TD)*. Values are given as mean ± standard error of the mean ($\bar{x} \pm SEM$)

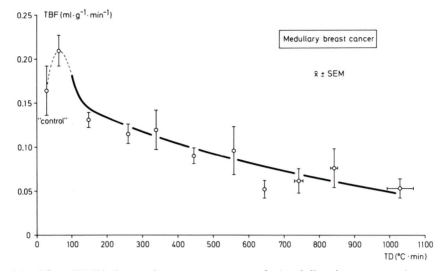

Fig. 4. Tumor blood flow *(TBF)* in human breast cancer xenografts (medullary breast cancers) as a function of thermal dose *(TD)*. Values are given as mean ± standard error of the mean ($\bar{x} \pm SEM$)

Pronounced intertumor variations in the flow changes upon heat treatment were detectable between individual xenografts. Thus, the biological behavior of individual tumors at elevated tissue temperatures can hardly be predicted. In some tumors blood flow even increased during hyperthermia although tumors of the same cell line, at the same implantation site, and of comparable size were treated with identical hyperthermia levels.

Determining the blood flow in the xenografts as a function of thermal dose (time integral of the tissue temperature elevation above the starting temperature during the treatment period) for the squamous cell breast carcinomas, showed that tumor blood flow decreases continuously with increasing thermal dose (Fig. 3). The continuous line in the figure was fitted using a least-squares method. The broken line indicates the trend at low thermal doses. Tumor blood flow in medullary tumor xenografts as a function of the thermal dose is shown in Fig. 4. A similar, but less pronounced decrease in flow was found with increasing thermal dose.

Red Blood Cell Flux in Tumors During Hyperthermia

Laser Doppler flowmetry has been used for noninvasive, direct, and continuous study of microcirculatory blood flow in normal skin and subepidermal tumor isotransplants (DS carcinosarcoma) during localized ultrasound hyperthermia in the rat. In normal rat skin, 40° C-hyperthermia only induced a marginal increase in the red blood cell (RBC) flux (Fig. 5). Significant increases occurred after 20 min at 42° C (Fig. 6) and after 4 min at 44° C. During 44° C-hyperthermia maximum fluxes were reached after 24 min. Thereafter, the flux declined and approached preheating levels after 60 min of hyperthermia (Fig. 7). In contrast, in subepidermal tumors 40° C-hyperthermia induced a slight decrease of average flux (Fig. 5). During 42° C-hyperthermia, a significant decrease was found after 40 min, flux reaching 60% of the initial value at the end of the heating period (Fig. 6). Following a transient increase in flux during the heating-up period, 44° C-hyper-

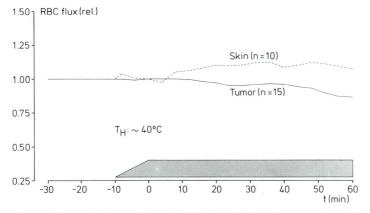

Fig. 5. Mean relative red blood cell flux *(RBC flux)* in normal rat skin and in subepidermal tumor isotransplants (carcinosarcoma) at the hind foot dorsum of Sprague–Dawley rats during 40° C-hyperthermia (*n*, number of measurements performed)

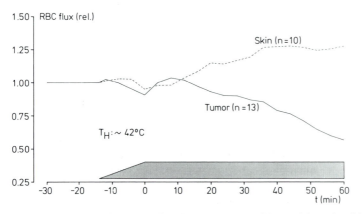

Fig. 6. Mean relative RBC flux in normal rat skin and in subepidermal tumors during 42° C-hyperthermia

thermia led to a significant impairment of flux after 24 min (Fig. 7). RBC flux finally reached 30% of the preheating value. A total shutdown of the RBC flux was observed in about 30% of the tumors at 44° C.

With elevated tissue temperatures, pronounced intertumor variations of the time- and temperature-dependent changes of RBC flux were observed. The heat-dependent flow patterns of some tumors were considerably different from those seen in most tumors at a given tissue temperature, although tumors of similar size were compared. Rhythmic oscillations of RBC flux were found in some subepidermal tumors (0.40 ± 0.05 cycles min^{-1}). Upon heating, these periodic variations slowed down significantly (0.20 ± 0.04 cycles min^{-1}), whereas in normal skin the frequency of the fluctuations increased. The mechanisms which lead to vascular stasis in tumors during hyperthermia are probably quite complex. As mentioned earlier, the results of the studies available are indicative of definite changes in tumor microcirculation that are reversible at low thermal doses and irreversible at higher temperatures with appropriate exposure times. In

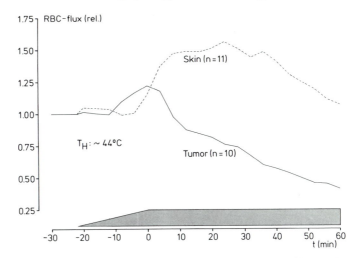

Fig. 7. Mean relative RBC fluxes in normal rat skin and in subepidermal tumors during 44° C-hyperthermia

the latter case coagulation necrosis is observed following treatment at therapeutic temperatures. The histopathological changes in the tumor microvasculature upon hyperthermia are identical in human and animal tumors, so that the mechanisms which lead to a breakdown of blood flow appear to be quite similar (for a review, see Vaupel and Kallinowski 1987).

Red Blood Cell Flux in Tumors During Hyperthermia Combined with Hyperglycemia

It has been suggested that certain agents could be used in conjunction with hyperthermia to improve cancer treatment if these substances decrease tumor blood flow and/or pH without modifying normal tissue response. One agent that has been studied extensively for this purpose is glucose. If this sugar is combined with hyperthermia the blood flow and tumor pH could be reduced to a greater extent during or prior to hyperthermia. Hence, the efficacy of the treatment could be improved (for a recent review see Ward and Jain 1987). Von Ardenne (1971) was the first to postulate that hyperglycemia may enhance the cytotoxic effects of heat on tumors. This rationale has led to many studies concerning the effect of combined hyperglycemia and hyperthermia on tumor blood flow and pH. However, in some of these investigations, glucose concentrations in the arterial blood far exceeded 25 mM (e.g., Calderwood and Dickson 1980; von Ardenne and Reitnauer 1980). This "excessive" hyperglycemia can initiate a series of deleterious pathological effects if maintained in patients for a longer period of time. For this reason, we initiated a study using subepidermal rat tumor isotransplants (carcinosarcoma) in the hind foot dorsum of Sprague-Dawley rats. In these experiments the effect of a "moderate" hyperglycemia (glucose concentrations 20–25 mM) on the RBC flux in tumors was investigated both during elevation of the blood glucose alone and in combination with hyperthermia.

Following the determination of baseline values, saline (12 ml kg^{-1}h^{-1} for control studies) or glucose (4.8 g kg^{-1}h^{-1}, 40% glucose solution) were given intravenously. In addition, a study was also performed to determine changes in the microcirculation following intravenous administration of galactose (4.8 g kg^{-1}h^{-1}, 40% galactose solution), a sugar not metabolized by the tumors studied.

In the control series (mean tumor tissue temperature at the site of laser Doppler flowmetry 34° C), the average RBC flux did not change during the observation period. Likewise, no alterations of the RBC flux were observed upon intravenous injection of glucose or galactose, although in the hyperglycemia experiments the blood glucose increased from 6.6±0.6 ($\bar{x}$ ±SEM) to 24.1±0.7 mM within 1 h. During the hyperglycemia experiments, the lactate concentration in the arterial blood was increased from 1.1±0.2 to 2.9±0.1 mM.

Heating the tumor tissue to 40° C was associated with only a marginal decline in the average RBC flux. Hyperglycemia (5.9±0.1 to 20.5±0.8 mM glucose; 1.1±0.1 to 2.7± 0.1 mM lactate) in combination with 40° C-hyperthermia also had little effect.

During localized 42° C-hyperthermia of the tumors, a significant decrease in flow was found. This decrease could not be intensified significantly by the additional intravenous injection of galactose. However, hyperglycemia (5.3±0.4 to 24.0±1.6 mM glucose; 1.0±0.1 to 3.1±0.3 mM lactate) in combination with 42° C-hyperthermia reduced the RBC flux to a greater extent than hyperthermia alone ($p < 0.005$; see Fig. 8).

The decrease in flow observed in the tumors with 44° C-hyperthermia could also be intensified by combining heat with hyperglycemia (6.0±0.2 to 25.0±1.5 mM glucose; 0.9±0.1 to 2.9±0.1 mM lactate). This additional flow reduction, however, was smaller than that observed during 42° C-hyperthermia combined with hyperglycemia.

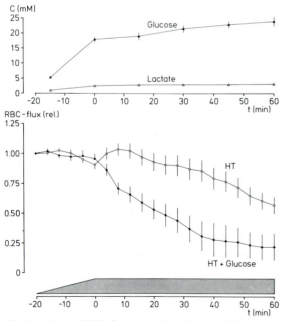

Fig. 8. Mean RBC flux in subepidermal DS-carcinosarcomas during 42° C-hyperthermia (*HT,* mean tumor wet weight 0.76±0.09 g, 13 tumors investigated), and during 42° C-hyperthermia in conjunction with hyperglycemia (*HT + Glucose,* mean tumor wet weight 0.77±0.08 g, 10 tumors investigated). Glucose and lactate concentrations *(C)* in the arterial blood are shown in the *upper panel.* Values are given as mean ± SEM. The heating period is marked by the *shaded area*

From these results one has to conclude that "moderate" hyperglycemia in conjunction with hyperthermia can decrease tumor blood flow to a greater extent than hyperthermia alone. This effect was most pronounced during 42° C-hyperthermia. For a severe flow decrease due to hyperthermia alone (e.g., at 44° C), the impact of combining hyperglycemia was to some extent reduced. When hyperthermia had no effect on tumor blood flow (e.g., at 40° C), the combination with "moderate" hyperglycemia showed no flow reduction.

Tumor Tissue Oxygenation in Hyperthermia

In general, the tissue oxygenation of isotransplanted rat tumors parallels the heat-induced flow changes (Vaupel et al. 1982b). If tumor heating induces only marginal alterations of tumor microcirculation, then tissue oxygenation also changes only marginally.

The same holds true for the nutrient supply and the removal of acidic waste products. This indicates that nutritive blood flow is the main parameter determining the micromilieu surrounding the tumor cells (Vaupel et al. 1983).

Tumor Tissue pH Distribution upon Hyperthermia

When the tissue pH distribution in subcutaneously growing Yoshida sarcomas with wet weights between 2 and 3 g was measured, the results clearly showed that the tissue pH ranges from 6.40 to 7.80, with an average of 6.89, under control conditions (Fig. 9). At 1 h after heating (44° C for 60 min), the pH distribution shifted to significantly lower values (average pH of 6.63), i.e., upon hyperthermia employing thermal doses of ca. 540° C min, the mean tumor pH dropped by about 0.25. From these results it can be concluded that the tissue pH is usually lower in rat tumors than in normal tissues (mean pH in muscle and subcutis ca. 7.35) and hyperthermia further decreases the tumor pH if appropriate thermal doses are applied. At 24 h after heating the tumor pH distribution has returned to its preheating condition, the average tissue pH 1 day after heating being 6.92. However, pH values higher than 7.2 are more frequent than with control conditions.

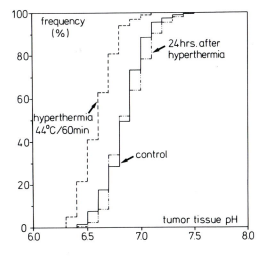

Fig. 9. Cumulative frequency distributions of intratumor pH in Yoshida sarcomas (wet weight 2–3 g) before (control, *continuous line*), immediately after *(broken line)*, and 24 h after *(dotted line)* localized ultrasound heating to 44° C for 60 min

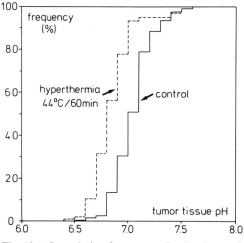

Fig. 10. Cumulative frequency distributions of intratumor pH in human squamous cell carcinomas of the breast xenografted subcutaneously into rnu/rnu rats before (control, *continuous line*) and immediately after *(broken line)* localized ultrasound heating to 44° C for 60 min

This is probably due to enlarged areas with necrotic debris following heat application (Vaupel 1982).

Heating the tumor tissue at different rates seems to induce different pH changes during 44° C-hyperthermia sustained for 60 min. Whereas heating rates $<0.7°$ C min^{-1} are followed by a mean pH decrease of approximately 0.4 after rapid heating ($>1.5°$ C min^{-1}) the average pH decrease is only 0.25. Up to now there have been no convincing experimental results indicating whether this difference is caused by the somewhat greater thermal dose applied during 44° C-hyperthermia for 60 min with a heating up rate $<0.7°$ C min^{-1} as compared with more rapidly heated tumors.

Although no pH changes upon hyperthermia were observed in the subcutis, in xenografts of squamous cell carcinomas of the breast heat treatment was followed by a significant shift of the pH frequency distribution to lower pH values (from 7.07 in control conditions to 6.91 after 44° C-hyperthermia for 60 min; see Fig. 10). Comparable shifts were also observed after 44° C-hyperthermia for 30 min and after 42° C for 60 min.

In xenografts of medullary breast carcinomas no significant pH shifts to lower values were found upon comparable thermal doses. However, there was a trend towards more acidic values with increasing thermal dose. The mean pH was 6.82 under control conditions and 6.72 after tissue temperatures of 44° C for 60 min. This slight pH shift to more acidic values in the medullary tumors coincides with a less pronounced flow decrease in this tumor type.

Conclusions

The hostile tumor microenvironment caused by an inadequate tumor microcirculation sensitizes tumor cells to heat. During heat treatment, a positive feedback mechanism is postulated, since adequate thermal doses further reduce tumor blood flow and thereby make the micromilieu more hostile. This pattern can be observed both in isotransplanted rat tumors and in xenotransplanted human breast cancers. Furthermore, from the results presented here it can be concluded that "moderate" hyperglycemia in conjunction with hyperthermia decreases tumor blood flow to a greater extent than hyperthermia alone.

References

Calderwood SK, Dickson JA (1980) Effect of hyperglycemia on blood flow, pH, and response to hyperthermia (42° C) of the Yoshida sarcoma in the rat. Cancer Res 40: 4728–4733

Gullino PM (1980) Influence of blood supply on thermal properties and metabolism of mammary carcinomas. Ann NY Acad Sci 335: 1–21

Gullino PM, Yi PN, Grantham FH (1978) Relationship between temperature and blood supply or consumption of oxygen and glucose by rat mammary carcinomas. J Natl Cancer Inst 60: 835–847

Jain RK, Ward-Hartley K (1984) Tumor blood flow – Characterization, modifications, and role in hyperthermia. IEEE Trans Sonics Ultrasonics SU-31: 504–526

Mueller-Klieser W, Vaupel P (1984) Effect of hyperthermia on tumor blood flow. Biorheology 21: 529–538

Reinhold HS, Endrich B (1986) Tumour microcirculation as a target for hyperthermia. Int J Hyperthermia 2: 111–137

Scheid P (1961) Funktionale Besonderheiten der Mikrozirkulation im Karzinom. Bibl Anat 1: 327–335

Song CW (1984) Effect of local hyperthermia on blood flow and microenvironment. Cancer Res (suppl) 44: 4721s–4730s

Vaupel P (1982) Einfluß einer lokalisierten Mikrowellenhyperthermie auf die pH-Verteilung in bösartigen Tumoren. Strahlentherapie 158: 168–173

Vaupel P, Kallinowski F (1987) Physiological effects of hyperthermia. Recent Results Cancer Res 104: 71–109

Vaupel P, Ostheimer K, Thome H (1977) Blood flow, vascular resistance, and oxygen consumption of malignant tumors during normothermia and hyperthermia. Microvasc Res 13: 272

Vaupel P, Frinak S, Mueller-Klieser W, Bicher HI (1982a) Impact of localized hyperthermia on the cellular microenvironment in solid tumors. Natl Cancer Inst Monogr 61: 207–209

Vaupel P, Otte J, Manz R (1982b) Oxygenation of malignant tumors after localized microwave hyperthermia. Radiat Environ Biophys 20: 289–300

Vaupel P, Mueller-Klieser W, Otte J, Manz R, Kallinowski F (1983) Blood flow, tissue oxygenation, and pH distribution in malignant tumors upon localized hyperthermia. Strahlentherapie 159: 73–81

Vaupel P, Kallinowski F, Kluge M, Egelhof E, Fortmeyer HP (1987a) Microcirculatory and pH alterations in isotransplanted rat and xenotransplanted human tumors associated with hyperthermia. Recent Results Cancer Res (in press)

Vaupel P, Fortmeyer HP, Runkel S, Kallinowski F (1987b) Blood flow, oxygen consumption, and tissue oxygenation of human breast cancer xenografts in nude rats. Cancer Res 47: 3496–3503

von Ardenne M (1971) Theoretische und experimentelle Grundlagen der Krebs-Mehrschritt-Therapie, 2nd edn. Verlag Volk und Gesundheit, Berlin

von Ardenne M, Reitnauer PG (1980) Selective occlusion of cancer tissue capillaries as the central mechanism of the cancer multistep therapy. Japan J Clin Oncol 10: 31–48

Ward KA, Jain RK (1987) Response of tumors to hyperglycemia: characterization, significance and role in hyperthermia. Int J Hyperthermia (in press)

Technical Aspects of Hyperthermia

P. F. Turner and T. Schaefermeyer

BSD Medical Corporation, 420 Chipeta Way, Salt Lake City, UT 84108, USA

Introduction

The history of hyperthermia is both ancient and extensive. Throughout the ages many varying techniques have been used to accomplish localized or whole-body heating of the human body. Other reviews of this subject have been previously published (Short and Turner 1980; Conway and Anderson 1986). The primary technical challenge in hyperthermia today is that of heating the whole tumor to adequate temperatures without overheating normal tissues. Secondary is the measurement of tissue temperatures, to serve as a guide in treatment and an assessment of how well the tumor is being treated. Currently, hyperthermia temperature measurement involves invasive placement of sensors into the tumor site and normal tissues. Little treatment improvement is possible, no matter how extensive the thermometry, if the method used is not capable of delivering the heat to the target site. The thrust of this paper will, therefore, be the technical aspects of heat delivery.

A major advancement in the delivery of hyperthermia was reported by Geyser (1916), who developed a system to make use of electromagnetics. Geyser described his system as operating at 3 MHz with capacitive-type electrode plates which were placed in contact with the tissue. He demonstrated successful high elevation of temperature in tumor sites using this radio frequency technique, which was similar to that used today. At that time the achievement of a 3 MHz frequency represented the state of the art. As technology grew and improved over the years the techniques used to generate hyperthermia became greatly diversified and improved. A review of these techniques and methods will serve as a useful background for those interested in hyperthermia treatments and studies.

Hyperthermia Modalities

The modalities of hyperthermia can be classified into five basic areas; whole-body hyperthermia, deep/regional hyperthermia, superficial hyperthermia, interstitial hyperthermia, and body orifice insertion hyperthermia.

Whole-body hyperthermia has as theoretical goal the elevation of the entire body to a uniform raised temperature. This is generally accomplished by thermal conduction or radiant light techniques. Often this includes thermal insulation or isolation of the body surfaces from surrounding air. The maximum temperature achievable with this technique is generally considered to be 41.8° C. At higher temperatures unacceptable toxicity to vital organs occurs. One technical difficulty of most whole-body hyperthermia techniques is that variations in temperature can still be detected at various points within the body even when great precautions are taken to minimize such differences. This technique is noted as having only moderate benefits, as the temperature level must remain low so as to protect the normal tissues. Interest in whole-body hyperthermia combined with chemotherapy is increasing, and many patients have whole-body mestatic disease.

Deep/regional hyperthermia implies depth greater than 5 cm and has generally been performed using electromagnetic fields. This technique attempts to place a large deep heating field at the tumor site but frequently heats other surrounding tissues as well. The challenge with this modality of hyperthermia is to obtain sufficient depth and sufficient localized heating in the full tumor volume with minimal heating of the surrounding tissue. This has presented a formidable challenge technically to physicists and engineers. Some research is under way with ultrasound in this area.

Superficial hyperthermia has gained much wider acceptance by clinicians. This modality involves the heating of tissues from the surface of the body down to as deep as 5 cm. Generally electromagnetic fields have been used to accomplish this type of hyperthermia. Usually tissue surfaces are not flat and thus do not match the common applicators. During treatment, the heating field can be altered by non-flat tissue surfaces. As in all types of hyperthermia, the variability of the blood flow within the treated region also contributes to the temperature variation within the tumor region.

Interstitial hyperthermia is also gaining increased acceptance. This treatment modality involves placement of heating devices directly into the tumor. Most interstitial hyperthermia has involved the electromagnetic heating technique. The popularity of this modality has increased as interstitial radiation therapy has become more widely practiced, as the inserted needles or catheters used for the interstitial radiation therapy implant provide entry or access portals for the hyperthermia applicators. The strong appeal of interstitial therapy is that the heating occurs primarily within the tumor volume itself because of the selective placement of the implant. This enables higher tumor temperatures with lower normal tissue temperatures. The procedure is very invasive, and until the practice of interstitial radiation became more widespread clinicians were not willing to accept such a level of invasion alone or with external beam radiation.

Hyperthermia has also been accomplished by inserting heating devices into natural body orifices containing malignant growths. Electromagnetics has been the most common method for introduction of hyperthermia in these areas. The rationale for this modality is much the same as in the interstitial techniques: even in deep sites, the heating can be localized to the tumor and heating of normal tissues can be avoided.

Methods To Induce Hyperthermia

There are four basic methods to induce hyperthermia: electromagnetic techniques, ultrasound, radiant light, and thermal conduction.

The electromagnetic techniques have been widely used and are by far the most popular. These methods have the ability to heat as deep as 20 cm. They are well tested, they provide moderate focusing, and there is only moderate interaction between differing tissues and the electromagnetic field, except for the excess heating of surface fat layers with capacitance techniques.

Ultrasound methods have the advantage of increased depth with more highly focused heating than electromagnetics. This precision can be very difficult to steer properly, since the heat patterns generally form a very oblique ellipsoid with the long axis along the depth direction.

The method of radiant light is very simple to use and devices are simple to build; however, the penetration depth is limited to approximately 3 mm, which is not useful for most localized tumor sites. Some surface evaporative cooling can be used to effect a slight depth improvement.

Thermal conduction also has limitations of about 2 mm into the tissue for substantial heating, but is also simple in principle. The method frequently requires contact of tissues to a hot surface and if contact is not uniform, heating is also not uniform.

The interaction of the electromagnetic fields with complex tissue structures is a difficult thing to understand. Only recently are three-dimensional electromagnetic numerical models being developed to mathematically describe these interactions. Some of the electromagnetic approaches result in high heating of surface fat layers, which can limit the penetration depth of therapeutic heating. This fat heating is observed with low-frequency capacitive heating and with frequencies over 1 GHz with thick fat layers (>2 cm). At the higher frequencies of electromagnetics, such as 915 MHz, penetration of therapeutic heating does not exceed 3 cm. Lower frequencies and larger applicators enable deeper heating with a radiating electromagnetic approach.

Ultrasound is often found to be too highly focused. Also, excess heating of underlying or involved bone frequently result in excessive bone temperatures and severe pain (Fessenden et al. 1984). (This causes the designers of ultrasound devices to defocus the ultrasound beam by means of mechanical motion, multiple transducer arrays or lenses; Lele 1982; Pounds 1984; Cain and Umemura 1986. The basic heating pattern of an ultrasound beam, which is long and narrow in shape, can be greatly modifidied by array designs. Cain (1986) has shown how ultrasound array heating patterns can be steered by using different phase and amplitudes. The problem for high-power ultrasound is finding tumor sites with a large enough beam entry portal which does not contain bone- or air-filled zones. This eliminates ultrasound in most thoracic sites and in many head, neck, abdominal, and pelvic sites. As more sophisticated ultrasound applicators are developed the use of this modality will increase.

Electromagnetic Techniques

Electromagnetic energy is introduced into the body through any one of three techniques; capacitive, inductive, or radiative (microwave). The term "electromagnetic" means that the energy field is oscillating between an electric and a magnetic field potential state, and all of the techniques contain both electric fields and magnetic fields. In practice it is possible to enhance one or the other field to make it the primary mechanism in heating the tissues.

The capacitive or contacting electrode method is a non-radiative method generally practiced below 30 MHz. This technique involves placing at least two electrodes at different points along or near the tissue to be heated (Fig. 1). A quasi-static electric (voltage) field is applied between these electrodes. "Quasi-static" means that the field alternates its potential in time between the electrodes but does not induce substantial propagation of energy away from the region of the electrodes. This technique is very simple to understand conceptually, as currents are induced in the tissue as a result of the voltage field or

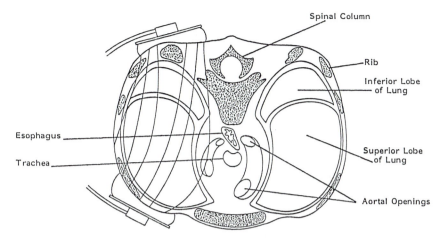

Spinal Column

Rib

Inferior Lobe
of Lung

Esophagus

Trachea

Superior Lobe
of Lung

Aortal Openings

Fig. 1. Various current paths for contact or capacitive electrode pair. Currents are perpendicular to tissue surface

potential between the two electrodes which alternates in time. This changing potential allows the energy to pass through air gaps or dielectric gaps sometimes placed between the electrode and the surface. The ability of both the capacitance and the contact electrodes to create an internal current within the tissue is what actually causes the tissues to heat. At these low frequencies, near 30 MHz and below, the dominant mechanism of heating in tissues results from conduction in current passing through the conductive tissues of the body. It is therefore possible approximately to model the current flow within the complex tissues of the body as a result of a three-dimensional resistive matrix, with different resistance of the tissues modifying the resistance of the equivalent resistor matrix in different zones. Therefore, when there is a fatty region or bone region in the localized area which has higher resistance, one can envision current flow being diverted into more conductive tissues which may lie parallel. However, when these conductive currents are forced to converge to narrow regions the current density increases. Also, when such currents must pass through a large high-resistance plane such as the superficial fat layer they induce higher power loss in that layer. This is reported to be typical of capacitance electrode techniques.

The inductive technique of hyperthermia was demonstrated in 1893 by d'Arsonval, who elevated the temperatures of animals by placing them within a solenoid helical coil through which a high-frequency current was passing. Other more recent researchers have also used the inductive heating method. These techniques have also primarily involved frequencies below 30 MHz, creating a quasi-static electromagnetic field which is non-radiative. With all of these devices, both an electric field and a magnetic field are produced. Low-frequency use of a coil reduces the electric field and increases the magnetic field-induced heating in the tissues. Normally these induced currents are parallel to the fat layer and penetrate without excessive fat heating (Fig. 2). However, most coils are also limited in penetration depth. Either the center of the body is not heated, or multiple heating zones are caused when a pair of coils are used with the magnetic field perpendicular to the surface of the tissue.

Radiative electromagnetic techniques have been used widely, generally at 27 MHz or above. These techniques involve the design of antenna devices to transmit or propagate

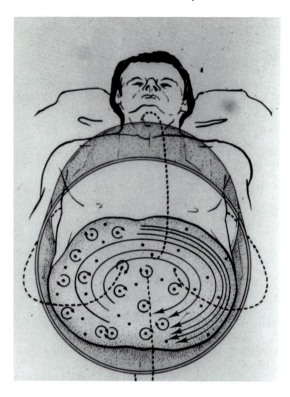

Fig. 2. Cylindrical current paths for a cylindrical coil which are inversely proportional to current path radius. Maximum occurs near surface and minimum in center. Currents are parallel to tissue surface

energy from the single antenna or an array into the tissues. A distinguishing factor of radiative methods is that two electrodes are not required to induce the fields and heating within the body. Thus the energy sources are independent of other sources. Examples of the devices are waveguides, dipoles, and spirals. When multiple antennas are used it is possible to achieve electronic focusing and heat-pattern steering with radiative arrays. Electronic steering of the heat pattern has not been generally used with inductive and capacitive methods; however, recent work reports steering of three or more electrodes using capacitive methods. By altering applicator setup, capacitive and inductive fields can be used to create a symmetrical or non-symmetrical heating field.

The achievable depth of heating by radiative methods is determined not only by frequency, but also by the size of the antenna used. Figure 3 shows the relationship between antenna size and frequency in effective penetration depth. For reference, the depth of a plane-wave field is shown. As frequency is decreased, the penetration depth begins to increase. This effect is reduced by restricting the size of the aperture or antenna at the lower frequency. Therefore, to increase the depth of penetration at lower frequency, it is necessary also to increase antenna size. This frequently causes undesirable or unacceptable limitations in clinical use of these lower frequencies. The penetration depth shown on the plot in Fig. 3 represents the $1/e$ penetration depth extending below the 1 cm depth position. Thus the penetration depth shown is that where the power has decreased to the level of approximately 13.5% of the power absorbed at 1 cm of depth.

As the frequency is lowered on an antenna of fixed size, the fields near the antenna become more dominated by the quasi-static fields and less of the energy is in the propagational or radiational field. This can cause increased amplitudes of the perpendicular

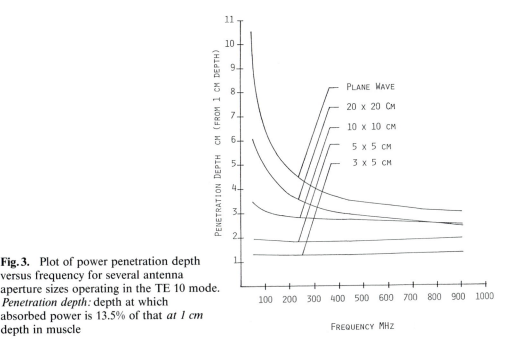

Fig. 3. Plot of power penetration depth versus frequency for several antenna aperture sizes operating in the TE 10 mode. *Penetration depth:* depth at which absorbed power is 13.5% of that *at 1 cm* depth in muscle

electric fields along the outer metallic edges of waveguide antenna and other types of antenna devices. These perpendicular fields can cause selective fat heating near the outer edges if a layer of superficial fat exists there. It has been found that increasing the space between the tissue and the applicator, or filling the space with a high dielectric (such as water), can reduce some of the effects of quasi-static fields by changing more of the energy from the quasi-static state into propagating energy. When antennas are operated in this low-frequency mode (where the antenna's field size is much less than a half-wavelength across the aperture), the strong quasi-static fields tend to act very much like the quasi-static capacitance fields of the electrode techniques. Figure 4 shows a comparative study of some of these deeper heating methods (Turner 1980). This plot shows that phased arrays provide improved deep power deposition.

For the capacitive methods, power absorption in superficial fat is about 10 times that in underlying tumor. This is not the case with the radiating applicators. The fat heating results from the electric fields between electrodes being perpendicular to the fat layer. In contrast, the fields of radiating applicators are primarily parallel to the surface fat layers. The heating results from both ionic conduction and vibration of the dipole molecules of water and proteins. The conduction current is produced by the drift of the free electrons in the material medium ($\bar{J}_C = \sigma|\bar{E}|$), while the displacement current (related to the historical term introduced by Maxwell) is proportional to the time rate of change of the electric field. In other words, the displacement current depends on the frequency of the electromagnetic wave. To illustrate this point, let us compare the magnitudes of the conduction current $\bar{J}_C$ and the displacement current $\bar{J}_D$ for the muscle tissues. The ratio between the conduction to the displacement current is given by:

$$\frac{|\bar{J}_C|}{|\bar{J}_D|} = \frac{\sigma}{2\pi f \varepsilon_0 \varepsilon'}$$

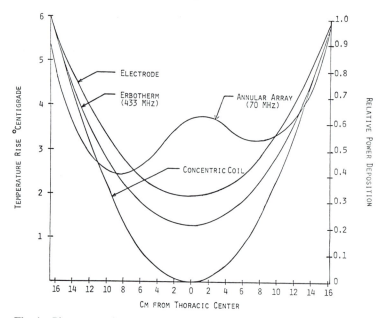

Fig. 4. Plot comparing the radial initial temperature rise for four deep heating methods in a cylinder of muscle-equivalent phantom. The central focusing of the annular array demonstrates superior central heating. (Turner 1980)

where σ is equivalent tissue conductivity accounting for losses of power in the tissue including both conduction and displacement losses, f is frequency, ε' is relative permittivity, and ε_0 is the free space permittivity. This current ratio for tissue with a high water content is 3.6 at 27 MHz, 2.2 at 100 MHz, 1.1 at 433 MHz, 0.5 at 915 MHz, and 0.35 at 2450 MHz. These values demonstrate that the displacement current is dominant for tissues only at higher frequencies. At lower frequencies the tissue heating phenomenon is dominated by the conduction current.

The absorption or heating patterns induced by the electromagnetic fields will be nonuniform and dependent on the dielectric properties of the tissues. The radiative absorption is high and the depth of penetration low in tissues of high water content such as muscle, brain tissue, internal organs, and skin, while the absorption is an order of magnitude lower in tissues of low water content such as fat and bone.

The patterns of the fields producing the heating are complex functions of the frequency, source configuration, tissue geometry, and dielectric properties of the tissues. The temperature patterns are further modified by the thermal properties of the tissues and neurocirculatory mechanisms.

The power absorbed by the tissue is usually quantified by the specific absorption rate (SAR), measured in units of watts per kilogram. For EM heating the SAR is related to the magnitude of the complex electric field in the tissue ($|\overline{E}|$) and tissue density ρ:

$$SAR = |\overline{E}|^2 \sigma / \rho$$

The initial rate of heating (R, in $^\circ$ C/s) of tissue for the first minute or so is related to SAR:

$$R \approx \frac{0.239 \times 10^{-3} \times (SAR)}{c}$$

where c is the tissue specific heat in kcal/kg ($^\circ$ C). In time the tissue temperatures are further modified by blood flow, thermal conduction, and metabolic heating. The required SAR for obtaining therapeutic temperatures is usually between 50 and 150 W/kg for moderately perfused tissues.

Phased Arrays

A great deal of interest has recently been aroused in the use of the microwave phased array. This employs the technique of coherent or synchronous phased arrays [either planar arrays parallel to the surface or cylindrical (annular) arrays surrounding the body] to obtain increased focusing in the target site. The special issue of the Microwave Theory and Techniques Journal of the IEEE published in May 1986 was dedicated to the topic of phased arrays for hyperthermia treatment of cancer (Lin 1986). In this journal many authors indicate strong physical evidence supporting the development of these phased arrays, both in high-frequency superficial devices and in low-frequency cross-fire devices (such as the annular phased array). It is expected that because of the only moderate interaction between tissues in these electromagnetic propagating fields, the microwave and radiofrequency phased array techniques will provide the most generally usable method of heating deep tumors.

The interstitial techniques for hyperthermia include microwave, antenna arrays, low-frequency local current fields which cause local current heating (LCF), and the heating of ferromagnetic seeds by the magnetic field. The most popular and best heating technique has been the microwave antenna arrays where a series of a small coaxial antennas are inserted into the tissue to heat the tumor (Mechling and Strohbehn 1986). The advantage of this technique is that the heating can be localized to the antenna and decreased heating of normal tissues along the insertion shaft. However, in some designs of the microwave interstitial antennas the heating pattern can be affected both by frequency changes and insertion depth changes. Normally these antennas are inserted into plastic catheters, typically 16 gauge. When these arrays are operated coherently or synchronously (where there relative phase is fixed between the applicators), there tends to be a coherent addition of the heating fields in tissues between the applicators. This greatly improves the uniformity of power deposition as compared to the non-synchronous operation mode, where antennas are operated either at slightly different frequency or sequentially turned on and off. It has generally been observed that using the synchronous or coherent type arrays allows fewer antennas to be inserted for the same type of heating field than does use of the non-synchronous mode.

LCF has the disadvantage of requiring more insertion sites or more needles than synchronous arrays in order to avoid cool zones between the inserted needles. The LCF technique generally uses metallic needles as electrodes. These are connected to a low-frequency generator (500 KHz). LCF electrodes can also be in the form of flexible metallized catheters to provide a flexible heating source. The advantage of this technique is that the exposed conductor length in the electrode can be modified by the addition of a dielectric coating of catheter insulation in areas of normal tissue. The length of metal in contact with tissue will limit the depth of heating. As with external capacitance electrodes, these invasive electrodes must be activated in groups or pairs, i.e., at least two electrodes are required to provide a current flow path between the electrodes. With recent developments and improvements in microwave interstitial antenna techniques, adjustable or variable heating pattern lengths are being developed. Therefore, most experts conclude that effective interstitial heating can be achieved with either technique (Stroh-behn 1984).

The ferromagnetic seeds method requires the surgical implantation of many ferro-magnetic seeds or wires into the tissues of the tumor. These seeds are heating by a strong magnetic field which is created at low frequency (100 KHz) with high-current magnetic coils. The magnetic materials are capable of absorbing power directly from the magnetic field. The human body is transparent to the magnetic field, but the alternating magnetic field is capable of inducing weakly coupled secondary currents (eddy currents) to flow within the conductive tissues of the body. Therefore, in the presence of a strong magnet-ic field it is possible for the ferromagnetic seeds to selectively heat and cause heating of the surrounding tissues by thermal conduction. Breznovich et al. (1984) and Stauffer et al. (1984) report the development of seeds having a Curie point in the therapeutic range. This means that the absorption characteristics of the ferromagnetic material decrease as the temperature of the seeds increases to the Curie temperature (42–50° C). It should be emphasized that the tissue heating resulting from the ferromagnetic seeds is via thermal conduction, as the power is not absorbed directly in the tissues of the body. This means that the seed temperatures will exceed the temperatures of the surrounding tissue and that there will be more temperature variation than with LCF and the microwave intersti-tial technique (which apply the power directly to the tissue). Researchers are now show-ing, however, that it is possible to achieve an acceptable temperature distribution in tu-mors by using the ferromagnetic seed technique with the seeds implanted at very close intervals, less than 1 cm (Mechling and Strohbehn 1986). Very little clinical testing has been done with these ferromagnetic seeds. There is still substantial work to be done on the development of biocompatible seeds, seeds designed for a sharper Curie transition point (changing from power-absorbing to non-power-absorbing), and the insertion tech-niques to place the seeds properly within the tissues.

Ultrasound

Ultrasound techniques involve the use of plane-wave single transducer applicators, me-chanically scanning transducer systems, or transducer focusing arrays. The heating pat-terns resulting from plane-wave transducer devices have been reported to be small in di-ameter and long and ellipsoidal in the depth plane. The increase in focal size and depth compared to a microwave applicator of the same size is a compelling reason to continue to study ultrasound. The plane-wave ultrasound applicator is a useful applicator for in-tact breast and certain non-bony soft tissue sites, but is not likely to be useful for a wide number of tumor sites. The mechanical scanning system designed by Lele (1982) pro-vides for heating of larger volumes than the single plane-wave transducers. In addition, the divergence caused beyond the focal zone of the multiple trajectories would reduce the excessive bone heating at deeper levels. Multiple transducer ultrasound arrays have been reported by Fessenden et al. (1984), Cain and Umemura (1986) and Foster et al. (1980). These are different approaches all attempting to improve the width as well as the depth of the heating field. These techniques promise a potential improvement over the electromagnetic techniques when they are used in areas having little underlying bone or no bone over the tumor site. Clinically it may be difficult to apply ultrasound to many sites of the body, such as the chest wall beyond the ribs, or along the portions of the ab-domen with air-filled regions. Air regions are not penetrated by the ultrasound signal. Even the pelvic region has a great deal of underlying bone which might limit the application of ultrasound. It is still uncertain to what extent the advanced ultrasound techniques will prove to be clinically applicable. A requirement with ultrasound is that if a water bolus medium is used to couple from the transducer to the tissue, the water must be de-gassed so as to not cause cavitation of air within the water that would disrupt the ultrasound beam and heating field.

It is possible to apply highly focused ultrasound as a form of non-invasive surgery or cautery for radical or selective destruction of tumor tissues. The small focal zone created by such an ultrasound device could cause very intense localized heating in deep target sites. Some researchers have proposed that application of this technique would be feasible (D. W. Pounds, personal communication). Tissues could be heated to temperatures well over 50° C for only a few seconds in small target regions and the focal beam could be moved or steered to cover various tumor sizes.

This would require very reliable energy targeting methods. The complications and difficulties which could occur using highly focused ultrasound suggest that caution and conservatism be exercised when considering such aggressive ultrasound methods. Targeting this highly focused ultrasound beam would require substantial imaging and a great deal of information to track the heating beam so as to spare normal critical tissue.

Conclusions

It has been reported that the focal size at depth of the microwave energy is much larger for substantial penetration than the focal size of ultrasound. This gives rise to the understanding that ultrasound can achieve deeper, more highly focused heating than electromagnetics. Often, however, the clinician's ability to precisely locate and direct energy to the tumor would be quite limited with a highly focused ultrasound device. Using electromagnetic methods, to achieve substantial heating depth, phased coherent focused arrays are required where the E fields of each antenna are also aligned with each other. The size of these phased arrays limits their implementation.

The ideal hyperthermia system and techniques do not exist and probably will not exist. We know that there will be deviations from the ideal treatment because of many complicated interactions of factors such as variable blood flow, variable energy coupling, nonuniformity of the power distributions, and various complex tissue interactions with the heating modality. In order to evaluate carefully the performance of these various techniques, either in phantoms or in the clinic, it is essential that temperature measurement in the heating field created be accurate and extensive. Such documentation will provide benchmarks of performance and quantified data to justify further clinical trials and improvements in hyperthermia equipment and techniques.

References

Brezovich IA, Atkinson WJ, Lilly MB (1984) Local hyperthermia with interstitial techniques. Cancer Res [Suppl] 44 (10): 4752s–4756s

Cain CA, Umemura SI (1986) Concentric-ring and sector-vortex phased array applicators for hyperthermia. IEEE Trans Microwave Theory Tech 34 (5): 542–551

Conway J, Anderson AP (1986) Electromagnetic techniques in hyperthermia. Clin Phys Physiol Meas 7 (4): 287–318

Fessenden P, Lee ER, Anderson TL, Strohbehn J, Meyer JL, Samulski TV, Marmor JB (1984) Experience with a multitransducer ultrasound system for local hyperthermia of deep tissues. IEEE Trans Biomed Eng 31 (1): 126–135

Foster FS, Hunt JW (1980) The focussing of ultrasound beams through human tissue. In: (eds) Acoustical imaging. vol 8. Plenum, New York, pp 709–718

Geyser AC (1916) The physics of the high frequency current. NY Med J: 891–893

Guy AW (1984) History of biological effects and medical applications of microwave energy. IEEE Trans Microwave Theory Tech 32 (9): 1182–1200

Guy AW, Lehmann JF, Stonebridge JB (1974) Therapeutic applications of electromagnetic power. IEEE Proc 62 (1): 55–75

Iskander (1981) Physical aspects and methods of hyperthermia production by RF currents and microwaves. AAPM Med Phys Monogr 8: 151–192

Johnson CC, Guy AW (1972) Nonionizing electromagnetic wave effects in biological materials and systems. IEEE Proc 60 (6): 692–716

Lele PP (1982) Local hyperthermia by ultrasound. In: Physical aspects of hyperthermia. Am Assoc Phys Med Monogr 8: 393–440

Lin JC (1986) Special issue on phased arrays for hyperthermia treatment of cancer. IEEE Trans Microwave Theory Tech 34 (5): 481–649

Mechling JA, Strohbehn JW (1986) A theoretical comparison of the temperature distributions produced by three interstitial hyperthermia systems. Int J Radiat Oncol Biol Phys 12: 2137–2149

Pounds DW (1984) Hyperthermia system. US patent 4 441 486

Short JG (1979) Hyperthermia and cancer: a brief review. BSD Medical, Salt Lake City

Short JG, Turner PF (1980) Physical hyperthermia and cancer therapy. IEEE Proc 68 (1): 133–142

Stauffer PR, Cetas TC, Jones RC (1984) Magnetic induction heating of ferromagnetic implants for inducing localized hyperthermia in deepseated tumors. IEEE Trans Biomed Eng 31 (2): 235–251

Steves RA, Paliwal BR (1986) Clinical and physical aspects of hyperthermia. 72nd Scientific Assembly of the Radiological Society of North America, Chicago

Strohbehn JW (1984) Summary of physical and technical studies. In: Overgaard J (ed) Hyperthermia oncology, vol 2. Taylor and Francis, London, pp 353–369

Turner PF (1980) Deep heating of cylindricals or elliptical tissue masses. Poster at International Hyperthermia Conference, Fort Collins, Colorado, June 1980

Advanced Technique in Localized Current Field Hyperthermia*

M. Weisser and P. Kneschaurek

Institut und Poliklinik für Strahlentherapie und Radiologische Onkologie,
Klinikum rechts der Isar, Technische Universität München, Ismaninger Straße 15,
8000 München 80, FRG

Introduction

Interstitial localized current field (LCF) hyperthermia has gained a good reputation as an excellent method for heating localized tumors (Joseph et al. 1981; Manning et al. 1982; Cosset et al. 1984; Goffinet et al. 1985; Vora et al. 1986; Linares et al. 1986; Emami and Perez 1985). Heating is performed by means of ohmic currents flowing through tissue between implanted electrodes. To gain a homogeneous temperature distribution throughout the tumor volume even if perfusion is not uniform it is necessary to continuously measure temperature inside the tumor at many points of interest. Temperature differences can be reduced by altering the specific absorption rate (SAR).

In this paper one such advanced technique will be described. Phantom measurements show the improvement in temperature distribution compared to simple sequencing of needle pairs.

Applicator Needles

We use hollow steel needles with an outer diameter of 1.6 mm. The inner diameter of about 1.2 mm allows the insertion of multithermistor sensors to continuously measure temperature also along the needle axis. Once the needles are placed the hyperthermia treatment may also be easily combined with interstitial irradiation by loading the needles with radioactive sources.

Hyperthermia System

The hyperthermia system is shown in Fig. 1. A detailed description is given by Kneschaurek and Weisser (1987a).

The amplified radio frequency (RF) signal is distributed to an array of up to 10 electrodes. "Hot", "cold" or "floating" condition can be selected for each electrode individually to allow a variety of different SAR. The applied RF voltage and the resulting RF

---

* This work was supported by the DFG, grant BR 678/3-1.

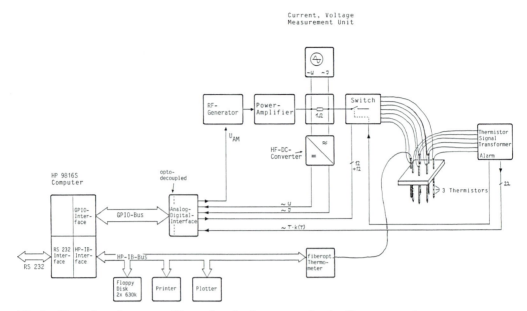

Fig. 1. Hyperthermia system. Electrode selection, power level adjustment and temperature measurement are performed automatically by computer. The different heating strategies are software-implemented. Parameters, e.g., reference temperature, may be varied interactively during treatment

current are measured. The voltage is kept constant independent of load by a feedback loop. Up to 22 temperatures can be measured. System control is done by computer with the possibility of interactively changing parameters, e.g. altering the reference temperature. Different heating strategies like sequential needle pair heating or coldest needle pair heating are software-implemented. The output power is regulated proportional to the difference between the actual temperature and a preset value.

Heating Strategies

Many different heating strategies can be devised (Astrahan and Norman 1982; Strohbehn 1983; Kneschaurek and Weisser 1987). One strategy commonly used with interstitial hyperthermia systems is sequential needle pair heating (SNPH). Heating is performed by sequentially connecting one needle pair after the other to the RF supply. The dwell time for each pair may be varied.

Coldest needle pair heating (CNPH) involves measuring the temperature of all the needles and then connecting only the two coldest needles to the RF source for a given dwell time. The rest of the electrodes are isolated. After the dwell time is over the cycle is repeated. A centrally located thermistor is used as a reference thermometer. The RF power level applied to the electrodes is regulated such that none of the heating electrodes gets hotter than a chosen temperature T-max and that in steady state the desired temperature T-soll is maintained at the reference thermometer.

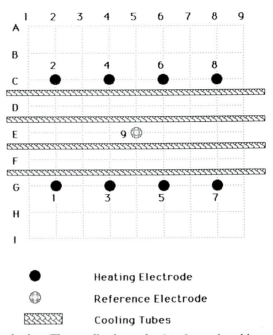

Fig. 2. Phantom: arrangement of electrodes and tubes. The needle electrodes 1 to 9 are placed in a sponge to reduce heat flow from convection. The sponge is placed in a saline solution of 37° C. By pumping the fluid of 37° C through the tubes blood flow is simulated

Measurements

All measurements were performed in a phantom. To avoid convection we used a sponge filled with saline solution of appropriate conductivity. Around the sponge the solution was held at 37° C. Blood flow was simulated by pumping cooling water through tubes placed inside the sponge. The arrangement of electrodes and tubes for the following measurements is shown in Fig. 2. The desired temperature T-soll of the reference electrode (no. 9, Fig. 2) was chosen as 43° C. T-max was set at 48° C. Measurements with and without flow were performed. The results will be discussed below.

No Flow

The time-temperature diagrams are shown in Fig. 3. Due to the high power deposition at the needle surface of the heated needles 1 to 8 compared to the center of the arrangement, the initial temperature rise is very steep. After a few seconds the maximum temperature T-max is reached at some of the heated needles. The applied power is then reduced so as not to exceed this temperature.

Near the end of the warm up phase, in SNPH electrode 1 is about 6° C colder than electrode 3. This can be explained by the localization of electrode 1 near the border of the sponge where the cooling water of 37° C is present. Electrode 2 is located a few millimeters further inside the sponge. Therefore the need for power of needles 1 and 2 is different, resulting in temperature differences that cannot be compensated by varying dwell times.

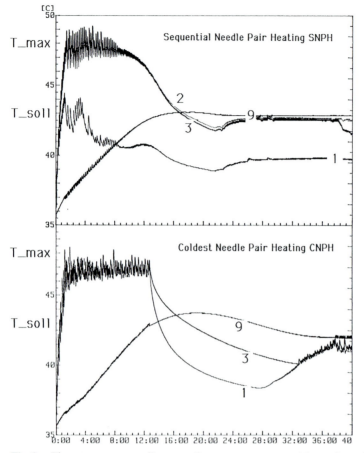

Fig. 3. Time-temperature diagrams for measurements without flow. Due to local differences in perfusion there are significant temperature differences among the heated needles 1 to 8 when using sequential needle pair heating. Coldest needle pair heating keeps the temperatures of the heated needles close together

In the case of CNPH this different need for power can be compensated by connecting needle 1 more often to the RF source than needle 2.

The temperature ripple in heating the coldest needles can be explained by the heating technique. It is as high as the measured temperature rise in the 1-s heating time.

After about 10 min the reference sensor reaches T-soll. The RF power is then shut off until the reference sensor again gets colder than T-soll. The more effective cooling of electrode 1 means that its temperature falls more quickly in the case of CNPH. When heating starts again after about 25 min with CNPH the temperature difference is compensated again. The temperature ripple is reduced because of the reduced power level to maintain steady-state temperatures. When SNPH is used a temperature difference of more than 3° C results in steady state.

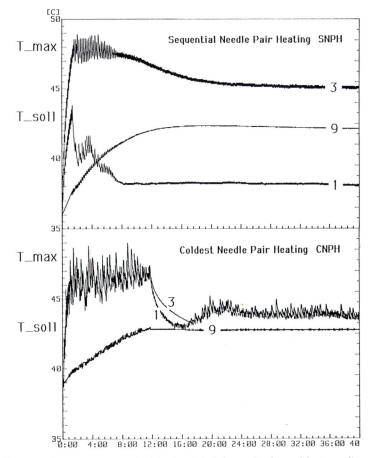

Fig. 4. Time-temperature diagrams for measurement with simulated flow. Again, coldest needle pair heating keeps temperatures close together

Simulation of Blood Flow

Figure 4 shows the situation with simulated blood flow. The warm-up phase is similar to the measurements without flow. Due to the cooling at the center of the phantom the overshoot observed at the reference sensor in the measurements without flow is now nearly compensated. Therefore the power is not switched off for such a long time, resulting in much smaller temperature differences when using CNPH at the end of this period.

During steady state, because of the even more effective cooling of electrode 1 the difference in temperature is increased. With SNPH it now amounts to 7° C between electrode 1 and electrode 3. With CNPH the electrode temperatures are kept close together. Both techniques require a higher power level to maintain steady-state temperatures when blood flow takes place. Therefore the temperature of the heating electrodes is increased.

Conclusion

With SNPH, heating is performed by sequentially connecting needle pairs to the RF supply. The dwell time for each pair can be set individually to compensate for temperature differences between the pairs. Temperature differences between the needles of one pair cannot be compensated with this technique. The improvement of the temperature distribution achieved with the new technique of CNPH is clearly demonstrated by the measurements.

References

Astrahan MA, Norman A (1982) A localized current field hyperthermia system for use with 192-iridium interstitial implants. Med Phys 9: 419–424

Cosset JM, Dutreix J, Dufour J, Janoray P, Damia E, Haie C, Clarke D (1984) Combined interstitial hyperthermia and brachytherapy: Institute Gustave Roussy technique and preliminary results. Int J Radiat Oncol Biol Phys 10: 307–312

Emami B, Perez CA (1985) Interstitial thermoradiotherapy – an overview. Endocuriether Hyperthermia Oncol 1: 34–40

Goffinet DR, Bagshaw MA, Lohrbach A, Sammushi TV, Fersenden P, Prionas SD, Lee E, Kapp DS, Herman TS (1985) The treatment of base tongue cancer and recurrent colorectal neoplasms by multiplexed ^{192}Ir flexible electrode interstitial rf-hyperthermia. Int J Radiat Oncol Biol Phys 11: 119

Joseph CD, Astrahan M, Lipset J, Archambau J, Forell B (1981) Interstitial hyperthermia and interstitial iridium-192 implantation: A technique and preliminary results. Int J Radiat Oncol Biol Phys 7: 827–833

Kneschaurek P, Weisser M (1987a) Computerized system for interstitial hyperthermia. Strahlentherapie 163: 154–163

Kneschaurek P, Weisser M (1987b) A new method for uniform heating in interstitial rf-hyperthermia. Radiology 161 (P): 304

Linares LA, Nori D, Brenner H, Shiu M, Ballon D, Anderson L, Alfieri A, Brennan M, Fucs Z, Hilaris B (1986) Interstitial hyperthermia and brachytherapy: A preliminary report. Endocuriether Hyperthermia Oncol 2: S39–S44

Manning MR, Cetas TC, Miller RC, Oleson JR, Connor WG, Gerner EW (1982) Clinical hyperthermia: results of a phase I trial employing hyperthermia alone or in combination with external beam or interstitial radiotherapy. Cancer 49: 205–216

Strohbehn JW (1983) Temperature distributions from interstitial RF electrode hyperthermia systems: theoretical predictions. Int J Radiat Oncol Biol Phys 4: 1655–1667

Vora N, Shaw S, Forell B, Desai K, Archambeau J, Penke R, Lipset J, Covell J (1986) Primary radiation combined with hyperthermia for advanced (stage III–IV) and inflammatory carcinoma of breast. Endocuriether Hyperthermia Oncol 2: 101–107

Weisser M, Kneschaurek P (1987) Kombination von interstitieller Hyperthermie mit Afterloadingtherapie hoher Dosisleistung. Strahlentherapie 163: 654–658

A 500-kHz Localized Current Field Hyperthermia System for Use with Ophthalmic Plaque Radiotherapy

M. Astrahan, P. Liggett, Z. Petrovich, and G. Luxton

Department of Radiation Oncology, Kenneth Norris Cancer Hospital and Research Institute, School of Medicine, University of Southern California, 1441 Eastlake Avenue, Los Angeles, CA 90033, USA

Introduction

Choroidal melanoma is the most common primary intraocular tumor in adults. Radiotherapy using an episcleral plaque containing an isotope such as ^{125}I has been demonstrated to be effective in the treatment of small choroidal melanomas (Brady et al. 1982; Lommatzsch 1983; Packer and Rotman 1980; Stallard 1966) and is being used with increasing frequency. This technique permits eradication of the tumor while preserving the eye. Although the initial tumor responses following plaque radiotherapy have been good, late complications are now being reported (Payne et al. 1986). These complications may include cataracts, vasculopathy of the retina and optic nerve, and neovascular glaucoma. The occurrence of late complications limits the radiation dose which may be delivered to larger tumors, particularly those in excess of 8 mm in height.

Hyperthermia enhances the effectiveness of radiotherapy for many tumors, particularly malignant melanoma. The thermal enhancement ratio for melanoma is reported as 2.0 (Overgaard 1986; Gillette 1984, Kim et al. 1982). In addition, hyperthermia may have its greatest therapeutic potential when used in conjunction with low dose-rate irradiation (Harisiadis et al. 1978). The combination of hyperthermia and radiotherapy in the treatment of choroidal melanoma is expected to improve local control of larger tumors and may eventually permit a reduction of radiation dose without compromising control of these tumors. Lowering the radiation dose may decrease the incidence of late complications.

Localized current field (LCF) heating was selected as a suitable means for producing hyperthermia from episcleral plaques. LCF heating results from the power dissipated by a 500-kHz electric current conducted through tissue between two or more metallic electrodes. The electrodes may be cylindrical (Doss and McCabe 1976; Astrahan and Norman 1982) or planar in form.

One application of planar electrodes suggested by Doss and McCabe (1976) involves the use of a small electrode which has been surgically implanted adjacent to a tumor, and a second electrode of much larger surface area placed elsewhere on the body. The difference in electrode size results in proportionally greater current density near the smaller electrode. The power dissipated at any point in the tissue by radio frequency (RF) current may be expressed as $P = J^2 r$, where P (W/cm^3) is the power density, J (A/cm^2) is the magnitude of the current density, and r (Ωcm) is the local tissue resistivity.

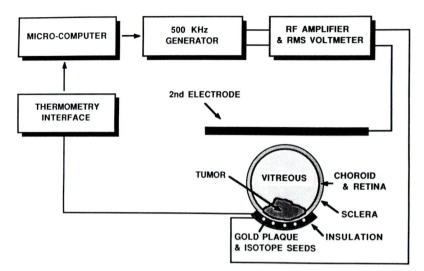

Fig. 1. The LCF ophthalmic hyperthermia system in diagrammatic form. A microcomputer controls a 500-kHz sine wave generator and an RF power amplifier. Temperature is regulated by feedback from microthermocouple sensors epoxied onto the concave surface of the plaque. The tissue volume adjacent to the concave surface of the plaque is selectively heated due to the increased current density relative to the second electrode

Since heating is proportional to J^2, significantly greater local heating occurs near the smaller electrode.

The principle of localized heating by the pairing of large and small planar electrodes may be applied to plaque radiotherapy by considering the episcleral plaque to be the small electrode. The RF current may be further localized to the concave surface of the plaque-electrode adjacent to the sclera by electrically insulating the "back" (convex surface) and sides of the plaque. This is illustrated diagrammatically in Fig. 1.

Plaque Design

The design of the plaque-electrode is illustrated in Fig. 2. The plaque is fundamentally a 2-mm-thick portion of a 25-mm-inner-diameter spherical shell. To construct the plaque, a wax model is created. Eyelets for suturing the plaque to the sclera and for connecting a power wire are added at the edges of the model. One-millimeter-deep slots for ^{125}I seeds and a central channel for a microthermocouple array are then added. The plaque is then cast in gold from the model using a "lost wax" technique. The high atomic number and high density of the gold provides radiation shielding to adjacent organs, and its high thermal conductivity helps to homogenize the temperature distribution.

To complete the plaque, radioisotope seeds are glued into the slots in the plaque, an insulated 30-gauge wire is soldered to one of the eyelets (to supply the RF current), and a microthermocouple array is epoxied into the central channel. The back and sides of the plaque are then painted with an electrically insulating varnish. Placing the seeds in slots insures that the plaque will make mechanically smooth and thus uniform electrical contact with the scleral surface. A cyanoacrylate adhesive is used to attach the seeds so that they may be removed from the plaque by soaking it in acetone.

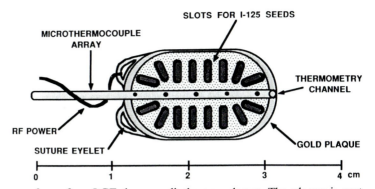

Fig. 2. View of the concave surface of an LCF thermoradiotherapy plaque. The plaque is cast from gold and contains slots into which ^{192}Ir or ^{125}I seeds may be glued. The high atomic number and high density of the gold provides radiation shielding to adjacent organs. The concave metallic surface of the plaque acts as one electrode of the LCF hyperthermia system. The high thermal conductivity of the gold acts to homogenize the heating characteristics of the plaque-electrode. A five-sensor microthermocouple array is epoxied into a channel which follows the long axis of the plaque surface. Power is supplied to the plaque by a small wire soldered to one of the suture eyelets. The convex surface of the plaque is coated with an electrically insulating varnish

Methods

The LCF system illustrated in Fig. 1 was used to evaluate the technique. A microcomputer acts as system controller for a 500-kHz function generator, an RF power amplifier, and an eight-channel microthermocouple thermometry system. The amplifier provides 50 dB power gain into a 50-ohm load over the range 20 kHz to 10 MHz, and is capable of delivering 100 watts to certain loads. In this design, as a safety factor, the maximum output power of the system was intentionally limited to under 30 watts. The performance of the ophthalmic LCF system was evaluated in tissue-like phantom material (Astrahan 1979a, b), in excised bovine eyes, and in vivo in rabbits.

Temperature distributions anterior to the concave surface of various plaques were measured under steady-state conditions. Measurements in rabbit and bovine eyes were made near the perpendicular central axis of the plaque. In phantoms, measurements were made on a 1 mm × 2 mm grid in the plane which intersects the central axis, and the long axis of the plaque. In the phantoms and the bovine eyes, the second electrode was placed about 30 mm directly above the plaque. In the rabbits, the second electrode was placed on the inner surface of an ear.

Results

In all cases (in phantom, in vitro, and in vivo) the temperatures measured on the plaque surface were slightly higher towards the edges of the plaque, but varied overall by not more than 0.5° C.

The steady-state temperature distribution produced by the plaque described in Fig. 2 was measured in a tissue-like phantom. This is illustrated in Fig. 3, together with the ef-

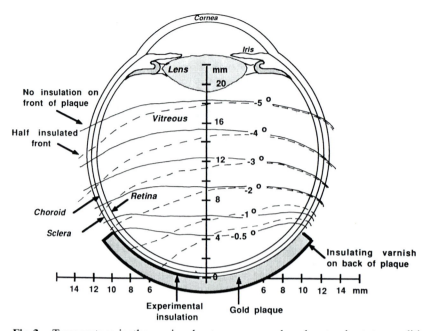

Fig. 3. Temperature isotherms in phantom measured under steady-state conditions in the plane containing the central axis and the long axis of the plaque surface. Isotherms are referenced to the average temperature of the plaque surface. The *solid lines* are for the conventional plaque. The *dashed lines* are for the same plaque in which half of the concave (front) surface has been electrically insulated. Decayed ^{125}I seeds were glued into the slots. The second electrode was placed about 30 mm directly above the plaque. The isotherms have been superimposed on a cross-section of a human eye to illustrate the scale

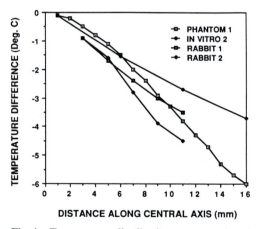

DISTANCE ALONG CENTRAL AXIS (mm)

Fig. 4. Temperature distributions measured under steady-state conditions along the central axis of the plaque in various phantoms, in excised bovine eyes, and in vivo in rabbit eyes. Temperatures are referenced to the average temperature of the plaque surface

fect of electrically insulating half of the plaque surface. Isotherms are referenced to the average temperature of the plaque surface. In this figure, measurements made in phantom are superimposed on a cross-section of a human eye to better illustrate the scale.

Central axis temperature distributions measured in phantoms, in excised bovine eyes, and in vivo in rabbit eyes are plotted in Fig. 4. Temperatures are referenced to the average temperature of the plaque surface. In each case the plaque temperature was raised to about 8° C above the ambient temperature of the object being heated.

Discussion

The uniformity of temperatures measured across the surface of the plaques suggests that a single thermocouple sensor may suffice for a clinical device. The best position for this sensor appears to be midway between the central axis and the edge of the plaque to which the power wire is attached. This would permit the plaque to hold more radioisotope, since the central channel would not have to extend the length of the plaque.

The ability of selective insulation to alter the temperature distribution is demonstrated in Fig. 3. This capability may prove useful when tumors are smaller than the plaque. It may also provide a temperature reduction of about 1–2° C to neighboring structures such as optic nerve. To be effective, however, the insulation must be applied to a fairly large portion of the surface. This is primarily due to the high thermal conductivity of the gold plaque itself. High thermal conductivity within the plaque is intended to reduce the intensity of "hot spots" near the edges of the plaque and to minimize the effects of imperfections on the concave surface. The most common surface imperfection is excessive isotope seed adhesive, which may inadvertently insulate small portions of the plaque surface.

Figure 4 suggests that phantom measurements may, in fact, be useful in predicting in vivo temperature distributions for the eye plaques. The phantom and in vitro temperatures along the central axis are generally within 1° C of the in vivo measurements, and the gradient of about −0.3° C/mm between 3 and 11 mm is similar in all cases. In addition, the temperature of the plaque surface is consistently the maximum temperature measured. This suggests that temperature measured on the plaque surface may alone provide sufficient information to permit safe clinical application.

In the normal rabbit eye, the steady-state central axis temperature distribution showed a 3.5–4.5° C temperature decrease at a height of 11 mm above the plaque surface, and less than 0.5° C variation across the plaque surface. This suggests that in the rabbit, heating the plaque to 45° C should result in therapeutic temperatures at distances up to 10 mm from the plaque surface.

The low cost and operational simplicity of this device should enable a rapid progression from the laboratory to clinical application. In addition, the simplicity of plaque-electrode construction makes a wide variety of custom shapes and sizes possible.

References

Astrahan MA (1979a) Hyperthermia phantom. Med Phys 6: 72
Astrahan MA (1979b) Concerning hyperthermia phantom. Med Phys 6: 235
Astrahan MA, Norman A (1982) A localized current field hyperthermia system for use with Ir-192 interstitial implants. Med Phys 9: 419–424

Brady LW, Shields JA, Augsburger JJ, Day JL (1982) Malignant intraocular tumors. Cancer 49: 578–585

Doss JD, McCabe CW (1976) A technique for localized heating in tissue: An adjunct to tumor therapy. Med Instrum 10: 16–20

Gillette EL (1984) Clinical use of thermal enhancement and therapeutic gain for hyperthermia combined with radiation or drugs. Cancer Res [Suppl] 44: 4836s–4841s

Harisiadis L, Sung D, Kessaris N, Hall EJ (1978) Hyperthermia and low dose-rate irradiation. Radiology 129: 195–198

Kim JH, Hahn EW, Ahmed SA (1982) Combination hyperthermia and radiation therapy for malignant melanoma. Cancer 50: 478–482

Lommatzsch PK (1983) B-irradiation of choroidal melanoma with Ru-106/Rh-106 applicators. Arch Ophthalmol 101: 713–717

Overgaard J (1986) The role of radiotherapy in recurrent and metastatic malignant melanoma: a clinical radiobiological study. Int J Oncol Biol Phys 12: 867–872

Packer S, Rotman M (1980) Radiotherapy of choroidal melanoma with iodine-125. Ophthalmology 87: 582–590

Payne DG, Simpson ER, Japp B, Fitzpatrick PJ, Gallie B, Palvin J (1986) Experience with plaque irradiation of choroidal melanomas (Abstr). Int J Oncol Biol Phys [Suppl] 12: 122

Stallard HB (1966) Radiotherapy for malignant melanoma of the choroid. Br J Ophthalmol 50: 147–155

Simulation of Interstitial Microwave Hyperthermia Using the Finite Element Method

K.-D. Linsmeier, M. Seebass, W. Hürter, F. Reinbold, W. Schlegel, and W. J. Lorenz

Institut für Nuklearmedizin, Deutsches Krebsforschungszentrum, Im Neuenheimer Feld 280, 6900 Heidelberg, FRG

Introduction

Interstitial stereotactic brain tumor therapy is the subject of a cooperation project between the Institute for Nuclear Medicine at the German Cancer Research Center and the Department of Neurosurgery of the University Clinic of Heidelberg. The treatment of brain tumors by surgery or radiotherapy involves a high risk of damage to healthy tissue resulting in permanent impediments or personality changes. Interstitial radiotherapy, in combination with stereotactic operation methods and computer-aided treatment planning, achieves effective treatment of brain tumors while minimizing damage to the surrounding healthy tissue.

Experimental and clinical studies have shown the possibility of enhancing the effectiveness of radiotherapy by hyperthermia treatment (Haveman 1986). The following methods of interstitial hyperthermia are suitable for combination with stereotactic brachytherapy: radio frequency (RF) needles, ferromagnetic seeds and microwave antennas. In brain tumor therapy the number of implants has to be as small as possible, and thus we regard microwave antennas as the most suitable applicators. Of great importance is their ability to deposit energy by interference within tissue regions located between several antennas but out of the range of any single one. Moreover the distribution of the tissue-dependent specific absorption rate (SAR) of such an antenna resembles an iodine seed dose distribution. Comparative calculations by Strohbehn and Mechling (1986) confirm our considerations.

Phase I studies at American clinics demonstrate the clinical safety of this method (Roberts et al. 1986; Salcman and Samaras 1983). Calculations of temperature distributions with given antenna configurations were included in the more recent of the two studies. The correspondence with data taken during the course of the treatment was said to be good.

Method

Like the calculation technique used in the study mentioned above, our simulations are founded on the so-called bio-heat transfer equation (BHTE), a balance equation. Heat delivering and dissipating processes are set off. As heat sources, metabolism, but also, and primarily absorption of energy from external sources like electromagnetic fields, are taken into account. As a transport mechanism, blood flow is far more important than heat conduction. However, the mathematical formulation of this balance results in a partial differential equation (Bowman 1982). The only way to solve such an expression under general conditions like inhomogeneities of tissue or asymmetrical antenna arrangements is to do it approximately by using numerical techniques on computers.

One necessary but very unfamiliar task in working with numerics is to "discretize" the region in which calculations are supposed to be done, i.e., to assign representative points or "nodes" in that region. This discretization allows the transfer of the original formulation of the model into a system of linear equations that can be solved by known methods.

One proven numerical technique to project a differential equation onto an equation system is the method of finite differences. This method was used for temperature calculations in a homogeneous tissue slice orthogonal to the antennas with a circular antenna configuration (Strohbehn et al. 1982). It approximates differential quotients by differences of the variables at the nodes, which form a rectangular grid. Such discretization is not well suited for inhomogeneous media. The representation of curvilinear contours in rectangular meshes necessitates a very high number of nodes and therefore places high demands on the storage capacity and computing velocity of the treatment planning computer.

Mathematically more fastidious, but also more flexible concerning discretization, is the finite element (FE) method. The region of interest is divided into geometrical elements like triangles, under consideration of given structures. Nodes are assigned on the elements, e.g., the triangle corners. The use of interpolating functions and the move from the differential to an equivalent integral expression are further characteristics of this method, making it a convenient tool in examination of transport phenomena (Vemuri and Karplus 1981). Reports of first applications in simulating the performance of various hyperthermia applicators have already been published (Strohbehn and Roemer 1984).

Results

First we constricted the calculations of temperature distributions on a plane orthogonal to the antenna axes and the state of thermal equilibrium. A program package has been developed which accomplishes grid generation, the FE method (Schwarz 1981) and the processing of the solution up to the imaging of isotherms.

Of special interest is the error of the numerical solution, e.g., its distance from the true solution. Using numerical methods presumes a convergence of the calculated approximation to the true solution. The finer the mesh is, i.e., the more it resembles the continuous reality, the more the approximate solution is thought to resemble the true one. On the other hand, one has to minimize computing costs: the finer the mesh, the bigger the amount of storage and the greater the computing time. Therefore the goal was an optimized grid with a prescribed error.

Thus we had to implement computer programs not only for the generation of basic grids but also for the two possibilities of refinement: global and local. "Global" means refinement of the whole mesh, "local" means making the grid finer at specific locations only.

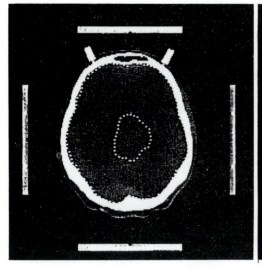

Fig. 1. Target volume and outer boundary

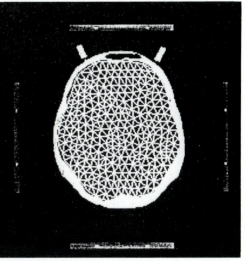

Fig. 2. Basic CAT grid

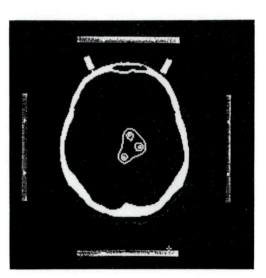

Fig. 3. Calculated isotherms:
38° C, 40° C, 42° C

First, we regarded it easier to handle symmetrical antenna arrangements and homogeneous conditions. With a prescribed error of 1/10° C the resulting optimal grid was characterized as follows: Because of a steep SAR gradient at the antennas, it was necessary to reach a maximal node distance of 0.07 cm by local refinement at the antenna location. For the rest of the grid 1 cm was found to be sufficient.

A program for semiautomatic discretization of irregular regions such as can be found in CAT scans was developed next (Lo 1985) (Figs. 1, 2). Convergence analyses gave the same results as the symmetrical and homogeneous cases. The isotherms calculated on a once global-refined and twice local-refined grid are shown in Fig. 3.

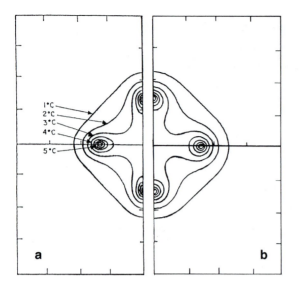

Fig. 4a, b. Temperature distribution calculated by (**a**) the finite difference method (Strohbehn et al. 1982) and (**b**) the finite element method (Linsmeier 1987) (four antennas in resting muscle)

All calculations have been done on a VAX 11/780. Using the optimal grid (2000 nodes, 2300 triangles), the computing time from the CAT grid generation up to the isotherm plotting was about 5 min per slice.

Discussion

Comparisons with the results of the calculations published by Strohbehn et al. (1982) confirm the reliability of the method and the programs (Fig. 4). The flexibility of the FE method in combination with the new technique of grid generation and refinement provide new potential for the use of simulations for radiotherapy and hyperthermia planning.

Inclusion of better blood flow models, measurement of blood flow by means of positron emission tomography and expansion of the calculations to three dimensions are our current ongoing projects.

In summary, we believe that simulations, using the method described, will be incorporated into the planning of combined brachytherapy and hyperthermia treatment in the near future.

References

Bowman HF (1982) Thermodynamics of tissue heating: modeling and measurements for temperature distributions. In: Nußbaum GH (ed) Physical aspects of hyperthermia. AAPM Medical Physics Monograph 68: 511–547

Haveman J (1986) Enhancement of radiation effects by hyperthermia. In: Anghileri LJ, Robert J (eds) Hyperthermia in cancer treatment I. CRC Press, Boca Raton, pp 169–182

Linsmeier KD (1987) Computersimulation einer interstitiellen Mikrowellenhyperthermie bei Hirntumoren mit der Methode der finiten Elemente. Inaugural dissertation. Dept of Clinical Medicine, University of Heidelberg

Lo SH (1985) A new mesh generation scheme for arbitrary planar domains. Int J Num Math Eng 21: 1403–1426

Roberts DW, Coughlin CT, Wong TZ, Fratkin JD, Double EB, Strohbehn JW (1986) Interstitial hyperthermia and iridium brachytherapy in treatment of malignant glioma: a phase-I clinical trial. J Neurosurg 64: 581–587

Salcman M, Samaras GM (1983) Interstitial microwave hyperthermia for brain tumors: results of a phase-I clinical trial. J Neurooncol 1: 225–236

Schwarz HR (1981) FORTRAN-Programme zur Methode der Finiten Elemente. Teubner, Stuttgart

Strohbehn JW, Mechling JA (1986) Interstitial techniques for clinical hyperthermia. In: Hand JW, James JR (eds) Physical techniques in clinical hyperthermia. Wiley, New York, pp 210–287

Strohbehn JW, Roemer RB (1984) A survey of computer simulations of hyperthermia treatments. IEEE Trans Biomed Eng 31 (1): 136–148

Strohbehn JW, Trembly BS, Double EB (1982) Blood flow effects on the temperature distributions from an invasive microwave antenna array used in cancer therapy. IEEE Trans Biomed Eng 29 (9): 649–661

Vemuri V, Karplus WJ (1981) Digital computer treatment of partial differential equations. Prentice Hall, Englewood Cliffs

Some Basic Effects in Cellular Thermobiology

H. Jung and E. Dikomey

Institut für Biophysik und Strahlenbiologie, Universität Hamburg, Martinistraße 52, 2000 Hamburg 20, FRG

The purpose of this introductory review is to provide a brief summary of some of our recent experimental work in the field of thermal biology that might facilitate the understanding of the biological rationale of combining hyperthermia and radiotherapy in the treatment of cancer. Emphasis is placed on some molecular and cellular aspects of cell killing by heat and of hyperthermic radiosensitization; these appear to be distinct rather than related phenomena.

Cell Killing by Heat Alone

The effects of hyperthermia on mammalian cells assayed in terms of cell killing leads to survival curves that are similar to those obtained with ionizing radiation. It has thus become customary in studies measuring cell killing by heat to analyze survival (S) curves by the multitarget, single-hit equation (see Dertinger and Jung 1970) $S = 1 - [1 - \exp(-t/t_0)]^n$ or by the linear-quadratic model, first proposed by Kellerer and Rossi (1971): $S = \exp(-\alpha t - \beta t^2)$. Either model may be applied to hyperthermic cell-killing data (Roti Roti and Henle 1980) but neither takes account of step-up or step-down heating phenomena.

As an alternative, a model has been developed (Jung 1986) according to which cellular inactivation by heat is a two-step process. In the first step, heating produces nonlethal lesions. In the second step, the nonlethal lesions are converted into lethal events. The conversion of one of the nonlethal lesions in a cell leads to cell death. Rigorous mathematical modeling leads to the following equation that describes the surviving fraction $S(t)$ of cells heated to a certain temperature for time t:

$$S(t) = \exp\{(p/c)[1 - ct - \exp(-ct)]\} \tag{1}$$

where p is the rate constant for the production of nonlethal lesions per cell and per unit of time, and c is the rate constant for the conversion of one nonlethal lesion into a lethal event per unit of time. The curve drawn in Fig. 1 through the 43° C data was calculated with $p(43° C) = 9.64 \pm 1.03$ h^{-1} and $c(43° C) = 0.49 \pm 0.065$ h^{-1}. That means that heating of CHO cells to 43° C produces an average of 9.6 nonlethal lesions per cell and per hour of treatment, and each nonlethal lesion has a 49% probability of being converted into a lethal event during 1 h of treatment at 43° C. When heating is performed at two different temperatures consecutively, i.e., when pretreatment at temperature T_1 for time t_1 is fol-

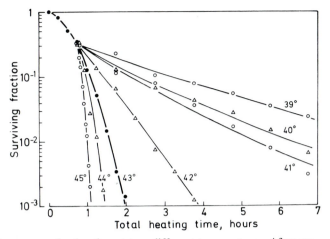

Fig. 1. Inactivation of CHO cells by consecutive heating at two different temperatures. After pretreatment at 43° C for 45 min, the cells were immediately exposed to the temperature indicated and then assayed for colony formation. ●, single heating at 43° C. The lines drawn were calculated from Eq. 2. Data from Jung (1986)

lowed by a graded exposure to temperature T for time t, the surviving fraction is given by the equation

$$S(t_1, t) = \exp\{(p_1/c_1)\exp(-ct)[1 - c_1 t_1 \exp(ct) - \exp(-c_1 t_1)] + (p/c)[1 - ct - \exp(-ct)]\} \qquad (2)$$

where p_1 and c_1 are the production rate and the conversion rate at the temperature of pretreatment T_1, and p and c are the corresponding values at the temperature of the second treatment T. This means that the same pair of parameters (p and c) that describe the survival curve after single heating to a certain temperature also apply to those curves obtained after step-up or step-down heating although the curves differ widely in shape, slope, and curvature (Fig. 1).

By fitting Eq. 2 to the experimental data of many heat survival curves, the values of p and c were determined for the temperature range 39°–45° C. In this range, the conversion rate c increases exponentially with temperature; the slope corresponds to an activation energy E_a of 86 ± 6 kcal/mol. The Arrhenius plot of the production rate p shows an inflection point at 42.5° C. Above that temperature, the E_a is 185 ± 14 kcal/mol; below, an E_a of 370 ± 30 kcal/mol was obtained (Jung 1986).

The proposed model makes it possible to describe the entire survival curve of heated cells by two parameters: these depend only on temperature but are the same for single heating as well as for step-up or step-down heating. In addition, it predicts that recovery phenomena are not involved in the occurrence of the shoulder in heat survival curves and, thus, points at the fundamental difference between cell killing by heat and by ionizing radiation.

Thermotolerance

Thermotolerance is a transient state of thermal resistance which develops in cells following prior heat treatment (for reviews, see Hahn 1982; Sciandra and Gerweck 1986).

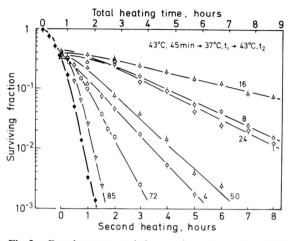

Fig. 2. Development and decay of acutely induced thermotolerance. After a priming heat treatment at 43° C for 45 min, the cells were incubated at 37° C for various periods of time (t_i ranging from 4 to 85 h, as indicated by the *curves*) before a second heat treatment at 43° C for the time t_2 was given. ●, single heating at 43° C. Data from Jung et al. (1986)

Thermotolerance can develop during heating for a few hours to temperatures below 42.5° C (chronic heating) or after a brief priming heat shock (acute heating) followed by a period of incubation at 37° C. Chronic heating causes a continuous decrease in heat sensitivity leading to biphasic cell survival curves (see Dewey et al. 1977).

The development and decay of acutely induced thermotolerance is shown in Fig. 2. After a priming heating at 43° C for 45 min, the cells were incubated at 37° C for various times before the second treatment was given. Maximum thermotolerance was observed after 16 h, when the cells were about 20 times more resistant than without pretreatment.

The maximum degree of thermotolerance as well as the time at which this maximum is reached depend on the severity of the conditioning heat treatment. The maximum thermotolerance ratio (i. e., the ratio of the slopes of the survival curves of control versus preheated cells; TTR) measured after heating for different times at 43° C increased linearly with decreasing log survival (Fig. 3). The pH during treatment was found to affect thermotolerance only to the extent that it modified survival after the priming heat treatment (Fig. 3). Furthermore, the lower the survival rate after the first heating, the later the peak level of thermotolerance occurred (Eickhoff and Dikomey 1984). Similar observations were reported by Henle et al. (1979) and Nielsen and Overgaard (1982).

The decay of thermotolerance occurs with exponential kinetics; it depends on treatment time, temperature, and the proliferative status of the cells. It was shown for CHO cells that the half-life of TTR decay at different pHs was proportional to the cell-number doubling time at the corresponding pH (Eickhoff and Dikomey 1984), and that the decay occurred more rapidly in proliferating than in plateau-phase cells (Gerweck and Delaney 1984).

It appears that the kinetics of thermotolerance induction are largely the same for cells in culture, tumors, and normal tissue. However, thermotolerance has generally been observed to decay faster in vitro than in vivo. For cell cultures heat sensitivity reached control values by 80–90 h after the priming heat treatment (Fig. 2; Henle and Leeper 1976), whereas it took 120–144 h before thermotolerance disappeared in normal tissue

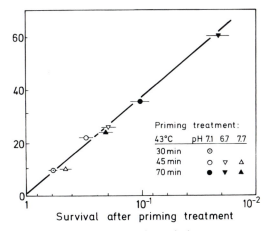

Fig. 3. Maximum thermotolerance ratio *(TTR$_{max}$)* plotted against survival after priming treatment. Thermotolerance was induced in CHO cells by heating to 43° C for various times and at different pHs followed by incubating at 37° C for variable times before thermal sensitivity was assayed by graded exposures to 43° C at the pH indicated. Data from Eickhoff and Dikomey (1984)

(Law 1981) or murine tumors (Maher et al. 1981; Kamura et al. 1982; Mooibroek et al. 1984). Whether this difference is due to different proliferation rates in vitro and in vivo is not yet clear.

Thermal Radiosensitization

The enhancement of cellular radiosensitivity by additional heat treatment is a well-documented phenomenon (for review see Dewey 1983). In general, the maximum cytotoxic effect is observed when radiation is applied simultaneously with heat. If radiation is administered before heat, the sensitization decreases steeply with increasing time at 37° C between the two modalities and disappears within 1–2 h. For heat followed by irradiation, it takes about 6–8 h before the enhanced sensitivity has disappeared (Li and Kal 1977; Mills and Meyn 1983).

When heat is applied immediately prior to X-rays, thermal radiosensitization increases with exposure time at a given temperature (Fig. 4). The same is found when the temperature is increased at a fixed duration of thermal exposure. The effect of thermotolerance on hyperthermic radiosensitization is shown in Fig. 4. Thermotolerance was induced in CHO cells by acute pretreatment at 43° C for 45 min followed by an incubation period of 16 h at 37° C, which leads to maximum development of heat resistance (see Fig. 2). When CHO cells were only irradiated at the time of maximal thermotolerance, the radiosensitivity was not different from that of normal cells (Jung et al. 1986; Dikomey and Jung 1987). However, thermotolerance was found to modify the effect of combined treatment, but only to the extent that thermotolerance enhanced cell survival after heating (Fig. 4). By contrast, the effect of step-down heating is different. If heat treatment at 43° C for 30 min is followed by a second exposure at 40° C for 5 h, the second treatment reduces survival by a factor of about 3, but does not further enhance radiosensitivity (Dikomey and Jung 1987).

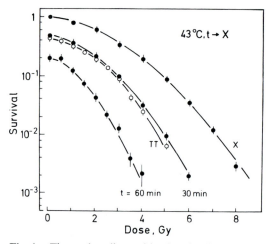

Fig. 4. Thermal radiosensitization in CHO cells. The cells were preheated to 43° C for 30 or 60 min, immediately followed by X-irradiation at 0° C. *X*, X-rays only; *TT*(○) effect of thermotolerance on thermal radiosensitization (43° C, 45 min→37° C, 16 h→43° C, 60 min→X-rays). Data from Dikomey and Jung (1987)

Loss of Polymerase β Activity

At the molecular and metabolic level, heat primarily induces conformational changes and destabilization of macromolecules or of multimolecular structures and leads to altered rates of metabolic reactions during heat treatment (Streffer 1985). Although many of these alterations have been measured so far, it is still not clear which of the molecular changes are involved in cell killing and in thermal radiosensitization. The polymerase β is of special interest in this respect, since it is known to be involved in DNA repair and, furthermore, appears to be especially sensitive to heat (Dube et al. 1977).

Heating of CHO cells at temperatures ranging from 40° to 46° C was shown to cause a reduction of polymerase β activity (Dikomey et al. 1987). As shown for heating to 43° C (Fig. 5), biphasic response curves were obtained for all temperatures tested, indicating that the heat sensitivity decreased with longer treatment times. The initial and final slopes were characterized by the t_0 values.

The activation energies were determined from the Arrhenius plot obtained for the initial and final slopes. The initial heat sensitivity of polymerase β is characterized by an E_a of 120 ± 10 kcal/mol for the entire temperature range, whereas the final slope shows an inflection point at 43° C with an E_a of 360 ± 40 kcal/mol below 43° C and an E_a of 130 ± 20 kcal/mol above 43° C. For cell killing, the E_a is 345 ± 20 or 147 ± 11 kcal/mol (Dikomey et al. 1984), and thus corresponds only to the final heat sensitivity of polymerase β. This indicates that the loss of polymerase β activity is not involved in heat-induced cell death.

It is interesting to note that the induction of thermotolerance affects the heat sensitivity of polymerase β. Making CHO cells thermotolerant either by an acute heat shock at 43° C for 45 min followed by incubation at 37° C for 16 h (Fig. 5) or by chronic treatment at 40° C for 16 h, resulted in a clear lowering of polymerase β heat sensitivity, by a factor of up to 5 (Dikomey et al. 1987). By contrast, if thermosensitization is induced by step-down heating, for instance by giving 43° C for 30 min followed by graded treat-

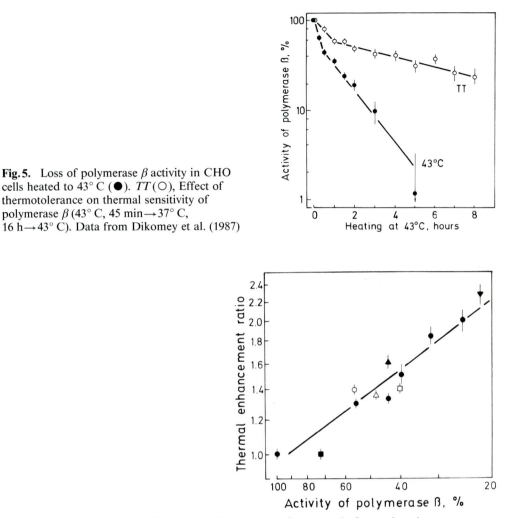

Fig. 5. Loss of polymerase β activity in CHO cells heated to 43° C (●). TT (○), Effect of thermotolerance on thermal sensitivity of polymerase β (43° C, 45 min→37° C, 16 h→43° C). Data from Dikomey et al. (1987)

Fig. 6. Correlation between the activity of polymerase β measured after various heat treatments, and thermal enhancement ratio determined after the same heat pretreatments on the 10% survival level of heat response curves normalized to 100% at zero radiation dose. *Filled symbols*, single heating at different temperatures for different times; *open symbols*, multiple heat treatments leading to acute thermotolerance (△), chronic thermotolerance (○), or thermosensitization (□). Data from Dikomey and Jung (1987)

ments at 40° C, polymerase activity is not measurably reduced by the second treatment. From this result we conclude that the heat sensitivity of cells is not correlated with polymerase β activity, as has previously been postulated by some other groups (Dewey and Esch 1982; Mivechi and Dewey 1984; Spiro et al. 1982). Our conclusion is further supported by the findings of Jorritsma et al. (1986) and Kampinga et al. (1985), who showed that in HeLa and EAT cells, fractionated hyperthermic treatments had different effects on the activity of polymerase β and on cell death.

Polymerase β activity measured after various single or multiple heat pretreatments is plotted in Fig. 6 against the thermal enhancement ratio (TER) as determined from several dose-response curves for heat combined with X-rays (e. g., Fig. 4). It becomes obvious that the radiosensitivity of preheated CHO cells correlates well with polymerase β activity at the beginning of irradiation (Dikomey and Jung 1987). A similar correlation was observed for CHO cells by Mivechi and Dewey (1985) and for HeLa cells by Jorritsma et al. (1986), indicating that hyperthermic radiosensitization might be associated with the transient loss of polymerase β activity.

Impairment of Repair of Radiation Damage

The experiments described in the preceding section have shown that heat causes damage to repair enzymes such as polymerase β. It thus appeared of interest to analyze the effect of heat on the kinetics of repair of DNA strand breaks.

After X-irradiation with 9 Gy (performed at $0°$ C to prevent repair during exposure), the number of DNA strand breaks per cell decreases with repair time at $37°$ C (Fig. 7). This decrease may be described by the sum of three exponential functions representing three classes of strand breaks (Dikomey and Franzke 1986). Of the radiation-induced strand breaks, 70% are repaired with a half-time of about 2 min (class I), 25% with a half-time of 20 min (class II) and 5% with a half-time of 170 min (class III).

When the cells are heated to $43°$ C for 60 min prior to irradiation the number of strand breaks produced (zero time interval) is only slightly higher than after irradiation alone, the difference being exclusively due to heat-induced breaks. This result indicates that the number of radiation-induced DNA strand breaks is not increased by additional heating.

It is obvious from Fig. 7 that after combined treatment, repair of DNA strand breaks is much slower than after X-irradiation alone. A closer analysis of such data measured after different temperatures and various exposure times showed that the proportions of the three classes of DNA strand breaks were altered by heat, and also that the half-times

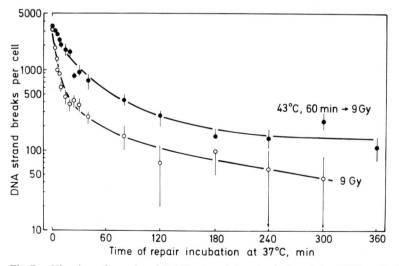

Fig. 7. Kinetics of repair of DNA strand breaks induced in CHO cells by a single dose of X-rays ($\bigcirc$) or by heating to $43°$ C for 60 min immediately followed by irradiation ($\bullet$). Data from Dikomey and Franzke (1986) and Franzke (unpublished)

of repair were prolonged, demonstrating that heat damage impairs the subsequent repair of radiation damage. As a consequence of decreased repair, the possibility of misrepair should be higher, leading to an increase in chromosomal aberrations (Dewey et al. 1978) and finally to the increased radiosensitivity that is generally observed in heated cells.

References

Dertinger H, Jung H (1970) Molecular radiation biology. Springer, Berlin Heidelberg New York

Dewey WC (1983) Hyperthermia in cancer therapy. In: Broerse JJ, Barendsen GW, Kal HB, van der Kogel AG (eds) Proceedings of the Seventh International Congress of Radiation Research. Martinus Nijhoff, Amsterdam, pp 517–527

Dewey WC, Esch JL (1982) Transient thermal tolerance: cell killing and polymerase activities. Radiat Res 92: 611–614

Dewey WC, Hopwood LE, Sapareto SA, Gerweck LE (1977) Cellular responses to combinations of hyperthermia and radiation. Radiology 123: 463–474

Dewey WC, Sapareto SA, Betten DA (1978) Hyperthermic radiosensitization of synchronous Chinese hamster cells: relationship between lethality and chromosomal aberrations. Radiat Res 76: 48–59

Dikomey E, Franzke J (1986) Three classes of DNA strand breaks induced by X-irradiation and internal β-rays. Int J Radiat Biol 50: 893–908

Dikomey E, Jung H (1987) Correlation between polymerase β activity and thermal radiosensitization in CHO cells. Recent Results Cancer Res (in press)

Dikomey E, Eickhoff J, Jung H (1984) Thermotolerance and thermosensitization in CHO and RIH cells: a comparative study. Int J Radiat Biol 46: 181–192

Dikomey E, Becker W, Wielckens K (1987) Reduction of DNA-polymerase β activity of CHO cells by single and combined heat treatments. Int J Radiat Biol (in press)

Dube DK, Seal G, Loeb LA (1977) Differential heat sensitivity of mammalian DNA polymerases. Biochem Biophys Res Commun 76: 483–487

Eickhoff J, Dikomey E (1984) Development and decay of acutely induced thermotolerance in CHO cells by different heat shocks at various external pH values. In: Overgaard J (ed) Hyperthermic oncology 1984. Taylor and Francis, London, pp 91–94

Gerweck LE, Delaney TF (1984) Persistence of thermotolerance in slowly proliferating plateau phase cells. Radiat Res 97: 365–372

Hahn GM (1982) Hyperthermia and cancer. Plenum, New York

Henle KJ, Leeper DB (1976) Interaction of hyperthermia and radiation in CHO cells: recovery kinetics. Radiat Res 66: 505–518

Henle KJ, Bitner AF, Dethlefsen LA (1979) Induction of thermotolerance by multiple heat fractions in Chinese hamster ovary cells. Cancer Res 39: 2486–2491

Jorritsma JBM, Burgman P, Kampinga HH, Konings AWT (1986) DNA polymerase activity in heat killing and hyperthermic radiosensitization of mammalian cells as observed after fractionated heat treatments. Radiat Res 105: 307–319

Jung H (1986) A generalized concept for cell killing by heat. Radiat Res 106: 56–72

Jung H, Dikomey E, Zywietz F (1986) Ausmaß und zeitliche Entwicklung der Thermoresistenz und deren Einfluß auf die Strahlenempfindlichkeit von soliden Transplantationstumoren. In: Streffer C, Herbst M, Schwabe H (eds) Lokale Hyperthermie. Deutscher Ärzte-Verlag, Cologne, pp 23–38

Kampinga HH, Jorritsma JBM, Konings AWT (1985) Heat-induced alterations in DNA polymerase activity of HeLa cells and of isolated nuclei. Relation to cell survival. Int J Radiat Biol 47: 29–40

Kamura T, Nielsen OS, Overgaard J (1982) Development of thermotolerance during fractionated hyperthermia in a solid tumor in vivo. Cancer Res 42: 1744–1748

Kellerer AM, Rossi HH (1971) RBE and the primary mechanism of radiation action. Radiat Res 47: 15–34

Law MP (1981) The induction of thermal resistance in the ear of the mouse by heating at temperatures ranging from 41.5° C to 45.5° C. Radiat Res 85: 126–134

Li GC, Kal HB (1977) Effect of hyperthermia on the radiation response of two mammalian cell lines. Eur J Cancer 13: 65–69

Maher J, Urano M, Rice L, Suit HD (1981) Thermal resistance in a spontaneous murine tumor. Br J Radiol 54: 1086–1090

Mills MD, Meyn RE (1983) Hyperthermic potentiation of nonrejoined DNA strand breaks following irradiation. Radiat Res 95: 327–338

Mivechi NF, Dewey WC (1985) DNA polymerase α and β activities during the cell cycle and their role in heat radiosensitization in Chinese hamster ovary cells. Radiat Res 103: 337–350

Mooibroek J, Zywietz F, Dikomey E, Jung H (1984) Thermotolerance kinetics and growth pattern changes in an experimental rat tumour (RIH) after hyperthermia. In: Overgaard J (ed) Hyperthermic oncology 1984. Taylor and Francis, London, pp 215–218

Nielsen OS, Overgaard J (1982) Influence of time and temperature on the kinetics of thermotolerance in L1A2 cells in vitro. Cancer Res 42: 4190–4196

Roti Roti JL, Henle KJ (1980) Comparison of two mathematical models for describing heat-induced cell killing. Radiat Res 81: 374–383

Sciandra JJ, Gerweck LE (1986) Thermotolerance in cells. In: Watmough DJ, Ross WM (eds) Hyperthermia. Blackie, Glasgow, pp 99–120

Spiro IJ, Denman DL, Dewey WC (1982) Effect of hyperthermia on CHO DNA polymerase α and β. Radiat Res 89: 134–149

Streffer C (1985) Metabolic changes during and after hyperthermia. Int J Hyperthermia 1: 305–319

Growth, Cell Proliferation and Morphological Alterations of a Mouse Mammary Carcinoma After Exposure to X-Rays and Hyperthermia

K. C. George, D. van Beuningen, and C. Streffer

Institut für Medizinische Strahlenbiologie, Universitätsklinikum Essen, Hufelandstraße 55, 4300 Essen 1, FRG

Introduction

There are numerous studies on the effects of radiation and hyperthermia on experimental animal tumours (Denekamp and Fowler 1977; Hahn 1982). Tumour regression is the criterion of response frequently used in most of these studies. However, a number of biological parameters can influence the effects of radiation and hyperthermic treatment on tumours, and only few of these studies consider simultaneously tumour growth, tumour cell proliferation and morphological alterations (Thomlinson and Craddock 1967; Hermens and Barendsen 1969; Kovacs et al. 1976; Nelson et al. 1976; Rowley et al. 1980). Therefore, it was of interest to study several such parameters on the same tumour system. Metabolic studies were carried out by our group with a C57 mouse mammary carcinoma. The tumours were irradiated with one of various doses of X-rays and heated to 43° C for 30 min following exposure to 10 Gy. Tumour growth, cell proliferation, formation of micronuclei and morphological changes in necrosis and density of small blood vessels were studied.

Materials and Methods

Animals, Tumour and Transplantation

The experiments were performed with 6- to 8-week-old C57 male mice weighing 18–26 g. An adenocarcinoma originally developed from a spontaneous mammary tumour of C57 mouse was used as the test system. The tumour was maintained by serial transplantation every 7 days. A cell suspension from the excised tumour tissue was prepared and 1.75×10^5 viable tumour cells in 0.3 ml NaCl solution were injected intramuscularly into the right hind leg of experimental animals.

Irradiation

Tumours were treated with radiation and/or hyperthermia on 6 days after tumour inoculation when they reached a volume of 1100 ± 290 mm³. Tumours were locally irradiated with single doses of 10, 20 or 30 Gy of X-rays at a dose rate of 100 R/min using a

Siemens Stabilipan X-ray machine (240 kVp, 15 mA, 0.5 mm Cu filter). Two tumours were irradiated simultaneously while the remainder of the body was shielded with lead.

Hyperthermic Treatment

Hyperthermia (43° C tumour core temperature) was locally administered for 30 min as a single dose using an ultrasonic machine (1.7 MHz). The tumour core and the rectal temperatures were measured by thermocouples implanted in the mice throughout the period of heat treatment. Control animals received similar treatment without heating.

For studying the combined effect of radiation and hyperthermia tumours were first irradiated with 10 Gy and then immediately (within 10 min) heated to 43° C for 30 min as described above.

Tumour Growth and Cell Proliferation

Tumour volume was determined from two diameters. For study of cell proliferation, tumours were excised at 6, 12, 24, 48, 72, 96 and 120 h after radiation and/or hyperthermic treatment, and DNA content was measured as described earlier (Streffer et al. 1984). Micronuclei were evaluated after staining the cell nuclei preparation with ethidium bromide.

Morphological Alterations

For estimating morphological changes in necrosis and in vascular density after various treatments, tumours were fixed in Bouin's fluid. Paraffin sections were prepared and stained with haematoxylin-eosin for studying necrosis and haematoxylin-Masson for blood vessels. The point count method was employed to determine the amounts of necrosis, tumour tissue and muscle present in the tumour.

Results and Discussion

Figure 1 shows the relative tumour volume after various treatments. From the originally measured tumour volume, the amount of muscle and of necrosis were deducted. Without this deduction, only a delay of tumour growth between 1–3.8 days was seen (data not shown). It can be seen that the effects of the treatments became more apparent when muscle and necrosis were deducted. The effect of the combined treatment with 10 Gy plus 43° C was equivalent to that obtained with 20 Gy. In the absence of a full dose-response curve, it is presently difficult to state whether there is any synergism between radiation and hyperthermia. In tumours exposed to 30 Gy, there was no growth up to 5 days after treatment.

Treatment of tumours with radiation and/or hyperthermia caused considerable perturbation of the cell cycle which lasted about 2 days. For instance, the well-known radiation-induced G_2 block of the cells was clearly observed at 12 h after irradiation, as indicated by an increase of the G_2-+ M-phase cells associated with a depletion of the G_1

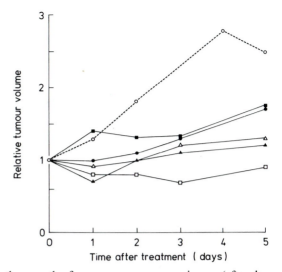

Fig. 1. Effect of X-rays and hyperthermia on the growth of mouse mammary carcinoma (after deducting the amounts of muscle and necrosis). ○, untreated; ●, 43° C (30 min tumour core temperature); ■, 10 Gy; △, 10 Gy + 43° C; ▲, 20 Gy; □, 30 Gy. In this and the following experiments tumours were treated on 6 days after tumour inoculation (volume 1100 ± 290 mm^3) and hyperthermia was administered within 10 min following irradiation with single dose of X-rays. Each point represents the mean value of two to three tumours

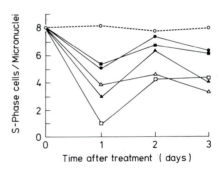

Fig. 2. Ratio of S-phase cells to micronuclei in mouse mammary carcinoma after treatment with X-rays and hyperthermia. Symbols as in Fig. 1. Each point is the mean value of 2–15 tumours

cells, whereas it was delayed when radiation was combined with heat (data not shown). A similar delay of G_2 block after combined treatment with radiation and hyperthermia was also observed in human melanoma cells in vitro, and this phenomenon is presently being investigated in detail. Twenty-four hours after various treatments, the proportion of S-phase cells decreased considerably, although the formation of micronuclei showed only a small increase (data not shown). However, the ratio of S-phase cells to micronuclei was significantly reduced during this period, as shown in Fig. 2. Since this ratio may reflect the cell turnover (Streffer et al. 1984), it might explain the growth delay induced by radiation and hyperthermia.

Figure 3 shows the changes in necrosis after various treatments. It can be seen that there is no significant increase of necrosis up to 2 days after any of these treatments. However, 5 days after exposure to 10 Gy plus heat, as well as to 30 Gy, the proportion of

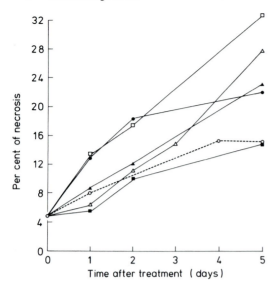

Fig. 3. Necrosis in mouse mammary carcinoma after exposure to X-rays and hyperthermia. Symbols as in Fig. 1. Each point represents the mean value of 2–3 tumours

necrosis was markedly enhanced. Although none of these treatments affected the density of small blood vessels such as capillaries and arteries in the tumour (data not shown), it may be pointed out that damage to tumour microvasculature as well as changes in the dynamics of blood flow may occur following radiation and hyperthermic treatments (Thomlinson and Craddock 1967; Nelson et al. 1976; Reinhold et al. 1982). Indeed, the increase of necrosis after treatment with 10 Gy plus heat and 30 Gy as observed in the present study may indicate such circulatory system inadequacies in clearing cell debris and is consistent with the observations of Hilmas and Gillette (1975) and Walter and Maurer-Schultze (1987).

Thus the present simultaneous study of tumour volume, morphology and cell proliferation of the mammary carcinoma demonstrates that the kinetics of cellular response to radiation and hyperthermic treatment occur well in advance of measurable volume changes. The absence of any remarkable reduction in the originally measured tumour volume, even after exposure to relatively large single doses of X-rays and to a combination of X-rays plus heat, indicates that tumour size alone is a poor criterion of the effects of radiation and hyperthermia. Other parameters, such as cell proliferation and morphological alterations, may influence the response of tumours to radiation and hyperthermic treatment.

Summary

A C57 mouse mammary carcinoma was irradiated with 10, 20 or 30 Gy of X-rays or heated to 43° C for 30 min preceded or not by exposure to 10 Gy. Tumour growth, cell proliferation kinetics, induction of micronuclei and morphological changes in necrosis and vascular density were simultaneously determined.

Treatment with radiation and/or hyperthermia produced only a delay in tumour growth of between 1 and 3.8 days. However, the effects of the treatments became more apparent when the amounts of muscle and necrosis were deducted from the originally measured tumour volume. Radiation-induced G_2 block of the cells was observed at 12 h after irradiation alone. After the combined treatment, however, the G_2 block was delayed beyond 12 h. Moreover, 24 h after the various treatments, the proportion of S-phase cells decreased considerably although the formation of micronuclei showed only a marginal increase. However, the ratio of S-phase cells to micronuclei was significantly reduced during this period. Whereas the amount of necrosis was markedly enhanced 5 days following treatment with 10 Gy plus heat, as well as after 30 Gy, no alterations in the density of small blood vessels could be observed during this period.

These results clearly demonstrate that the apparent changes in tumour volume after X-rays and hyperthermia do not truly reflect the response of the constituent cells and that there are many other factors, for instance cell proliferation and morphological alterations, that influence the effects of radiation and hyperthermia on tumours.

References

Denekamp J, Fowler JF (1977) Cell proliferation kinetics and radiation therapy. In: Becker FF (ed) Cancer: A comprehensive treatise, vol. 6. Plenum, New York, pp 101–137
Hahn GM (1982) Hyperthermia and cancer. Plenum, New York
Hermens AF, Barendsen GW (1978) The proliferation status and clonogenic capacity of tumour cells in a transplantable rhabdomyosarcoma of the rat before and after irradiation with 800 rad of X-rays. Cell Tissue Kinet 11: 83
Hilmas DE, Gillette EL (1975) Microvasculature of C3H/Bi mouse mammary tumours after X-irradiation. Radiat Res 61: 128–143
Kovacs CJ, Hopkins HA, Evans MJ, Looney WB (1976) Changes in cellularity induced by radiation in a solid tumour. Int J Radiat Biol 30: 101–113
Nelson JSR, Carpenter RE, Burboraw RE (1976) Mechanisms underlying reduced growth rate in C3H BA mammary adenocarcinoma recurring after single doses of X-rays or fast neutrons. Cancer Res 36: 524–531
Reinhold HS, Wike-Hooley JC, van den Berg AP, van den Berg Blok A (1982) Environmental factors, blood flow and microcirculation. In: Overgaard J (ed) Hyperthermic oncology. Taylor and Francis, London, pp 41–52
Rowley R, Hopkins HA, Bestill WL, Retnour ER, Looney WB (1980) Response and recovery kinetics of a solid tumour after irradiation. Br J Cancer 42: 586–595
Streffer C, van Beuningen D, Bamberg M, Eigler FW, Gross E, Schabronath J (1984) An approach to the individualization of cancer therapy – determination of DNA, SH groups and micronuclei. Strahlentherapie 160: 661–666
Thomlinson RH, Craddock EA (1967) The gross response of an experimental tumour to single doses of X-rays. Br J Cancer 21: 108–123
Walter J, Maurer-Schultze B (1987) Regrowth, tumour cell proliferation and morphological alterations of the adenocarcinoma EO771 following a single dose of 30 Gy ^{60}Co γ-rays. Strahlentherapie (in press)

Analysis of Results in Neck Node Metastases from Tumors of the Head and Neck

G. Arcangeli

Istituto Medico e di Ricerca Scientifica, Via S. Stefano Rotando 6, 00184 Roma, Italy

Introduction

The clinical experience accumulated in recent years has clearly shown the potential of using hyperthermia in addition to radiotherapy (Dethlefsen and Dewey 1982; Overgaard 1984 and 1985; Storm 1983). However, the information on specific tumor site and histology, as well as on the duration of the heat-induced improvement in radiation response, is limited (Perez et al. 1983; Arcangeli et al. 1985; Dewhirst and Sim 1984) and more careful investigation is warranted.

Materials and Methods

This study was carried out on 38 patients with a total of 81 multiple neck node metastases from squamous cell carcinoma of the head and neck treated between 1977 and 1982.

The details of heat and radiation treatment have been described elsewhere (Arcangeli et al. 1980, 1983 a, b). Briefly, radiation was given with 6 MeV photons or with electrons of various energies by means of linear accelerators.

Several heat devices were used, including a 27-MHz diathermic generator and a microwave generator with a frequency ranging from 300 to 3000 MHz (mainly employed at frequencies around 500 and 2450 MHz). The temperature was monitored at regular intervals, with the power off, by inserting an 18-gauge constantan-copper thermocouple wire inside the plastic lumen of a standard intracatheter probe previously placed at the base of the lesion. Radiation was delivered as three fractions/day of 1.5–2 Gy each, with 4-h intervals between fractions, to a total dose of 60 Gy; hyperthermia at 42.5° C for 45 min, was applied every other day, immediately after the second daily fraction of radiation, for a total of seven treatments. Heating time was the effective time at the treatment temperatures.

Before 1983, multisensor probes were not available in our institution and therefore the temperature was measured only at one point at the base of the lesion. Because of poor matching, it was not always possible to achieve the prescribed temperature. In this case, treatments of 45 min at the maximum tolerable temperature were delivered, and the temperature data were converted to equivalent minutes at 42.5° C (Eq. 42.5) using an empirical isoeffect relationship recently employed by several investigators (Sapareto

1982; Field and Morris 1983) based on 1° C being equivalent to a factor of 2 in heating time above the transition point (i.e., 42.5° C in our cases), and to a factor of 4 below. Although we do not believe that this is the best way to quantify thermal dose, we think that this relationship can be used to compare responses to different time-temperature combinations. For this purpose, Eq. 42.5 was simply derived from the average of temperatures taken every 5 min during the whole period of heating.

Tumor response was simply recorded as failure or success (i.e., total disappearance of lesion) at the end of treatment or soon after. Recurrences were defined as the reappearance of tumor following complete response.

Statistical significance was tested by means of the two-tailed Fisher exact test (Sokal and Rohlf 1969).

The influence of the treatment modality on the time from beginning of treatment to first detection of recurrence was also analyzed. The response duration was evaluated by means of the Kaplan and Meier (1958) product limit method. The difference between the curves was tested by means of the Mantel-Haenszel test (Mantel 1966). Every patient had at least two lesions, each one assigned to a specific treatment modality. Patients who died without suffering the event of interest (i.e. local recurrence of the lesion treated with the specific modality under consideration) were designated recurrence-free.

Results

The immediate overall complete response rates were 79% (30/38) and 42% (18/43) respectively for the lesions treated with combined modality and radiotherapy alone. The difference was statistically significant ($p < 0.05$).

A general idea of response duration is given in Fig. 1 which shows an actuarial analysis of local control according to the Kaplan and Meier (1958) product limit method. The maximum follow-up was 28 months. At 2 years, local control was 58% and 14% respectively in the combined modality and radiotherapy alone groups.

The estimated local control distribution as a function of tumor volume in the two treatment arms is shown in Fig. 2. The response duration seems to be significantly dependent upon tumor volume in both arms. However, the volume effect was more pro-

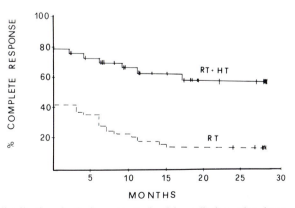

Fig. 1. Estimated tumor local control distribution in lesions treated with radiation plus heat *(RT + HT)* and radiation alone *(RT)* according to the Kaplan and Meier product limit method. The difference between the curves is statistically significant *(p < 0.05)*. (From Arcangeli et al. 1985)

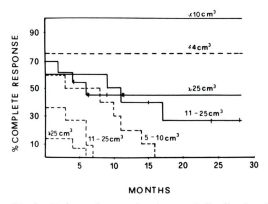

Fig. 2. Estimated tumor local control distribution for different ranges of tumor volumes in the two treatment arms. ----, radiation alone; ——, radiation + heat. The difference between the two arms was statistically significant for the curves corresponding to each class of volume ($p < 0.05$). (From Arcangeli et al. 1985)

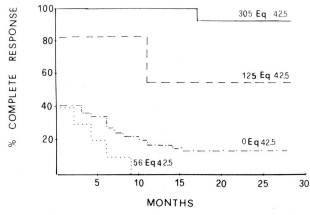

Fig. 3. Estimated tumor local control distribution for different Eq 42.5 values. The lesions that received 0 Eq 42.5 are those treated with radiotherapy alone. The difference between this latter curve and the curves at 125 and 305 Eq 42.5 was statistically significant ($p < 0.05$). (From Arcangeli et al. 1985)

nounced in the radiation alone arm than in the combined modality arm. For lesions of small volume the response duration was similar in both arms, in that there was no recurrence during the follow-up period. However, in lesions of larger volumes local control after radiation alone was 0% at 15, 7 and 6 months for increasing volumes. In the combined modality arm, 27% of lesions 11–25 cm³ in size and 45% of those over 25 cm³ were still controlled at 28 months. The difference between the two treatment arms was statistically significant for the curves corresponding to each class of volumes ($p < 0.05$).

Response duration was also analyzed as a function of the thermal dose (Fig. 3). The curve at 0 Eq 42.5 is that obtained after radiation alone and is derived from Fig. 1. After 28 months, 56% and 92% of the lesions treated with 125 and 305 Eq 42.5 respectively were still controlled, in comparison with 14% of the lesions in the group treated with 0

Eq 42.5. The curve at 0 Eq 42.5 was not statistically different from that at 56 Eq 42.5; however, when the former was compared with the curves at 125 and 305 Eq 42.5, the difference was statistically significant ($p < 0.05$), clearly indicating the presence of a threshold dose above which response duration increases on increasing the thermal dose.

In this study, the addition of heat did not result in any enhancement of early or late radiation effects on normal skin and subcutaneous tissue. The percentage of acute skin reaction and of late fibrosis was similar in both treatment arms. However, thermal damage (blisters) was seen in eight patients.

Discussion

Our findings on neck node metastases from head and neck cancer show that both immediate response and response duration are enhanced by the addition of heat. However, there are some important variables, such as tumor volume and "isoeffect thermal dose," which can influence local tumor control.

The volume effect was more pronounced for the lesions treated with radiation alone than for those treated with the combined modality: this is clearly seen on looking at the estimated local control distribution, which appeared to be less influenced by tumor volume when radiation was given in combination with heat than when it was given without heat.

Tumor response has been positively correlated with mean temperature, lowest daily average temperature and minimum temperature (Dewhirst and Sim 1984). In this study, we have also correlated response duration with the isoeffect thermal dose by converting the recorded temperature into equivalent minutes at 42.5° C (Sapareto 1982; Field and Morris 1983). It is beyond the purpose of this report to discuss the value of using the equivalent time as the method of measurement of the thermal dose. We have used it essentially to compare treatments at different temperatures. One could argue about the legitimacy of this when hyperthermia is combined with radiation. Nevertheless, the estimated local control distributions show that the duration of response is better in the lesions that received more Eq 42.5. These results are in agreement with other clinical findings (Perez et al. 1983) and with those of a randomized trial on pet animal tumors (Dewhirst and Sim 1984).

Despite the technical difficulties, there must be further and intensified efforts to develop an effective approach to treatment of radioresistant tumors using combined heat radiation.

References

Arcangeli G, Barni E, Cividalli A, et al. (1980) Effectiveness of microwave hyperthermia combined with ionizing radiation: clinical results on neck node metastases. Int J Radiat Oncol Biol Phys 6: 143–148

Arcangeli G, Cividalli A, Lovisolo G, Nervi C (1983a) Clinical results after different protocols of combined local heat and radiation. Strahlentherapie 159: 82–89

Arcangeli G, Cividalli A, Creton G, et al. (1983b) Tumor control and therapeutic gain with different schedules of combined radiotherapy and local hyperthermia in human cancer. Int J Radiat Oncol Biol Phys 9: 1125–1134

Arcangeli G, Arcangeli GC, Guerra A, et al. (1985) Tumor response to heat and radiation: prognostic variables in the treatment of neck node metastases from head and neck cancer. Int J Hyperthermia 1: 207–217

Dethlefsen LA, Dewey WC (1982) Proceedings of the Third International Symposium on Cancer Therapy by Hyperthermia, Drugs and Radiation. Natl Cancer Inst Monogr 61

Dewhirst MW, Sim DA (1984) The utility of thermal dose as a predictor of tumor and normal tissue responses to combined radiation and hyperthermia. Cancer Res 44: 4772s–4780s

Field SB, Morris CC (1983) The relationship between heating time and temperature: its relevance to clinical hyperthermia. Radiother Oncol 1: 179–186

Kaplan EL, Meier P (1958) Non-parametric estimation from incomplete observation. J Am Stat Assoc 53: 457–481

Mantel N (1966) Evaluation of survival data and two new rank order statistics arising in its consideration. Cancer Chemother Rep 50: 163–170

Overgaard J (1984) Rationale and problems in the design of clinical studies. In: Overgaard J (ed) Hyperthermic oncology, vol 2. Taylor and Francis, London, pp 325–338

Overgaard J (ed) (1985) Hyperthermic oncology, vols 1, 2. Taylor and Francis, London

Perez CA, Nussbaum G, von Gerichten D (1983) Clinical results of irradiation combined with local hyperthermia. Cancer 52: 1597–1603

Sapareto S (1982) The biology of hyperthermia in vitro. In: Nussbaum GH (ed) Physical aspects of hyperthermia. American Institute of Physics, New York, pp 1–19

Sokal RR, Rohlf FS (1969) Analysis of frequencies. In: Sokal RR, Rohlf FS (eds) Biometry. Freeman, San Francisco, pp 549–620

Storm FK (1983) Hyperthermia in cancer therapy. Hall, Boston

Two Versus Six Hyperthermia Treatments in Combination with Radical Irradiation for Fixed Metastatic Neck Nodes: Progress Report*

R. Valdagni**

Centro Oncologico, Istituti Ospedalieri, 38100 Trento, Italy

Introduction

Local hyperthermia (HT) has been demonstrated to be an effective adjunct to ionizing radiation (RT) in cancer therapy. Several clinical trials employing RT plus HT (Overgaard 1984) have shown that HT can improve local control rates of both primary and recurrent tumors. However, the precise role of some hyperthermia parameters in clinical RT and HT studies is still not well defined. In an attempt to substantiate and clarify the results of our previous experience with different numbers of heat sessions (Valdagni et al. 1986a), a randomized clinical trial for N_3 neck nodes (UICC 1973) comparing radical irradiation plus two heat treatments versus radical irradiation plus six heat sessions was activated in September 1985. This study is part of a randomized protocol comparing radical irradiation alone versus radical irradiation plus heat (Valdagni et al. 1986b).

Methods and Materials

Treatment Protocol

In September 1985, a prospective clinical trial employing radical irradiation versus radical irradiation plus local microwave hyperthermia for fixed and inoperable N_3 (TNM-UICC) metastatic neck nodes was initiated (Valdagni et al. 1986b). Nodes in the combined treatment arm were further randomized to receive either two or six HT treatments. Heat sessions were delivered after RT treatment, two times a week with an interval of at least 72 h (e.g., Monday and Thursday or Tuesday and Friday; Table 1). Multiple N_3 neck nodes in the same patient were individually randomized and treated as separate lesions.

---

* This work was partly supported by Provincia Autonoma di Trento and Fondazione per La Ricerca sul Cancro "L. Pece".

** The author wishes to thank Dr. Fei Fei Liu for reviewing the manuscript and Drs. M. Amichetti and G. Pani for their helpful clinical collaboration. He also thanks Marge Keskin for skillful secretarial assistance.

Table 1. Treatment protocol

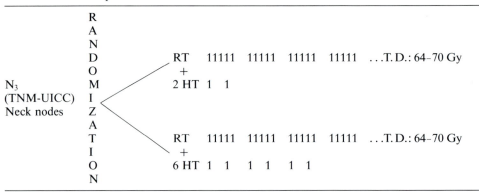

	R A N D O M I Z A T I O N						
N₃ (TNM-UICC) Neck nodes		RT 11111 11111 11111 11111 ...T.D.: 64–70 Gy + 2 HT 1 1					
		RT 11111 11111 11111 11111 ...T.D.: 64–70 Gy + 6 HT 1 1 1 1 1 1					

_T. D._, total dose of irradiation.

Patient Selection and Characteristics

Patients who satisfied the following criteria were included in this trial: pathologically proven squamous cell carcinoma; fixed N_3 (TNM-UICC) neck lymph nodes with the following nodal size characteristics: diameter ≤ 7 cm, depth ≤ 5 cm; staging of the primary as T_{1-3} (TNM-UICC) or of unknown origin; Karnofsky performance status $\geq 60\%$; a life expectancy of more than 3 months.

To be eligible for randomization, patients could not have received prior radiation to the neck regions or previous chemotherapy. Pretreatment workup included a complete history, physical examination with a detailed ear-nose-throat assessment, and blood tests. Computed tomography or ultrasonography of nodes was advised but not considered mandatory.

As of December 1986, 19 patients with a total of 21 nodes were entered into this study. Seventeen nodes (in 15 patients) were evaluable at the 3-month follow-up time after completion of therapy: seven of these nodes had been treated with a total of two heat sessions and 10 nodes with six heat sessions. Patient and nodal characteristics in each

Table 2. Patient and nodal characteristics

	RT+2 HT	RT+6 HT
Sex: M/F	5/2	9/1
Age range/median age (years)	53–76/66	48–65/58
Maximum nodal diameter		
3.5–3.9 cm	14% (1)	40% (4)
4.0–5.9 cm	43% (3)	20% (2)
6.0–7.0 cm	43% (3)	40% (4)
Site of primary		
Oral cavity	3	2
Oropharynx	–	1
Hypopharynx	–	3
Larynx	3	2
Unknown	1	2

arm are outlined in Table 2. The male/female ratio and the median age are higher in the RT+2 HT treatment arm than in the RT+6 HT treatment arm. Larger nodes are equally represented in both arms (43% and 40% of nodes greater than 6.0 cm). The sites of primary in the RT+2 HT treatment arms were the following: oral cavity (3), larynx (3), unknown (1); in the RT+6 HT treatment arm primary tumors were located in oral cavity (2), oropharynx (1), hypopharynx (3), larynx (2), unknown (2).

Radiation Therapy

The primary tumor and node(s) were treated using either 6- or 12-MeV linear accelerators (photon or electron beam) or a cobalt-60 unit with parallel opposed portals and "enface" portals for the supraclavicular regions. Conventional fractionation irradiation (2–2.5 Gy/fraction) was delivered 5 times a week for a total tumor dose of 64–70 Gy.

Hyperthermia and Thermometry

Heat treatments are delivered using the BSD-1000 units (BSD Medical Corporation, Salt Lake City, Utah, USA) and all nodes were treated with the same MA-150 applicator operating at a frequency of 280–300 MHz. Surface cooling was occasionally activated in order to prevent damage to uninvolved skin. Heat was delivered within 20–25 min after irradiation for a total duration of 30 min after the measured minimal intratumoral temperature reached the desired level of 42.5° C. Thermometry was performed with the use of Bowman thermistors. A minimum of five intratumoral sites had constant temperature monitoring, and no less than three surface thermistors were used for each treatment. Within the target region, thermal probes were preferentially positioned at tumor/normal tissue interfaces. Probes were always withdrawn after the end of each heat treatment.

Response Evaluation

Clinical response is determined on the following WHO/UICC criteria: complete response, disappearance of all palpable nodal disease; partial response, reduction in the volume of the node of greater than 50%; no response, reduction of less than 50%; progression of disease, increase in the measured volume of the node. Tumor volume (V) was calculated from the measurement of two orthogonal diameters, according to the formula:

$$V = \frac{4}{3}\pi a b^2$$

where a is the maximum radius of node and b is the radius orthogonal to it.

Acute side effects of the treatments were assessed in the following sites: (a) neck regions receiving RT+2 HT, (b) neck regions receiving RT+6 HT, and (c) contralateral neck regions which received only RT. The assessment of acute toxicity was based on the WHO/UICC cutaneous grading system (Miller et al. 1981) with the addition of "blister" to the grade 3 category.

Results and Discussion

Preliminary analysis of the clinical response in 17 evaluable nodes was performed 3 months after completion of combined therapy (Table 3). RT+2 HT resulted in six (85.7%) complete responses with one (14.3%) progressive disease; RT+6 HT resulted in eight (80%) complete responses, one (10%) partial response, and one (10%) progressive disease.

The two arms are relatively well balanced for most pretreatment and treatment variables (Tables 2, 3), such as age, site of primaries, and especially median maximum nodal diameter (4.86 cm for the 2 HT arm versus 4.90 cm for the 6 HT arm). With regard to radiation parameters, nodes heated twice received an average total RT dose of 67.46 Gy, which is comparable to an average total RT dose of 68.05 Gy delivered in the arm with six heat treatments.

Analyzing the hyperthermia treatment parameters, the average total maximum "thermal dose" (total max. $\overline{T.D.}$), expressed in equivalent minutes of 42.5° C (Eq 42.5) (Sapareto 1982), is 186.7 for the 2 HT arm versus 508.4 for the 6 HT arm; total min. $\overline{T.D.}$ is 14.5 and 82.2 respectively. In general, nodes in the 6 HT arm seem to have achieved a slightly higher temperature ($T°$) and this is corroborated by a difference between the two arms of 0.7° C in the average minimum temperatures recorded in all treatments.

To date, two failures have occurred in the 17 nodes, one in each treatment arm, and both of these metastatic lymph nodes were partially tucked under the mandible.

Both of these lymph nodes had several intratumoral thermal probes; one demonstrated "poor" heating, and the other appeared to have been "satisfactorily" heated. The former node achieved a total max T.D. of 9.3; total min T.D. was 0; average $T°$ max in all treatments was 40.75° C, and average $T°$ min was <40° C in all treatments. The latter node achieved a total max. T.D. of 53.92; total min T.D. was 5.6; average $T°$ max in all treatments was 42.3° C, and average $T°$ min in all treatments was 41.3° C. It might be presumed that there were shifts in the electromagnetic field and subsequent isotherm lines immediately adjacent to the overlying mandible, but this cannot be proven because the thermistors were not specifically placed under the bone.

It should be pointed out that the "poorly" heated node was paired with a larger node in the same patient; the latter achieved a higher intratumoral $T°$ and at 3 months evalua-

Table 3. Clinical results and treatment parameters

	RT+2 HT	RT+6 HT
Complete response % (n)	85.7% (6)	80% (8)
Partial response % (n)	0% (0)	10% (1)
Progressive disease % (n)	14.3% (1)	10% (1)
Average total RT dose	67.46 Gy	68.05 Gy
Average maximum nodal diameter	4.86 cm	4.90 cm
Total max $\overline{T.D.}$ (Eq 42.5)	186.7	508.4
Total min $\overline{T.D.}$ (Eq 42.5)	14.5	82.2
Total min, max $\overline{T.D.}$ (Eq 42.5):	Average of total minimum/maximum "thermal dose" expressed in equivalent minutes at 42.5° C	

tion it exhibited a complete response. One patient with a 7×7 cm node died 4 months after the completion of therapy of an intercurrent illness; his lymph node demonstrated continual regression, with 80% reduction of its original volume at time of death.

Acute side effects have been analyzed in the neck regions which received RT+2 HT or RT+6 HT and in the contralateral neck which received RT. Except for one blister noted in a patient with skin involvement (RT+6 HT arm), no significant difference in acute local toxicities was observed in the two arms of this trial. Acute side effects were also comparable between RT only and RT+(2 or 6) HT neck contralateral regions.

To date our clinical results are in agreement with that of the Stanford randomized clinical trial of two versus six heat treatments plus irradiation in a variety of tumors, which is showing no benefit of six heat treatments over two in terms of local control rates (Kapp et al. 1987). Table 4 summarizes clinical response percentages observed in clinical trials employing different numbers of total hyperthermic sessions. Tumor responses do not differ significantly according to the number of heat treatments in the studies by Hiraoka et al. (1984), by Arcangeli et al. (1987) with melanoma metastases, and by Kapp et al. (1987). Our previous nonrandomized report (Valdagni et al. 1986a) suggested that there was no correlation between the number of heat treatments given and the clinical response. The lower complete response rate of 50% observed with the six heat treatments group was influenced by the presence of larger lymph nodes. A benefit of four heat treatments over eight such treatments was shown by Alexander et al. (1987) in a preliminary analysis of a randomized clinical trial, but different RT doses per fraction were used (2 Gy vs 4 Gy), making early evaluation difficult.

Obviously more data are needed to substantiate these preliminary findings. Optimization of known hyperthermia parameters is needed to increase the clinical efficacy of combined treatment, and thus to avoid undesired overheating of normal tissues, and, last but not least, to contain the increased cost of heat treatments.

Table 4. Clinical response as a function of number of total and weekly heat sessions

		No. of total heat treatments	No. of weekly heat treatments	Complete response (%)
Arcangeli et al. (1984)	– NR –	5	1	64
various tumors		10	2	78
Hiraoka et al. (1984)	– NR –	2–7	2	50
various tumors		8–12	2	53
Valdagni et al. (1986)	– NR –	<6	2/3	83
SCC neck nodes[a]		6	2/3	50
		>6	2/3	71
Alexander et al. (1987)	– R –	4	1	42
various tumors		8	2	21
Arcangeli et al. (1987)	– NR –	5	2	75
melanoma		8	2	77
Kapp et al. (1987) –	– R –	2	1	68
various tumors		6	2	63
Valdagni (this study) 1987	– R –	2	2	85
SCC neck nodes[a]		6	2	80

NR, nonrandomized trial; R, randomized trial; SCC, squamous cell carcinoma.
[a] Radical irradiation plus local hyperthermia.

References

Alexander GA, Moylan DJ, Waterman FM, Nerlinger RE, Leeper DB (1987) Randomized trial of 1 vs. 2 adjuvant hyperthermia treatments in patients with superficial metastases (Abstr). 35th Annual Meeting of the Radiation Research Society, February 21–26, Atlanta

Arcangeli G, Nervi C, Cividalli A, Lovisolo GA (1984) Problem of sequence and fractionation in the clinical application of combined heat and radiation. Cancer Res 44: 4857s–4863s

Arcangeli G, Benassi M, Cividalli A, Lovisolo G, Mauro F (1987) Radiotherapy and hyperthermia: analysis of clinical results and identification of prognostic variables. Cancer, (submitted)

Hiraoka M, Shiken J, Dodo Y, Ono K, Takahashi M, Nishida H, Abe M (1984) Clinical results of radiofrequency hyperthermia combined with radiation in the treatment of radioresistant cancer. Cancer 54: 2898–2904

Kapp DS, Samulsky TV, Fessenden PF, Bagshaw MA, Lee ER, Lohrbach AW, Cox RS (1987) Prognostic significance of tumor volume on response following local-regional hyperthermia (Ht) and radiation therapy (RT) (Abstr). 35th Annual Meeting of the Radiation Research Society, February 21–26, Atlanta

Miller AB, Hoogstraten B, Staquet M, Winkler A (1981) Reporting results of cancer treatment. Cancer 47: 207–214

Overgaard J (1984) Rationale and problems in the design of clinical studies. In: Overgaard J (ed) Hyperthermic oncology. Taylor and Francis, London, pp 325–338

Sapareto S (1982) The biology of hyperthermia in vitro. In: Nussbaum GH (ed) Physical aspects of hyperthermia. American Institute of Physics, New York, pp 1–19

UICC (1973) T. N. M. classification of malignant tumors. UICC, Geneva

Valdagni R, Kapp DS, Valdagni C (1986a) N_3 (TNM-UICC) metastatic neck nodes managed by combined radiation therapy and hyperthermia. Clinical results and analysis of treatment parameters. Int J Hyperthermia 2: 189–200

Valdagni R, Amichetti M, Pani G, Ambrosini G (1986b) A prospective randomized clinical trial of radical irradiation alone versus radical irradiation combined with hyperthermia in the treatment of N_3 metastatic neck nodes: preliminary report (Abstr). Int J Hyperthermia 4: 411

First Results After Hyperthermia Treatment with the BSD System in Essen*

M. Molls[1], W. Baumhoer[2], H. J. Feldmann[1], R. D. Müller[1], and H. Sack[1]

[1] Abteilung für Strahlentherapie, Universitätsklinikum, Westdeutsches Tumorzentrum, Hufelandstraße 55, 4300 Essen, FRG
[2] Abteilung für Medizinische Röntgenphysik, Universitätsklinikum, Westdeutsches Tumorzentrum, Hufelandstraße 55, 4300 Essen, FRG

Introduction

Up to 1982, 126 patients were treated with hyperthermia alone or in combination with either radiotherapy or chemotherapy at the Department of Radiotherapy at the University Hospital in Essen. We used conventional generators (13, 27, 434, and 1540 MHz). The combination of radiation plus heat was most favorable (Gerhard and Scherer 1983).

Because of our own and others' encouraging observations (for review see Molls and Scherer 1987), a new hyperthermia facility was established in our institution. We received substantial support from the *Deutsche Krebshilfe* founded by Dr. Mildred Scheel. The new hyperthermia section was opened on 25 and 26 April 1986, with an international symposium, the first European BSD users' conference (proceedings in *Strahlentherapie und Onkologie* 163 (4), 1987).

Treatment of patients started in September 1986, after different applicators of our BSD-1000 system had been tested in phantoms. The patients suffered from locally advanced or metastasized cancer, and treatment was usually performed with a palliative intention. The tumors were situated on the surface or in half-depth. Our aim was to acquire clinical experience of heat supply and intratumor thermometry with the BSD system. The hyperthermia treatments were planned and performed in such a manner that the extent of side effects such as pains or burns could be expected to be low.

Material and Methods

Hyperthermia treatments were performed with the BSD-1000 system. Three radiative contact applicators were used. We heated superficial tumors with the MA 151 (620 MHz), or, in most cases, the MA 150 (450 MHz) applicator. Tumors in half-depth were treated with the Dual Horn MA 201 (180 MHz) applicator. The applicators have a water bolus attached that is variable in volume. The bolus improves thermal coupling and controls surface temperature. The bolus temperatures ranged from 20° to 40° C.

Tumor temperatures were measured by thermistors (Bowman probes) introduced into the tumor along blind-ended Teflon catheters which had been inserted under local anesthesia. In the patients with metastases of the os sacrum (Table 4), the tumors pene-

* This work was supported by the Deutsche Krebshilfe.

trated through the bone, and so the insertion of catheters could be carried out without complications. The position of the catheters was controlled by computed tomography, by radioscopy, or by clinical means. Temperatures were determined in the centers of the tumors and continously monitored during treatment. A young girl (G. V., Table 4) and a patient of 71 years (B. G., Table 4) refused the invasive procedure and thermometry was not performed. Hyperthermia treatments lasted 45 min, and started as soon as possible after irradiation (within 20 min).

Radiotherapy was planned according to tumor site and dimensions and especially to previous therapy. We irradiated with four or five fractions weekly up to a total dose of 18–70 Gy. In most cases the single doses were 2 Gy. We irradiated with photons except for one patient (D. O., Table 2) who received electrons.

From September 1986 to April 1987, 28 patients received 273 hyperthermia treatments, usually to palliate symptoms. Hyperthermia was applied when tumors recurred, when they had not responded to earlier radiotherapy, or when they were expected to be radioresistent. Observations on 18 patients have been evaluated and reported. Ten patients are either still undergoing treatment, or treatment was not completed because of progress of disease or on request of the patient. With regard to histology and tumor site, our group of patients was very heterogeneous, as is shown in Table 1. Tumor responses were graded as "complete response" (CR: total disappearance of the tumor), "partial response" (PR: at least 50% decrease in tumor area or volume), or "no change" (NC: less than 50% decrease or less than 25% increase).

Results

The results are shown in Tables 2–4. The highest temperatures of the individual tumors during hyperthermia treatment ranged from 40.7° to 46.5° C. The average time per treatment with a tumor temperature higher than 41° C was longest in the patient with a large malignant melanoma in the paravertebral soft tissue (W. K., Table 4). In one case (L. H., Table 2) with a cervical lymph node metastasis we never obtained temperatures above 41° C.

In four patients (22%) the combined treatment led to PR, and in a further four it led to CR. NC was found in six patients (33%). The tumor response could not be judged in four cases (22%) because of a short follow-up or death of the patient. An unexpected result was obtained in the patient with a sternal metastasis of a breast carcinoma (M. C.,

Table 1. Tumor characteristics

n	Tumor	Site	Histology
7	Head and neck	Cervical lymph nodes, skin metastases	Squamous cell carcinoma
3	Rectal	Sacral metastases penetrating through bone	Adenocarcinoma
2	Breast	Supraclavicular lymph node, sternal metastasis penetrating through bone	Intraductal, undifferentiated carcinoma
2	Skin	Paravertebral soft tissue, chest wall skin	Malignant melanoma
1	Esophageal	Supraclavicular lymph node	Squamous cell carcinoma
1	Kidney	Metastasis at L5/S1 infiltrating the bone	Adenocarcinoma
1	Soft tissue sarcoma	Lower leg	Liposarcoma
1	Neuroectodermal	Paravertebral at L4/S1	Askin tumor

Table 2. Treatment with the MA 150 applicator

	D.O. ♂, 58 years, melanoma, chest wall	G.W. ♂, 45 years, tongue carcinoma, supraclavicular node	B.W. ♂, 41 years, esophageal carcinoma, supraclavicular node	S.A. ♀, 62 years, breast carcinoma, supraclavicular node	C.W. ♂, 78 years, glottic carcinoma, skin metastasis	L.H. ♂, 60 years, floor of mouth carcinoma, cervical node	R.R. ♂, 43 years, tonsillar carcinoma, cervical node	B.B. ♂, 57 years, hypopharyngeal carcinoma, cervical node	H.K. ♂, 52 years, tonsillar carcinoma, cervical node
Largest tumor ∅ (cm)	5; 8[a]	6	4	5	2	3	4	6	9
Dose (Gy)	48	56	49	40	36	30	40	50	70
Fractionation (Gy)	4×3	5×2	4×2.5; 3×3	5×2	5×2	5×2	5×2	5×2	5×2
No. of hyperthermia treatments	16	27	13	11	13	9	13	20	20
Highest temp. (°C)	43.1	46.5	43.7	43.1	42.5	40.7	43.1	43.0	42.4
Average time (min) at temp.>41°C	18	19	21	38	20	–	20	11	25
Response	PR	PR	NC	CR	CR	PR	NC	NC	PR

[a] 2 melanoma nodules treated.

Table 3. Treatment with the MA 151 applicator

	D.K. ♂, 45 years hypopharyngeal carcinoma, cervical node
Largest tumor ∅ (cm)	4
Dose (Gy)	18
Fractionation (Gy)	5 × 1,8
No. of hyperthermia treatments	10
Highest temp. (° C)	43.1
Average time (min) at temp. > 41° C	20
Response	NC

Table 4) which had penetrated through the bone. We used the Dual Horn MA 201 applicator and the treatment resulted in histologically confirmed CR (follow up of 4 months) as well as pain relief, although the radiation dose was very low (20 Gy). Due to previous therapy we had to limit the irradiation.

The most impressive effect was seen in a patient with a large malignant melanoma (5 cm × 20 cm) in the paravertebral soft tissue (W. K., Table 4, Fig. 1). The tumor completely disappeared after a radiation dose of 46 Gy and 15 hyperthermia treatments (Fig. 1). Due to the proximity of the spinal cord we did not give a higher radiation dose. At present CR has persisted for 3 months, and as the patient initially suffered from very severe pain, the complete relief from pain was an excellent palliative result.

Complete or partial relief from pain was achieved in 8 out of 10 patients showing the symptom at the beginning of therapy. As mentioned above, the result was excellent in patient W. K. (Table 4) and also in patients B. G. (Table 4) and S. A. (Table 2). In one case the treatment improved the ability to walk (S. M., Table 4) and in another, to swallow (G. W., Table 2).

No acute or severe treatment-related complications were found, although the following side effects occurred: pain during hyperthermia treatment (4/18), erythemas (6/18), and small epitheliolytic lesions (2/18).

Discussion

When considering our first clinical results after hyperthermia with the BSD system, we are aware of the fact that the average follow-up was short (2.5 months). Thus, in patient D. O. (Table 2), in whom treatment has now been finished, a further shrinkage of both melanoma nodules can be expected. Furthermore, the heat doses (average time per treatment with a tumor temperature higher than 41° C) were relatively low. This was due to our intention of strictly avoiding severe side effects in a phase during which we were acquiring our fist clinical experiences with the BSD system. Now, after hyperthermia (including thermometry) has been established as an additional therapy in our department of radiation therapy, we have prolonged the single hyperthermia treatments.

Although the heat doses might have been relatively low in our small series, the CR of a large malignant melanoma after a radiation dose of 46 Gy combined with hyperthermia was a very impressive effect. The observation is consistent with the results of syste-

Table 4. Treatment with the Dual Horn MA 201 applicator

	W.K. ♂, 57 years, melanoma, paravertebral	G.V. ♀, 11 years, Askin Tumor, L4/S1	T.W. ♂, 27 years, liposarcoma, lower leg	M.C. ♀, 45 years, breast carcinoma, sternal metastasis	S.M. ♀, 56 years, rectal carcinoma, sacral metastasis	H.F. ♂, 63 years, rectal carcinoma, sacral metastasis	R.W. ♂, 58 years, rectal carcinoma, sacral metastasis	B.G. ♂, 71 years, renal carcinoma, metastasis L5/S1
Largest tumor Ø (cm)	20	3	Microscopic	4	4	4	5	6
Dose (Gy)	46	40	50	20	20	20	30	50
Fractionation (Gy)	4×2	5×2	5×2	5×2	5×2	5×2	5×2	5×2
No. of hyperthermia treatments	15	19	18	10	10	7	15	16
Highest temp. (°C)	43.5	n.d.	43.0	41.7	42.6	42.4	43.1	n.d.
Average time (min) at temp. > 41°C	35	n.d.	20	15	20	10	21	n.d.
Response	CR	CT to be performed	Short follow-up	CR	NC	Patient died	NC	CT to be performed

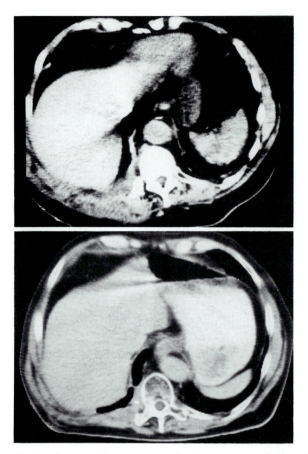

Fig. 1. Computed tomograms of a large paravertebral malignant melanoma (length 20 cm) before and after therapy. The tumor showed complete regression (W. K., Table 4)

matic, clinical studies showing that thermal enhancement of the radiation effect is most pronounced in malignant melanomas (Overgaard 1981).

A further interesting, unexpected tumor response was the CR of a sternal breast carcinoma metastasis after combining a very low radiation dose (20 Gy) with hyperthermia. In a relatively large group of breast carcinoma patients receiving radiation doses of only 12–25 Gy, a response rate (PR and CR) of about 80% has been found. (Van der Zee et al. 1986). Obviously, a significant thermal enhancement of the radiation effect can also be obtained in breast carcinoma. Thus, when palliative treatment of breast cracinoma is needed but the radiation dose must be limited, the combination of irradiation plus heat is a promising treatment possibility.

It is known that irradiation combined with hyperthermia is an effective treatment of neck lymph nodes in head and neck cancers (Arcangeli et al. 1985; Valdagni et al. 1986). Our own less favorable results might be explained by the limitation of the applied radiation dose (D. K., Table 3) and by the aggressive biological behavior of very advanced tumors in two patients who died soon after therapy (B. B. and R. R., Table 2).

When treating metastases of rectal carcinomas that penetrated through the os sacrum, we were aware of the fact that the comparatively small presacral parts of these tumors could not be heated sufficiently (temperatures reported in Table 4 were measured in the central region of the tumors, at a depth of about 4 cm). However, when we started the treatments, our goals were PR and palliation. Partial relief from very severe pain could be achieved (S. M. and R. W., Table 4), but PR did not occur.

Because of earlier treatment, we only irradiated the sacral metastases using low doses (Table 4). This might have been the most important reason for the NC result demonstrated by computed tomography in patients S. M. and R. W. (Table 4). Sacral recurrences of rectum carcinomas require relatively high radiation doses of about 50 Gy (James and Pointon 1985), and doses higher than 20–30 Gy must apparently be given with the combined treatment. Thus, the radiosensitizing effect of additional hyperthermia (Streffer and van Beuningen 1987), can be expected to enhance tumor regression of recurrent rectum carcinomas.

References

Arcangeli G, Arcangeli G, Guerra A, Lovisolo G, Cividalli A, Marino C, Mauro F (1985) Tumor response to heat and radiation: prognostic variables in the treatment of neck node metastases from head and neck cancer. Int J Hypertherm 1: 207–217

Gerhard H, Scherer E (1983) Lokalhyperthermie in der Behandlung fortgeschrittener Tumoren: kurative und palliative Wirkungen. Strahlentherapie 159: 521–531

James RD, Pointon RCS (1985) Gastrointestinal tract. In: Easson C, Pointon RCS (eds) The radiotherapy of malignant disease. Springer, Berlin Heidelberg New York, pp 429–435

Molls M, Scherer E (1987) The combination of hyperthermia and radiation: clinical investigations. In: Streffer C (ed) Hyperthermia and the therapy of malignant tumors. Springer, Berlin Heidelberg New York, pp 110–135

Overgaard J (1981) Fractionated radiation and hyperthermia: experimental and clinical studies. Cancer 48: 1116–1123

Streffer C, van Beuningen D (1987) The biological basis for tumor therapy by hyperthermia and radiation. In: Streffer C (ed) Hyperthermia and the therapy of malignant tumors. Springer, Berlin Heidelberg New York, pp 24–70

Valdagni R, Kapp DS, Valdagni C (1986) N 3 (TNM-UICC) metastatic neck nodes managed by combined radiation therapy and hyperthermia: clinical results and analysis of treatment parameters. Int J Hyperthermia 2: 189–200

Van der Zee J, Van Putten WLJ, Van den Berg AP, Van Rhoon GC, Wike Hooley JL, Broekmeyer-Reurink MP, Reinhold HS (1986) Retrospective analysis of the response of tumors in patients treated with a combination of radiotherapy and hyperthermia. Int J Hyperthermia 2: 337–349

Interstitial Radiotherapy Combined with Interstitial Hyperthermia in the Management of Recurrent Tumors

Z. Petrovich, K. Lam, M. Astrahan, G. Luxton, and B. Langholz

Department of Radiation Oncology and Preventive Medicine, University of Southern California School of Medicine, 1441 Eastlake Avenue, Los Angeles, CA 90033, USA

Introduction

Properly administered radiotherapy is an effective treatment for early tumors, resulting in over 90% local control for T1 and T2 lesions (Marcial and Pajak 1983). This incidence of local control decreases sharply with increase in tumor volume. Presence of metastatic tumor in the regional lymph nodes decreases the locoregional control for each T stage by approximately 50% (Marcial and Pajak 1985). In head and neck tumors, persistent or recurrent locoregional disease is the most frequent site of first failure (Marcial et al. 1983) and is responsible for patient death. Locoregional control has a profound importance on survival in tumors including malignant melanoma (Overgaard et al. 1968; Overgaard 1986). Similar problems, perhaps with a higher incidence of locoregional failure, are seen in surgically treated head and neck tumors (Strong 1983).

Interstitial hyperthermia (HT) has been reported to be an effective therapy for control of local recurrence and/or persistent tumors (Cosset et al. 1984; Emami and Perez 1985; Vora et al. 1982; Aristazabal and Oleson 1984; Puthawala et al. 1985). The role of HT with irradiation (RT) in increasing the incidence of locoregional control for T3 and T4 lesions has not been defined. The present report describer the University of Southern California experience with interstitial microwave HT and 192 Ir RT for recurrent and/or persistent malignant tumors.

Materials and Methods

From 1984 through 1986, 27 patients with 31 recurrent and/or persistent tumors were entered into a phase I/II study of interstitial thermoradiotherapy. Figure 1 shows the study schema. Basic eligibility requirements were for previously treated patients with biopsy confirmed, symptomatic recurrent tumors accessible for interstitial RT with microwave HT. Patients with metastatic disease were included if projected survival was more than 3 months. The upper respiratory and digestive tract was represented by 14 (52%) patients with 15 sites treated. Breast sites were seen in six (22%) patients, the cervix and vagina in four (15%), and three (11%) had other tumors. Histologically, squamous cell carcinomas predominated with 19 (61%) sites; adenocarcinoma was next in frequency with seven (23%) sites, malignant melanoma followed with four (13%), and 1 (3%) had abdominal wall sarcoma. There were 16 female and 11 male patients. Their average age

Recent Results in Cancer Research, Vol. 107
© Springer-Verlag Berlin · Heidelberg 1988

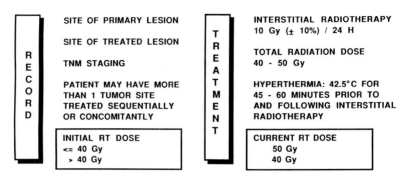

Fig. 1. Study schema

was 62 years. Follow-up ranged from 3 to 24 months with an average of 7 months. Prior treatment consisted of RT alone in eight (26%) sites, RT plus surgery (S) in 11 (35%), chemotherapy (CT) in four (13%), and RT-S-CT combination in eight (26%). Prior radiotherapy consisted of teletherapy with an average tumor dose of 56 Gy. In addition, four sites were treated by means of interstitial therapy, with an average dose of 25 Gy. Tumor volume prior to thermoradiotherapy was < 50 cm^3 in 13 (42%), 50–100 in 11 (35%), 100–150 cm^3 in three (10%), and > 150 cm^3 in four (13%). Of the 31 sites treated, eight (26%) had persistent tumor. Recurrent tumor < 12 months since diagnosis was seen in six (19%), from 12 to 24 months in 10 (32%), and seven (23%) had late recurrence (Table 1). Interstitial HT was administered with 915 MHz microwave in 18 patients using a BSD-300 system (BSD Medical Corporation, Salt Lake City, Utah), with the remaining nine patients receiving 630 MHz using a BSD-1000 apparatus. The patients were to receive two HT treatments of 45–60 min duration at 42.5° C. Thermal dose (TD) was defined as the time in minutes at 42.5° C. Temperature was increased slowly over a period of 10–15 min; faster increases in temperature resulted in greater patient discomfort. Patients were carefully monitored during administration of HT. Vital signs were checked every few minutes. No analgesia was used during HT sessions. In case of local discomfort, power was reduced irrespective of temperature reading. Multipoint thermometry was used. This included thermometry built into an antenna and separate thermistors ("Bowman probes"). A total of 58 HT treatments were given, with the majority (87%) receiving two sessions while the remainder (13%) had one session. In these four sites receiving a single treatment, erroneous temperature readings were noted and the second treatment was not given. This problem and its prevention are discussed elsewhere (Astrahan et al. 1987). TD < 60 was given to seven (23%), TD 60–90 to 14 (45%), and in 10 (32%) sites TD was > 90.

Interstitial RT was given with ^{192}Ir. In 11 sites the total dose ranged from 28 to 50 Gy with an average of 35 Gy, and in 20 sites it was over 40 Gy. The dose of RT depended on prior therapy (Fig. 1). Under general anesthesia, plastic catheters were introduced into the tumor through 17-gauge hollow steel needles and secured to skin with plastic buttons (Syed and Feder 1978). In five patients, the template technique was used (Puthawala et al. 1982). An attempt was made to distribute catheters evenly through the volume of interest. Typically, they were spaced 15 mm apart in one or more planes separated again by 15 mm. An additional two to three catheters were introduced to allow for more thermometry points. Response was classified into: complete (CR) with total tumor regression for 30 days or longer, partial (PR) with tumor decrease $> 50\% < 100\%$, and minimal (MR) with tumor regression $< 50\%$.

Results

In all 31 treated sites, tumor regression was noted. CR was found in 19 (61%) sites in 18 patients, with no recurrence during the period of observation. PR was obtained in 11 (36%) sites, and one (3%) had a lesser degree of tumor regression. Subjective improvement was reported by all patients. Decrease in local pain and bleeding and healing of a tumor-produced ulcer were the most frequent benefits. Tumors larger than 150 cm³ in volume did not respond as well to therapy as the smaller tumors. Of the such sites with these larger tumors, none showed CR. This difference was significant ($p < 0.02$, Fisher's exact test). Radiation dose played an important role in tumor regression. Tumors receiving ≤ 30 Gy exhibited 20% CR, those with 30–40 Gy displayed 60% CR, and the group with 40–50 Gy showed 82% CR ($p < 0.02$). Figure 2 shows the response relationship by the radiotherapy dose. No similar dose dependence was noted with the increase in TD. The sites receiving TD < 60 had 86% CR, those with TD 60–90 displayed 43% CR, and those with TD > 90 experienced 70% CR (Fig. 3). No significant difference in the incidence of CR was observed among different sites, and there was also no difference in CR rates by histological diagnosis. As expected, recurrent adenocarcinoma of the breast had a somewhat greater incidence of CR (NS). It is of interest to note that patients whose tumor recurrence was diagnosed after a longer period of time had a more

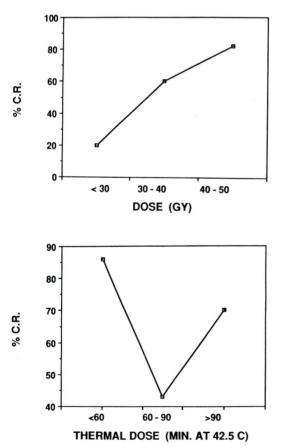

Fig. 2. Importance of radiation dose in the incidence of complete response

Fig. 3. Lack of correlation between complete response and thermal dose

Table 1. Treatment results by time of tumor recurrence from initial diagnosis

	CR	PR	MR	Total
Persistent	5 (62%)	3 (38%)	0	8
<12 months	2 (33%)	3 (50%)	1 (17%)	6
12–24 months	6 (60%)	4 (40%)	0	10
>24 months	6 (86%)	1 (14%)	0	7
Total	19 (61%)	11 (36%)	1 (17%)	31 (100%)

CR, complete response; PR, partial response; MR, minimal response.

favorable treatment outcome than those whose tumors recurred sooner following initial treatment ($p < 0.05$). Details of this difference in tumor response are shown in Table 1. CR was an important event, since eight of 18 patients who had this status remained free of all evidence of tumor for the duration of follow-up. Among the 11 patients with PR, one persistent tumor which had been unresectable following primary radiotherapy became resectable and the patient is alive with no evidence of tumor 15 months after therapy. Treatment complications of importance were seen in one patient who developed an area of skin necrosis 8 weeks following therapy and expired with metastatic disease 5 weeks later. A few patients had severe erythema and moist desquamation which required no treatment. Bleeding following removal of catheters from the lower neck was seen in two patients. It responded well to compression and skin sutures.

Discussion

Interstitial HT in combination with ^{192}Ir implantation offers good palliation to patients with recurrent tumors. This treatment is effective and well tolerated by patients, and has a low incidence of complications. A major drawback of interstitial therapy is the requirement for general anesthesia and a few days of hospitalization. Every effort was made to treat these patients with superficial HT and external beam radiotherapy. In fact, none of them were suitable candidates for this alternative, less invasive way of administering hyperthermia. The requirements for sites accessible to interstitial implantation and relatively small tumor volume limit the application of this technique to carefully selected patients. Recent technological advances have helped to optimize this treatment and made it easier and safer. Hyperthermia antennae with heating up to 6 cm in length, and effective heating to the tip of the applicator, permit treatment of larger volumes. Additionally, built-in thermometry reduces the number of catheters required for implants. It is apparent that randomized trials are necessary to establish the role of HT-RT combination in the treatment of recurrent tumors, and also as a primary treatment for locally advanced tumors.

Summary

During a 2-year period 31 sites in 27 patients with recurrent, previously treated tumors received a combination of interstitial ^{192}Ir radiation (RT) with microwave hyperthermia (HT). Head and neck sites were treated most frequently (48%), with breast and vagina each representing about 20% of sites. Complete response (CR) with no local recurrence was obtained in 61%, partial response (PR) in 11 (36%), and one (3%) had < 50% tumor regression. Of these patients, nine had no evidence of tumor following HT-RT therapy, 8/18 in the CR group and 1/11 in the PR group. Significant factors influencing CR were: radiation dose, tumor volume and duration of tumor control following the initial therapy ($p < 0.02$). Treated site, histology and thermal dose were not significant factors influencing tumor regression. Complications of significance developed in one patient who had local skin necrosis.

Interstitial HT-RT combination provides an effective palliative therapy for recurrent and/or persistent tumors. Randomized trials are necessary to assess the effectiveness of this combination as a component part of primary management of selected tumors.

References

Aristazabal SA, Oleson JR (1984) Combined interstitial irradiation and localized current field hyperthermia: results and conclusions from clinical studies. Cancer Res [Suppl] 44: 4757s-4760s

Astrahan M, Luxton G, Petrovich Z (1987) Thermometry characteristics of the BSD interstitial hyperthermia applicator. Endocuriether Hyperthermia Oncol, (in press)

Cosset JM, Dutriex J, Dufour J, Janoray P, Damia E, Haie C, Clark D (1984) Combined interstitial hyperthermia and brachytherapy: Institute Gustave Roussy technique and preliminary results. Int J Radiat Oncol Biol Phys 10: 307-317

Emami B, Perez CA (1985) Interstitial thermoradiotherapy: an overview. Endocuriether Hyperthermia Oncol 1: 35-40

Marcial VA, Pajak TF (1983) Radiation therapy alone or in combination with surgery in head and neck cancer. Cancer 55: 2259-2265

Marcial VA, Amato DA, Pajak TF (1983) Patterns of failure after treatment for cancer of upper respiratory and digestive tracts: a Radiation Oncology Group report. Cancer Treat Symp 2: 33-40

Overgaard J (1986) The role of radiotherapy in recurrent and metastatic melanoma: a clinical radiobiological study. Int J Rad Biol Phys 12: 867-872

Overgaard J, Overgaard M, Hansen VP, van der Maase K (1968) Some factors of importance in the radiation treatment of malignant melanoma. Radiother Oncol 5: 183-192

Puthawala AA, Syed AMN, Flemming P, Disaia PJ (1982) Reirradiation with interstitial implant for recurrent pelvic malignancies. Cancer 50: 2810-2814

Puthawala AA, Syed AMN, Sheikh KAMA, Rafie S, McNamaram CS (1985) Interstitial hyperthermia for recurrent malignancies. Cancer 1: 125-131

Strong EW (1983) Sites of treatment failure in head and neck cancer. Cancer Treat Symp 2: 5-20

Syed AMN, Feder BH (1978) Technique of afterloading interstitial implants. Radiol Clin 46: 458-475

Vora N, Farell B, Joseph C, Lipsett J, Archambeau JO (1982) Interstitial implant with interstitial hyperthermia. Cancer 50: 2518-2523.

Localized, Non-Invasive Deep Microwave Hyperthermia for the Treatment of Prostatic Tumors: The First 5 Years*

A. Yerushalmi

Department of Cell Biology, The Weizmann Institute of Science, Rehovot 76100, Israel

Introduction

The clinical value of local hyperthermia (LH) combined with radiation therapy (RT) is well documented in the treatment of a variety of human tumors (Arcangeli et al. 1983; Corry et al. 1980; Overgaard 1981; Scott et al. 1983; U et al. 1980; Yerushalmi et al. 1986). Clinical treatment of benign tumors with LH has not been reported in the literature except for the trials with patients suffering from benign prostatic hyperplasia carried out by our group (Yerushalmi et al. 1985).

The LH + RT studies reported in the literature have featured variations in hyperthermia equipment, protocols of LH, RT, and LH + RT, duration of LH, sequencing between LH and RT, tumor sites, and histologies. The normally advanced stage of the disease on admittance to LH treatment permits only short follow-up periods; no long-term response information is available from most clinical reports.

In the present paper, I review our development of clinical hyperthermia technology, adapted to the anatomy of the treated tumor site, and our results of combined RT + LH or LH alone on the treatment of the prostate, in prostatic carcinoma and benign prostatic hyperplasia. I summarize our group's findings (Yerushalmi et al. 1985, 1986) regarding the initial response and response duration and the acute and long-term side effects of hyperthermia in a patient population homogeneous with respect to tumor site and treated with the same machine according to the same LH protocol.

Technology

The treatment of deep-seated tumors involves yet unsolved technological problems of safe heat delivery to tumors situated deep in the body. The present state of clinical hyperthermia technology for deep-seated tumors does not guarantee complete avoidance of unwanted heating of non-invaded, unmonitored sites in the applicator to tumor pathway.

To circumvent the problems involved in the treatment of deep-seated tumors, I have chosen to treat a tumor in the vicinity of a natural cavity. This permits implementation of

* This work was supported by the Gulton Foundation, New Jersey, USA.

my strategy of non-invasive heating of a deep-seated tumor with minimal exposure of normal tissues to large microwave fields.

The main factor guiding the design of such a technology is the utilization of the natural cavity to afford close proximity of the applicator to the tumor, so that low microwave power will be sufficient to reach hyperthermic temperatures in the targeted volume, and concomitantly minimize normal tissue heating in the pathway from applicator to tumor.

I have selected the prostate as the first site for intracavitary hyperthermia because it is the site of two diseases in the male: prostatic carcinoma (PCA) and benign prostatic hyperplasia (BPH). Adenocarcinoma of the prostate is the second most frequently occurring tumor in men over 55 years of age, and is the third cause of death from malignant tumors in males after lung cancer and colorectal cancer (Grayhack 1980). BPH is the most common growth occurring in elderly men (Harbitz and Haugen 1972).

At the present time, there is no satisfactory medical management for BPH and the only effective treatment is surgery (Walsh 1979). However, because of the advanced age of most patients with BPH, surgery is contraindicated in a significant number of cases.

Radiation therapy is applied to patients with disease localized to the pelvis or for palliation at a more advanced stage. It is accompanied by complications of various degrees (Hill et al. 1974; Dewit et al. 1983) but is useful for pain relief (Benson et al. 1981).

The importance of a non-invasive method for the treatment of BPH or, in combination with RT, for PCA is self-evident.

The hyperthermia system developed for the treatment of prostatic tumors and the method of clinical application have been described in detail elsewhere (Yerushalmi 1986). Briefly, LH at 42.5–43° C was delivered using a transrectal applicator operated at 2.45 GHz, equipped with an appropriate cooling system to avoid overheating of the rectal mucosa and rectal wall. The applicator is positioned in the rectum: maximal heat energy can thus be aimed towards the prostatic mass, permitting heating of the tissue comprising the prostatic tumor with minimal exposure of normal tissues to large microwave fields. Thus, a deep-seated tumor can be heated without the need for the large electromagnetic fields required if treatment is performed using external antennae.

Feasibility Study of the Technology

The clinical applicability of such a technology depends on the tolerance of normal tissues in the vicinity of the treated tumor for hyperthermic temperatures. To test the feasibility of such a method, we investigated the effects of localized deep microwave hyperthermia to the normal rabbit prostate (Yerushalmi et al. 1983). The results indicated that our LH technology can safely be applied for the clinical treatment of prostatic tumors.

Material and Methods

Treatment and Protocol

LH alone was given to BPH patients twice weekly in 12–16 sessions, each lasting 1 h.

In PCA patients, pelvic irradiation was delivered using parallel opposed portals, AP and PA, 15×15 cm, at 100 cm focus-skin distance. A total dose of 30 Gy was delivered in 3 weeks, in five daily fractions per week of 2 Gy each. LH treatment was applied twice a week, 1–2 h post irradiation. After a 3- to 4-week rest period, further irradiation doses to the pelvis were administered to a total of 50 Gy without further LH treatment. A boost to the prostate of 10 Gy was then given in 1 week (five fractions) by two 120° lateral arcs (Yerushalmi et al. 1986).

.   In both protocols, treatments were given on an outpatient basis, without sedation or anesthesia.

Patients

Thirty-three patients with PCA and 67 patients with BPH were treated. In the PCA group, four were treated with heat alone, 27 underwent RT + LH and two were non-evaluable (Table 1). In the BPH group, 78 patients were referred to treatment, but six did not complete treatment and five were non-evaluable. Of the evaluable patients, 20 carried an indwelling catheter prior to treatment and 47 were without a catheter (Table 2).

Results

Prostatic Carcinoma

During the initial phase of the study four patients were treated with LH alone. Three of these patients showed objective local tumor regression, but relapsed within 6 months and were referred to RT treatment. Also, two protocols were examined. In the first protocol the patients underwent two courses of RT + LH, while the second protocol consisted of a first course of RT + LH and a second course of RT alone. Since the results of these two protocols were similar, we adopted the second protocol.

As can be seen from Table 1, 18 complete local responses (CR) and nine partial local responses (PR) were achieved in 27 patients treated with reduced RT + LH. Forty-three months after treatment, 11 of these 27 patients were alive with no evidence of disease. A further 11 were alive with disease 20–55 months after treatment. Thus 22/27 were alive with or without disease after 20–55 months follow-up. As can be seen from Table 1, there were no correlations between local recurrence, initial CR or PR response, and metastatic spread.

Table 1.   Prostatic carcinoma: clinical status of patients treated with RT + LH

Patients (n)	Local response		Local recurrence	Status (months after treatment)	Distant metastases	
	Initial	Duration (months)			+	–
11	CR	18–43	–	Alive, no disease 43	–	11
4	CR	6–20	2	Alive, disease 20–38	2	2
3	CR	12–24	3	Dead after 36–45	2	1
2	PR	4– 5	–	Dead after 4–5[a]	–	2
7	PR	7–38	3	Alive, disease 7–55	2	5

Two patients at stage B of the disease, 26 at stage C, and three at stage D were treated. Four underwent LH alone and 27 were treated with RT + LH. Radiation dose and schedule were as detailed in the Treatment and Protocol section. LH was applied twice a week, 1–2 h post irradiation.
[a] Died of unrelated disease.

None of the treated patients experienced thermal damage or burns. No reactions of the rectal mucosa, such as erythema, edema, or ulceration, were observed. The combined treatment was generally well tolerated, and the addition of HT to RT made no impact on the complications caused by RT alone.

Benign Prostatic Hyperplasia

Seventy-eight poor operative risk patients with BPH were treated with LH alone according to the protocol detailed above. Twenty-four were with and 54 without an indwelling catheter.

The results of treatment as detailed in Table 2 were derived from a scoring system designed for evaluation of urological symptoms prior to and after treatment (Yerushalmi et al. 1985). In the catheter-carrying group, 75% of 20 evaluable patients were freed of their catheter and resumed satisfactory voiding. Seventy percent of the treated patients were free of their catheter up to 51 months after treatment, and only 5% regressed to their pretreatment condition and required repeat catheterization.

Eighty-four percent of 47 evaluable patients who initially had severe prostatic symptoms were improved after LH. Relief from prostatic symptoms was maintained in 81% of the treated patients during long-term follow-up until 56 months after treatment. Only 7% regressed to their pretreatment condition. It should be noted that there were cases in which no improvement was noted after LH but urological symptomatology improved a few weeks later.

In both groups, there were no acute or late side effects of LH and treatment was generally well tolerated.

Table 2. Benign prostatic hyperplasia: treatment with LH alone

Clinical response in patients without a catheter: patients relieved of their initial symptoms (%)[a]

After treatment			*Clinical assessment up to 56 months after treatment*			
Marked improvement	Improvement	Failed	Same/better than after treatment	Worse than after treatment but satisfactory	Regressed to pretreatment condition	Failed
37	47	16	54	27	7	12

Total improvement: 84%   Total improvement: 81%

Clinical response in catheter-carrying patients: patients free of catheter, resumed voiding (%)[b]

After treatment		*Clinical assessment up to 51 months after treatment*		
Free of catheter	With catheter	Same/better than after treatment	Worse than after treatment but catheter-free	With catheter
75	25	60	10	30

Total improvement: 75%   Total improvement: 70%

[a] Fifty-four patients, seven non-evaluable. Number of sessions 8–17, mean 12.
[b] Twenty-four patients, four non-evaluable. Catheter carried for 2–12 months prior to LH.

Discussion

This study is unique with respect to the opportunity it affords to compare data obtained from treatment of two diseases of the same organ in two patient populations. One disease was treated with LH alone, while the second was treated with RT to a total dose of 60 Gy and LH. The clinical situation offered the following characteristics: the same anatomical site, the same applicator, and the same LH treatment protocol. In both patient populations, treated with LH alone for BPH or with RT+LH for PCA, there was no acute or long-term thermal damage, attesting to the safety of the hyperthermia technology employed.

In the PCA patients, LH did not increase radiation morbidity beyond that expected from RT alone (Yerushalmi et al. 1986). No correlations were found between local control, recurrence of the disease, and appearance of metastases. From the short relief obtained in PCA patients treated with LH alone, it seems that patients who have exhausted all conventional standard procedures, or patients with a short life expectancy, may benefit from heat alone as a palliative treatment.

In the BPH patients, both catheter-carrying and without catheters, the freedom from catheterization and the symptomatic relief obtained after treatment in 70%–80% of cases was maintained on follow-up.

The results indicate that LH is an effective modality for the treatment of BPH and an adjuvant to RT in the treatment of PCA.

Hyperthermia treatment is currently at its stage of maturation. It is now recognized that the anatomical site to be treated dictates the technology to be used, the type of applicator, and the method and frequency of treatment.

The design of specific applicators based on anatomical requirements will enable the administration of the minimal energy needed to elevate the tumor temperature to the required degree with minimal heating of surrounding, non-invaded tissues. The clinical results achieved and the safety and ease of application of the technology used in this project support this approach.

References

Arcangeli G, Cividalli A, Creton G, Lovisolo G, Mauro F, Nervi C (1983) Tumor control and therapeutic gain with different schedules of combined radiotherapy and local hyperthermia in human cancer. Int J Radiat Oncol Biol Phys 9: 1125–1134

Benson RC, Hasan SM, Jones AG, Schlise S (1981) External beam radiotherapy for palliation of pain from metastatic carcinoma of the prostate. J Urol 127: 69–71

Corry P, Barlogie B, Spanos W, Armour E, Barkley H, Gonzales M (1980) Approaches to clinical applications of nonionizing and ionizing radiations. In: Meyn RE, Withers HR (eds) Radiation biology in cancer research. Raven, New York, pp 637–644

Dewit L, Kian K, van der Schueren E (1983) Acute side effects and late complications after radiotherapy of localized carcinoma of the prostate. Cancer Treat Rev 10: 78–81

Grayhack JT (1980) Prostate cancer (Editorial). J Urol 123: 716

Harbitz TB, Haugen OA (1972) Histology of the prostate in elderly men. Acta Pathol Microbiol Scand [A] 80: 756–759

Hill DR, Crews QE, Walsh PC (1974) Prostate carcinoma: Radiation treatment of the primary and regional lymphatics. Cancer 34: 156–159

Overgaard J (1981) Fractionated radiation and hyperthermia: Experimental and clinical studies. Cancer 48: 1116–1123

Scott RS, Johnson RJR, Kowal H, Krishnamsetty RM, Story K, Clay L (1983) Hyperthermia in combination with radiotherapy. A review of 5 years experience in the treatment of superficial tumors. Int J Radiat Oncol Biol Phys 9: 1327–1333

U R, Noell KT, Woodward KT, Worde BT, Fishburn RI, Miller LS (1980) Microwave-induced local hyperthermia in combination with radiotherapy of human malignant tumors. Cancer 45: 638–646

Walsh PC (1979) Benign prostatic hyperplasia. In: Harrison JH, Gittes RF, Perlmutter AD, Starney TA, Walsh PC (eds) Campbell's urology. Saunders, Philadelphia, p 949

Yerushalmi A (1986) Hyperthermia treatment of prostate tumors. In: Anghileri LJ, Robert J (eds) Hyperthermia in cancer treatment, vol 3. CRC, Boca Raton, pp 89–103

Yerushalmi A, Shpirer Z, Hod I, Gotteffeld F, Bass DD (1983) Normal tissue response to localized deep microwave hyperthermia in the rabbit's prostate. Int J Radiat Oncol Biol Phys 9: 77–82

Yerushalmi A, Fishelovitz Y, Singer D, Reiner I, Arielly J, Abramovici I, Katsnelson R, Levy E, Shani A (1985) Localized deep microwave hyperthermia (LDMW) in the treatment of poor operative risk patients with benign prostatic hyperplasia. J Urol 133: 873–876

Yerushalmi A, Shani A, Fishelovitz Y, Arielly J, Singer D, Levy E, Katsnelson R, Rakowsky E, Stein JA (1986) Local microwave hyperthermia in the treatment of carcinoma of the prostate. Oncology 43: 209–305

External Microwave Hyperthermia Combined with Radiation Therapy for Extensive Superficial Chest Wall Recurrences

M. H. Seegenschmiedt, L. W. Brady, and G. Rossmeissl

Department of Radiation Oncology and Nuclear Medicine, Hahnemann University, Philadelphia, PA 19102, USA

Introduction

Extensive superficial chest wall malignancies after surgery, full-dose radiation and/or chemotherapy cause significant morbidity with impaired quality of life. Local tumor control at the chest wall is important for those patients, even if no survival benefit is achieved.

In 1986 Perez et al. reported on the combined use of radiation therapy and local hyperthermia for chest wall recurrences. They achieved complete tumor response rates of 65% for lesions > 3 cm and 80% for lesions < 3 cm in diameter.

From January 1986, 17 patients with extensive superficial chest wall recurrences were treated with combined radiation therapy and external microwave hyperthermia at Hahnemann University in Philadelphia. For this analysis, 145 hyperthermia treatment sessions for 27 hyperthermia fields in 13 patients were evaluated.

Materials and Methods

Patients

All 13 patients had a Karnofsky score > 60, life expectancy > 3 months, and disease to the chest wall which could be encompassed with radiation and hyperthermia fields. Patient characteristics are as follows:

Age:	33–82 years (mean 56.5 years)
Histology:	11/13: inferior ductal carcinoma
Previous surgery:	13/13:
	3 excision biopsy/partial mastectomy;
	10 radical modified mastectomy
Previous radiotherapy:	9/13: 32–90 Gy (mean 56 Gy)
Previous chemotherapy:	8/13: CMF/adriamycin/vincristine
Involved area:	50–450 cm^2 (mean 120 cm^2)
	4/13: < 100 cm^2
	9/13: > 100 cm^2

Other workup:	6/13: no metastasis
	7/13: bone/brain/liver metastases
Concomitant therapy:	6/13 antiestrogens: tamoxifen/methylestrogen
	5/13 chemotherapy: CMF/adriamycin/vincristine

Radiation Therapy

Radiation doses ranged from 16 to 50.5 Gy (mean 37.5 Gy). Electrons (6–15 MeV), sometimes mixed with photons (10 MV), were used encompassing a 3-cm margin around involved areas.

Hyperthermia

To avoid possible thermotolerance (Henle and Dethlefsen 1978) hyperthermia was delivered at 72- to 96-h intervals within 30 min after radiation therapy for a total treatment time of 60 min. A 915-MHz microwave generator and dielectrically filled waveguide applicators measuring 15×15 cm or 10×10 cm were used (Mark IV/Mark IX, Clini-Therm). Thorough coupling was performed to avoid air gaps. Skin "cooling" was kept at 37.5° C to 42.5° C.

Nine patients with extensive disease were treated with two or more microwave applicators to include all involved areas and a 3-cm safety margin in the applicator fields. Applicators were either abutted to each other and moved during simultaneous treatments or overlapped about 2.5 cm for sequential treatments in a junction strip area 10–15 cm in length.

Thermometry

Up to 24 temperature points were recorded using a semiconductor/fiberoptic thermometry system (Thermo Sentry 1200, Clini-Therm). Temperature probes were inserted in teflon catheters placed in the center and the periphery of each applicator field. Temperatures were also recorded in the junction strip area.

Maximum and minimum temperatures (TMAX/TMIN) per hyperthermia field were assessed and calculated in minutes equivalent 43° C (T43 Eq) using the "thermal dose" formula of Sapareto and Dewey (1984). For each hyperthermia field total time and mean time in T43 Eq minutes were calculated.

Clinical Tumor Response

At follow-up after 1 month complete response (CR) was seen in 13 of 27 hyperthermia fields (48%) and partial response (PR; > 50% reduction in tumor volume) in a further 13 fields (48%). The one remaining field (4%) was without response. At 3 months one patient (four fields) had died from systemic disease with PR in three of four fields. Four more CRs were noted, yielding a result of 17/23 (74%) CR and 6/23 (26%) PR. The two patients with longest follow-up (at 12 months) still had complete response in all five hyperthermia fields.

For analysis, tumor response per treatment field was related to total and mean minimum tumor temperature (TMIN43 Eq), which was demonstrated to be a powerful prognostic factor for tumor response in both in vivo animal studies (Dewhirst et al. 1984) and human studies (Oleson et al. 1984).

Treatment Complications

In 11 of 27 hyperthermia fields (41%) acute complications were observed: three (11%) subcutaneous necrosis, eight (30%) superficial blisters.

For analysis, complication rate was related to total and mean maximum tumor temperature (TMAX43 Eq).

Results

A broad range of maximum and minimum tumor temperatures was recorded in each treatment field (TMAX/TMIN) and in the junction strip area (JMAX/JMIN) (Fig. 1 a–d). Junction temperatures were lower than field temperatures.

Thermal dose (T43 Eq) per treatment field showed good relation to complication rate and CR:

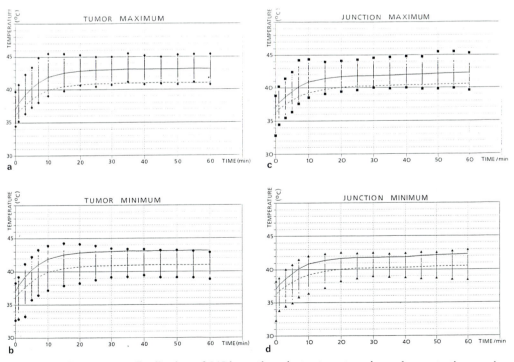

Fig. 1 a–d. Temperature distribution of 145 hyperthermia treatment sessions, demonstrating maximum and minimum tumor temperature (**a, b**) and maximum and minimum junction temperature (**c, d**) ranges

1. Complication rate was related to total and mean maximum tumor temperatures, TMAX43 Eq (Table 1)
2. CR at 1 month was related to total and mean minimum tumor temperatures, TMIN43 Eq (Table 2)
3. CR at 1 and 3 months was related to radiation dose (Table 3)
4. CR at 1 and 3 months was related to involved tumor area (Table 4)

Table 1. Complications *(COMPL)* related to TMAX43 Eq

COMPL at 1 month (%)	Total TMAX43 Eq	Mean TMAX43 Eq
11/27 (41%)	>300 min 10/13 (77%) COMPL	>60 min 9/9 (100%) COMPL
	<300 min 1/14 (7%) COMPL	<60 min 2/18 (11%) COMPL

Table 2. Complete response *(CR)* related to TMIN43 Eq

CR at 1 month (%)	Total TMIN43 Eq	Mean TMIN43 Eq
13/27 (48%)	>35 min 12/14 (86%) CR <35 min 1/13 (8%) CR	>7.5 min 11/12 (92%) CR <7.5 min 2/15 (13%) CR

Table 3. Complete response *(CR)* related to radiation dose *(cGy)*

	<2000 cGy	2000–4000 cGy	>4000 cGy
CR at 1 month 13/27 (48%)	1/7 (14%)	3/8 (38%)	9/12 (75%)
CR at 3 months 17/23 (74%)	1/3 (33%)[a]	5/8 (63%)	11/12 (92%)

[a] One patient with four hyperthermia fields (three PR, one no change) died.

Table 4. Complete response *(CR)* related to tumor area

	Tumor area >100 cm²	Tumor area <100 cm²
CR at 1 month 13/27 (48%)	3/13 (23%) CR	10/14 (71%) CR
CR at 3 months 17/23 (74%)	4/9 (44%) CR[a]	13/14 (93%) CR

[a] One patient with four hyperthermia fields (three PR, one no change) died.

Conclusions

1. The treatment of extensive superficial chest wall recurrences with combined radiation therapy and local hyperthermia is, providing certain criteria are observed, technically feasible and clinically successful.
2. Total and mean thermal dose (T43 Eq) are useful parameters to estimate complication rate (TMAX43 Eq) and tumor response rate (TMIN43 Eq).
3. Higher minimum tumor temperature, higher radiation dose and smaller involved tumor area yield better response rates.
4. CR rates after 1 month and 3 months follow-up were 48% and 74% respectively

Summary

From January 1986 to January 1987 at Hahnemann University Hospital, Philadelphia, 13 patients were treated with combined external hyperthermia and radiation therapy to extensive chest wall recurrences from carcinoma of the breast. A total of 145 hyperthermia sessions for 27 different hyperthermia fields were analyzed in this study.

Mixed photon (10 MV) or electron beam (6–15 MeV) radiation therapy (16–50.5 Gy, two to five weekly fractions of 1.5–2.0 Gy) was given followed within 30 min by external 915-MHz microwave hyperthermia (two weekly fractions for 60 min at 41–44° C).

Temperature recordings were made at up to 24 points in the center and periphery and at junctions of fields. The thermal dose concept of Sapareto (T43 Eq) was used to evaluate the relationship between maximum or minimum temperature (TMAX/TMIN) and rates of complication and of CR for each treatment field.

Our results indicate that total or mean TMAX43 Eq per hyperthermia treatment field is related to complication rate, whereas total or mean TMIN43 Eq is related to CR rate. Also, higher radiation dose and smaller involved tumor area are related to higher CR rates.

At 1 month follow-up 13/27 (48%) fields showed CR. At 3 months, four more CRs were noted, yielding 17/23 (74%) CR.

References

Dewhirst MW, Sim DA, Sapareto SA, Connor WG (1984) Importance of minimum tumor temperature in determining early and long-term response of spontaneous canine and feline tumors to heat and radiation. Cancer Res 44: 43–50
Henle KJ, Dethlefsen LA (1978) Heat fractionation and thermotolerance. A review. Cancer Res 38: 1838–1851
Oleson JR, Sim DA, Manning MR (1984) Analysis of prognostic variables in hyperthermia treatment of 163 patients. Int J Radiat Oncol Biol Phys 10: 2331–2339
Perez CA, Kuske RP, Emami B (1986) Irradiation alone or combined with hyperthermia in the treatment of recurrent carcinoma of the breast in the chest wall. A non randomized comparison. Int J Hyperthermia 2 (2): 179–188
Sapareto SA, Dewey WC (1984) Thermal dose determination in cancer therapy. Int J Radiat Oncol Biol Phys 10: 787–800

Microwave-Induced Hyperthermia and Radiotherapy in Human Superficial Tumours: Clinical Results with a Comparative Study of Combined Treatment Versus Radiotherapy Alone

C. E. Lindholm, E. Kjellen, P. Nilsson, T. Landberg, and B. Persson

Departments of Oncology, University Hospitals, 21401 Malmö and 22185 Lund, Sweden

Introduction

Encouraged by radiobiological and preliminary clinical reports on hyperthermia, we started in August 1980 to evaluate hyperthermia as an adjunct to radiation therapy in human superficial malignant tumours. Preliminary results were reported earlier (Lindholm et al. 1982). We now present an updated report including patients referred to and accepted for hyperthermia treatment up to November 1984.

Methods and Material

From the outset, we used equipment for 2450-MHz or 915-MHz microwave heating developed by members of the team. Details have been described elsewhere (Nilsson et al. 1982; Nilsson and Persson 1985). From 1983 onward, additional similar equipment (Lund Science AB, Lund, Sweden) was also used. When using 915 MHz, the microwave power was coupled from the applicator to the patient by a flexible plastic bag through which temperature-controlled deionized water was circulated. Thus the skin of the patient could be actively heated or cooled.

Polyethylene-coated thermistor probes (Yellow Springs Inc., Model 511) with an outer diameter of 0.6 mm were used for temperature measurements. When used invasively the probes were inserted into the tissue by means of intravenous cannulas (Venflon, 18 gauge) from which the metallic needles were removed after insertion, leaving in the tissue only the plastic cannulas with the thermistor probes. Cannulas and probes were removed after every heat session. Forty-five patients aged 22–94 years (median 68 years) with in total 98 superficial malignant tumours refractory to established treatment modalities were accepted for the study. The patients were anticipated to have a life expectancy of at least 3 months. The tumours, estimated to be growing no deeper than 3 cm under the skin surface, were either locoregional recurrences of primary tumours or metastatic lesions. This was verified by microscopic examinations of fine-needle aspiration and/or biopsy specimens.

In all, seven patients (13 tumours) were excluded, leaving 38 patients with 85 tumours for evaluation.

Fifty-seven tumours received radiotherapy combined with hyperthermia and 28 received radiotherapy alone (Table 1). In 18 patients with 2–10 tumour sites each, a total

Day	1	2	3	4	5	6	7	8	9	10	11	12
Radiotherapy (3.00 Gy/fraction)	x	x	x	x	x			x	x	x	x	x
Hyperthermia (41–45° C for 45 min, either 0.5–1.5 h or 3–4 h after radiotherapy)		x			x				x			x

Fig. 1. Treatment schedule for combined hyperthermia and radiotherapy

of 56 tumours underwent evaluation whereby 28 tumours received combined treatment and 28 tumours radiotherapy alone, the patient being his or her own control. This part of the study is henceforth referred to as "the comparative study" (Table 1). In all but four patients in this comparative study the smallest lesion was treated with radiotherapy alone while the combined modality was used for the larger lesion.

The majority of the tumours, 78% (66 lesions), were adenocarcinomas and 14% (12 lesions) were squamous cell carcinomas. More than half, 53% (45 lesions), were mammary adenocarcinomas. There were four anaplastic carcinomas, two soft tissue sarcomas and one malignant melanoma. The tumour regions had often been treated earlier with radiotherapy and/or hormonal therapy and/or chemotherapy.

Hyperthermia was delivered with 2450-MHz microwaves to 33 lesions and with 915-MHz microwaves to 24 lesions. The treatment schedule is described in Fig. 1. Two different time intervals between radiotherapy and hyperthermia were used, either 0.5–1.5 h or 3–4 h.

All except five tumours received the low-dose radiotherapy that was considered possible in these situations (3.00 Gy × 10 = 30.0 Gy or less during 2 weeks). The five exceptions received 39.0–51.0 Gy because they had had no or less previous radiotherapy in these regions. In all patients in the comparative study, however, the absorbed doses were the same in the two compared tumour regions.

The results of treatment with respect to response rates were evaluated by inspection, palpation, caliper measurements in two dimensions and photos with measuring scales. Registration of response to treatment was made at regular intervals, with the first follow-up occurring 3–4 weeks after completion of treatment. Tumour response was reported according to the following definitions; complete response (CR), 100% local tumour regression; partial response (PR), ≥ 50% local tumour regression; no response (NR), < 50% local tumour regression or < 25% increase in local tumour size; progressive disease (PD), ≥ 25% increase in the local tumour size or appearance of new local lesions (progression of disease during the treatment period was found in only one single tumour treated with radiotherapy alone). For tumour response evaluation two observations showing continuing response at least 1 month apart were required.

Results

The response rates in terms of CR, PR and NR for all tumours treated are given in Table 1. In 57 tumours treated with combined hyperthermia and radiotherapy, CR and PR rates 26 (46%) and 17 (30%) respectively, giving a total response rate (CR + PR) of 75% for the combined treatment. The median observed duration of response was 8 months and 4 months for CR and PR respectively (range 1–38 months). Analysis of

28 tumours treated with combined modality in the comparative study showed similar rates and durations of responses – 16 (57%) CR and nine (32%) PR, together 89%.

The range of response duration was 1–23 months (median 9 months and 4 months for CR and PR respectively) (Table 2).

In 28 tumours treated with radiotherapy alone, seven (25%) CR and seven (25%) PR were registered, giving a total response (CR+PR) of 50%. The median observed duration was 16 months and 4 months for CR and PR respectively (range 1–22 months). The difference in both CR and total (CR+PR) response rates between the two treatment modalities in the comparative study ($p=0.0027$ and $p=0.0039$ respectively, McNemar's test) is significant in favour of the combined treatment (Table 2).

Half of the tumours in our material were local recurrences or subcutaneous and/or cutaneous metastases from breast cancers. The response in all breast cancer tumours to combined treatment was 18 (64%) CR and eight (29%) PR, a total of 26 (93%) respond-

Table 1. Evaluable tumour material according to number of superficial tumours per patient in relation to treatment and response (total 85 tumours in 38 patients)

Number of tumours per patient	Number of patients	Combined treatment				Radiotherapy alone			
		Number of tumours	Response			Number of tumours	Response		
			CR	PR	NR		CR	PR	NR
1	18	18	6	6	6	0	–	–	–
2	10	{4	–	1	3	8a	–	2	6
		8a}	4	3	1				
3	3	{3	1	–	2	3a	1	1	1
		3a}	1	–	2				
4	4	{4	3	1	–	6a	1	2	3
		6a}	3	3	–				
6	2	6a	3	3	–	6a	–	2	4
10	1	5a	5	–	–	5a	5	–	–
	38	57	26	17	14	28	7	7	14

CR, complete response; PR, partial response; NR, no response.
a Comparative study.

Table 2. Response rates in the comparative study (56 tumours in 18 patients with 2–10 superficial tumours each)

	Number of tumours	Complete response (CR)	Partial response (PR)	CR+PR	Duration of response (months)	
					Range	Median
Hyperthermia + radiotherapy	28	16 (57%)	9 (32%)	25 (89%)	1–23	9 (CR) 4 (PR)
Radiotherapy alone	28	$p=0.0027$ 7 (25%)	n.s. 7 (25%)	$p=0.0039$ 14 (50%)	1–22	16 (CR) 4 (PR)

McNemar's test used for statistical evaluation.

Table 3. Skin reactions in 57 tumour regions treated with combined hyperthermia and radiotherapy. The numbers of different reactions are tabulated according to a scoring modified after the WHO (1979)

Skin reaction	2450 MHz without water bolus	915 MHz with water bolus
Grade 1		
a) No visible reaction	–	–
b) Minimal erythema	9	17
c) Marked erythema	7	3
Grade 2		
a) Erythema with slight desquamation	2	–
b) Dry desquamation	–	1
Grade 3		
a) Desquamation with blisters	7	1
b) Moist desquamation	–	–
Grade 4		
a) Small necrosis or ulceration	3	1
b) Massive ulceration	5	1
	33	24
Subcutaneous fat tissue necroses	3	1

ers among a total of 28 breast cancer tumours in 16 patients. Eleven of these patients had more than one tumour and could be used for a separate controlled study including 34 breast cancer tumours, 17 treated with the combined modality and 17 with radiotherapy alone. The difference in both CR and total (CR + PR) response rates between the two treatment modalities for breast cancer tumours is significant in favour of the combined treatment ($p = 0.0253$ and $p = 0.0082$ respectively, McNemar's test).

Cutaneous ulcerations (grade 4; Table 3) of up to a few centimetres in size were seen in skin uninvolved by tumour and were probably due to local overheating. Such "hot spot" effects were usually predictable immediately after removal of the applicator by the appearance of slight blisters on the skin. Their healing took a long time, in some cases 5 months. The scoring of skin reactions in relation to type of heating system and number of regions heated is listed in Table 3, which also includes information on the number of subcutaneous fat tissue necroses.

Discussion

Our results show that the response rate in certain recurrent tumours will be higher after combined treatment with hyperthermia and radiotherapy than after radiotherapy alone. The combination of hyperthermia with a small or moderate dose of radiotherapy may now be regarded as a useful method for palliation of superficial recurrences in sites where moderate to high doses of radiotherapy have been given earlier. The situation may, due to biological and technical factors, be different for primary tumours than for deep-seated tumours. Continued studies of relatively superficial tumours are still warranted to investigate some specific factors which may be important for the effect of combined hyperthermia and radiotherapy. Such factors include the size of the tumours, the condition of the non-tumour tissue in the anticipated treatment region and thermal distribution in the tumour related to treatment time factors necessary to reach CR. To determine the maximum possible temperature that can be given for a certain time without causing damage to normal tissue seems to be of growing importance as technology for deep hyperthermia improves. The practical importance of thermotolerance (Overgaard and Nielsen 1983) as an essential factor in the clinic should also be investigated further.

References

Lindholm CE, Kjellen E, Landberg T, Nilsson P, Persson B (1982) Microwave-induced hyperthermia and ionizing radiation: preliminary clinical results. Acta Radiol [Oncol] 21: 241–254
Nilsson P, Persson B (1985) Computer controlled microwave system for clinical hyperthermia. Phys Med Biol 30: 283–292
Nilsson P, Persson B, Kjellen E, Lindholm CE, Landberg T (1982) Technique for microwave-induced hyperthermia in superficial human tumours. Acta Radiol [Oncol] 21: 235–240
Overgaard J, Nielsen OS (1983) The importance of thermotolerance for the clinical treatment with hyperthermia. Radiother Oncol 1: 167–178

Hyperthermia and Chemotherapy: Biological and Clinical Studies

Interaction of Hyperthermia and Chemotherapy*

O. Dahl

Department of Oncology, Haukeland Hospital, University of Bergen, 5021 Bergen, Norway

Introduction

Although hyperthermia alone has limited antitumor effects at clinically tolerable doses, the method is currently being widely studied in combination with radiation, with promising results. A large amount of experimental and clinical data supports this combination. Much less is known about the possibilities of combining hyperthermia and chemotherapeutic drugs. This paper reports experimental data from cell culture and animals. The pharmacological effects of hyperthermia and known mechanisms for the interaction of heat and drugs will be presented. A better understanding of these processes is necessary for rational clinical application of the combined modality approach.

Experimental Studies

Alkylating Agents

The alkylating agent thio-TEPA decreased the survival of cultured Chinese hamster ovary cells (CHO) proportionally to the temperature rise between 35 and 42° C with activation energies as expected for alkylation (Johnson and Pavelec 1973). Similar increased cell killing at elevated temperatures has been demonstrated in several tumour lines in vitro for melphalan, chlorambucil, and mitomycin C (see Hahn 1982; Dahl 1986).

In rats bearing Yoshida sarcoma, Suzuki (1967) demonstrated a proportional increased effect by combining local hyperthermia above 40° C for 30 min with nitrogen mustard. Similar increased activity at elevated temperatures have been reported in many tumours in different animal species given thio-TEPA, melphalan, chlorambucil, mitomycin C and cyclophosphamide (see Dahl 1986). In our neurogenic rat tumour where local hyperthermia alone (44.0° C for 1 h) did not induce tumour regression, seven of eight complete tumour regressions were observed (four tumours were cured) when hyperthermia was combined with cyclophosphamide (200 mg/kg) which alone only resulted in transient partial tumour regressions (Dahl and Mella 1983). When the increase of antitumoural effects was compared with the increase of normal tissue side effects (bone

---

* Supported by the Norwegian Cancer Society.

marrow) in mice treated with systemically administered chemotherapeutics and whole-body hyperthermia (41° C for 45 min), only melphalan resulted in a therapeutic gain, in contrast to chlorambucil, cyclophosphamide, cisplatin. CCNU and BCNU (Honess and Bleehen 1985).

Nitrosoureas

Hahn (1978) demonstrated increased killing at higher temperatures in cultured Chinese hamster cells when the concentration of BCNU, CCNU and methyl-CCNU was correct-ed for the increased hydrolytic decay at elevated temperatures. The calculated activation energies suggested alkylation as the underlying mechanism. Enhancement of effect by combining hyperthermia and BCNU has been confirmed in Chinese hamster cells (Her-man 1983) and with CCNU in B16 melanoma and Lewis lung cancer treated in vivo but assayed as colonies in agar (Joiner et al. 1982). It is also of interest that BCNU, which is only mildly pH dependent at 37° C, exhibits higher cytotoxicity at pH 6.5–7.0 at 43° C than at normal pH (Hahn and Shiu 1983). This is a pH which is normally found in hu-man tumours and which is further lowered during hyperthermia. The pH effect is, how-ever, lower in cells chronically adapted to acidic conditions (Hahn and Shiu 1986).

Local hyperthermia (about 41.6° C) increased the cure rate of BCNU (20 mg/kg) in mice with EMT6 tumours (Twentyman et al. 1978), as did the same drug concentration in our rat glioma tumour at 44° C for 60 min (Dahl and Mella 1982). A synergistic effect of the combination of hyperthermia and ACNU in mouse melanoma tumour has also been reported (Yamada et al. 1984). However, no potentiation was found when a human colonic xenograft was treated by local hyperthermia (42.5° C) and methyl-CCNU in nude mice (Osieka et al. 1978). Systemic hyperthermia (41° C for 45 min) combined with BCNU (10–40 mg/kg) or CCNU (5–15 mg/kg) did not increase the therapeutic ratio in mice bearing the nitrosourea-resistant RIF-1 fibrosarcoma and the more sensitive KHT sarcoma (Honess and Bleehen 1982), possibly related to an increased effect on bone marrow (O'Donnel et al. 1979).

Cisplatin

Cisplatin interacts with DNA leading to intrastrand crosslinkage. Reduced survival was found in human colon cancer cells in culture (Barlogie et al. 1980) and in Chinese hamster cells (Fisher and Hahn 1982) treated with cisplatin at higher temperatures. This is confirmed in a human melanoma line in culture (Zupi et al. 1984) as well as in normal mouse bone marrow and P388 mouse leukemia cells treated in vivo and assayed in cul-ture (Alberts et al. 1980). Cisplatin-resistant cell lines can be rendered susceptible to combined treatment by hyperthermia in the therapeutic range (41–43° C) (Murthy et al. 1984; Wallner et al. 1986). This emphasizes the potential role of hyperthermia as a mo-dality to overcome drug resistance.

In a mouse mammary tumour, increased effect of combined therapy was observed (Douple et al. 1982). In investigations into the combination of local hyperthermia (44° C) and cisplatin (3 mg/kg) in rat glioma tumours (Mella and Dahl 1985), the best tumor delay occurred when both were given simultaneously, suggesting a true interac-tion. We have recently demonstrated increased renal damage caused by cisplatin due to low-grade systemic hyperthermia (about 41° C) during local hyperthermia (44° C for 60 min) (Mella et al. 1987). Despite the lack of general therapeutic gain, a useful local tumour effect was obtained. Hyperthermia can clinically be given to local areas where a greater antitumour effect is needed, thus serving as a targeting procedure for systemically administered drugs. The experiments further imply that chemotherapeutic agents can have severe systemic side effects when combined with whole-body hyperthermia.

Antibiotics

Anthracyclines

Combination of doxorubicin and hyperthermia in culture (see Dahl 1986) may result in increased cell killing, no change or even reduced survival at 41–43° C. Sensitive cells may turn resistant during prolonged heating (Hahn and Strande 1976; Dahl 1982). In animals the same discrepancy exists. Increased growth delay was observed on combined treatment in some tumours (Marmor et al. 1979; Dahl 1983), while only additional effect on survival, and no increased growth delay, occurred in P388 leukemia (van der Linden et al. 1984); or an increased effect occurred only at extreme doses (Overgaard 1976). Only few cures have been observed in animals after single combined treatments. The moderate effect may be related to development of thermotolerance or to the fact that thermotolerant cells may also acquire resistance to doxorubicin (Hahn and Strande 1976; Morgan et al. 1979; Dahl 1982). However, repeated low doses of doxorubicin combined with hyperthermia in metastatic head and neck cancers increased the response rate from 4/11 after drug alone to 9/10 in the combined modality group (Arcangeli et al. 1980). The related drugs mitoxantrone (Wang et al. 1984), and aclacinomycin A (Mizuno et al. 1980), but not daunorubicin, were more effective at elevated temperatures in vitro, as was 4-epidoxorubicin in vivo (Dahl 1983).

Bleomycin

Bleomycin, a complex antibiotic drug, has been investigated in many studies using different cell lines. Bleomycin has only slightly enhanced effect at 40–41° C, but above 42–43° C its effect is markedly potentiated, showing a threshold effect (Hahn 1982). The survival curve for bleomycin tends to have a biphasic form. Hyperthermia seems especially to sensitize the presumed resistant cell population for combined treatment. Many animal experiments have confirmed the efficacy of combined modality therapy with bleomycin and hyperthermia, usually giving tumour regressions, but rarely cures (Magin et al. 1979; von Szczepansky and Trott 1981; Dahl 1986). The pilot study by Arcangeli et al. (1980), however, again suggests a substantial advantage of the combination in man: objective responses in 11/11 against 6/11 after bleomycin alone. Due to lack of bone marrow toxicity, bleomycin can be combined with hyperthermia at the schedule most suited for hyperthermia alone. It must, however, be combined with local hyperthermia where temperatures above the threshold (42–43° C) can be achieved.

Actinomycin D

Actinomycin D, like hyperthermia, interacts with RNA synthesis. In CHO cells in culture a superadditive killing was observed for short hyperthermic treatments, but the effect was then reduced, probably related to development of resistance (Donaldson et al.

1978). In leukemic cells (Giovanella et al. 1970; Mizuno et al. 1980) and in a murine fibrosarcoma (Yerushalmi 1978), increased effects have been observed at 42–43° C but not
in EMT6 cells (Har-Kedar 1975).

Antimetabolites and Vinca Alcaloids

The antimetabolites methotrexate, fluorouracil and PALA have slightly increased effects
at elevated temperatures in some cell lines, but as a whole these drugs seems not to be
good candidates for combination with hyperthermia (see Dahl 1986). Herman et al.
(1981) reported increased cell killing at 43° C with combined treatment in Chinese hamster cells, possibly related to reduced synthesis of dihydrofolate reductase. Neither vincristine nor vinblastine exhibited an increased effect on leukemic cells when used at
41–42° C. No data for the latter drugs above 43° C have been published.

Pharmacological Changes Due to Hyperthermia

Physicochemical Changes

An increased hydrolytic degradation of the nitrosoureas BCNU, CCNU and methyl-
CCNU (Hahn 1978), melphalan (Honess et al. 1985) and chlorambucil (Eksborg and
Ehrsson 1985) has been found even at mild hyperthermia. As a general rule the aqueous
solubility of drugs increases with increasing temperature (Ballard 1974). In culture, special factors like binding to culture flasks and release of drugs from increasing numbers
of dying cells may contribute to the observed effects.

Enzymatic Changes

Most enzymatic reaction rates are temperature-dependent. Increased metabolic activation or inactivation may occur, and at high tempratures the reaction may stop. It is therefore not easy to make general rules. Cyclophosphamide which is normally metabolized
to active alkylating derivatives by microsomal oxidases in the liver, showed less alkylating activity when incubated with liver slices (Clawson et al. 1981) or by perfusion of rat
liver (Skibba 1982) at temperatures 41–44° C. In contrast, the rate of production of alkylating species from mitomycin C was increased at similar temperatures, 41–43° C, during
anoxic conditions (Teicher et al. 1981). It is also possible that drug metabolism within
the target cell may be changed, as doxorubicin produced more metabolites (aglycones)
at elevated temperatures (Magin et al. 1980; Dodion et al. 1986). Lack of effect of
fluorouracil may be due to impaired incorporation of the drugs into nucleotides. The alkylation reaction is also influenced by temperature.

Blood Flow

In vivo chemotherapeutic agents have to be transported to the target cells by the blood
flowing through the delicate vascular network within tumours. Generally there is increased flow in normal tissues at temperatures of 43–45° C, whereas there is only a transient increase, followed by shutdown of blood flow due to stasis and haemorrhages, in

most tumours (Reinhold and Endrich 1986). If the flow is stopped by hyperthermia, systemically administered chemotherapeutics cannot reach the tumour cells, but drug already present will also be trapped within the tumour. Increased cell killing was observed in the tumour periphery when cisplatin and hyperthermia were combined (Mella 1985). This might imply spatial cooperation, hyperthermia killing the centrally located tumor cells growing under poor nutrition conditions while the chemotherapeutic agent has increased effect in vascularized areas where the proliferating cells are insufficiently heated to be killed by hyperthermia alone and are also not completely eradicated by the drug alone.

Protein Binding

Only the unbound, diffusable fraction of drugs is able to react with target cells. With higher temperatures the various bonds between drugs and proteins tends to be weakened, leaving a higher free fraction (Ballard 1974). It is known that such protein binding takes place for several chemotherapeutic drugs, such as nitrosoureas (Oliverio 1973; Hahn 1978), methotrexate (Warren and Bender 1977) and cisplatin (Hecquet et al. 1985). Hyperthermia (42° C) increased the rate of metabolism or aquation of cisplatin leading to *increased* serum protein binding in vitro but also enhanced tissue extraction in dogs treated by whole body hyperthermia (Riviere et al. 1986).

Pharmacokinetics

Methothrexate serum concentration is now routinely checked during high-dose therapy. On perfusion of liver in rats and rabbits, an increased plasma half-life of doxorubicin was observed due to reduced biliary excretion (Mimnaugh et al. 1978, Skibba 1982) at temperatures 41–42.3° C. Systemic heat (41° C for 45 min) caused higher plasma and tumour melphalan concentrations than in unheated animals, but the effect was greater in plasma than in tumour (Honess et al. 1985). In dogs, systemic hyperthermia at 42° C significantly increased clearance, volume of distribution and $T_{\frac{1}{2}}$ of free ultrafiltrable cisplatin and the parent drug (Riviere et al. 1986). In man no change in half-life of total cisplatin was seen on infusing the drug during whole-body hyperthermia (Gerad et al. 1983), but enhanced nephrotoxicity was observed. Peak unchanged cyclophosphamide and serum alkylating activity was unchanged, but urinary excretion of unchanged parent drug increased and alkylating activity decreased when measurements were performed at normothermia and after whole-body hyperthermia at 41.8° C in two patients (Ostrow et al. 1982). Our knowledge in this field is still only fragmentary, but further studies should be encouraged to examine the interaction of whole-body hyperthermia and chemotherapeutics.

Mechanisms of Interaction

Drug Uptake

Cellular *membrane damage* is probably one of the main causes of cell killing due to hyperthermia alone (see Wallach 1977). Based upon this fact and reports of increased drug uptake at elevated temperatures, many authors refer to increased uptake of chemother-

Table 1. Thermal influence on measured intracellular concentrations of different chemotherapeutic agents. Although an increased uptake is found for many cell types and drugs, increased intracellular drug does not always lead to enhanced cell killing

Drug	Cell line	Temperature (°C)	Time (min)	Uptake	Cytotoxic effect	Author	Year
Doxorubicin	Leukemia	39.5	120	↑	– –	Block et al.	1975
	Chinese hamster	43	11–50	↑30↓	↑30↓	Hahn and Strande	1976
	P388 leukemia	43	60	↑	↑	van der Linden et al.	1984
	MAM 16/C	43	60	– –	– –	van der Linden et al.	1984
	L1210 leukemia	41.5	60	↑	?	Klein et al.	1977
	Ehrlish ascites cells	40–43	240	↑30↓	↑	Yamane et al.	1984
	CHO sensitive	40–50	15	↑	↑	Bates and Mackillop	1986
	CHO resistant	40–50	15	– –	– –	Bates and Mackillop	1986
	Mamma tumor[a]	43	60	– –	↑	Magin et al.	1980
	Rat liver slices	43	0–90	– –	?	Dodion et al.	1986
Daunorubicin	L1210 leukemia	41.5	60	–/↑	?	Klein et al.	1977
Mitoxan-throne	Human colon cancer	42	60	↑	↑	Wang et al.	1984
Actinomy-cin D	Chinese hamster cells	43	120	↑30↓	↑30↓	Donaldson et al.	1978
Bleomycin	Chinese hamster cells	43	60	↓	↑	Braun and Hahn	1975
	HeLa	43	90	↓	↑	Hassanzadeh and Chapman	1982
	Rat rhabdomyo-sarcoma[a]	43	90	– –	– –	Lin et al.	1983
Melphalan	Mouse RIF-1[a]	41	45	↑	↑	Honess et al.	1985
	CHO	40–45	60	↑	↑	Bates and Mackillop	1987
Cisplatin	Chinese hamster	43	0–60	↑	↑	Wallner et al.	1986
	Mouse leukemia[a]	42.3	30	↑	↑	Alberts et al.	1980
Methyl-CCNU	Colon xenografts[a]	42.5	60	– –	– –	Osieka et al.	1978
Metho-trexate	CHO resistant	43	0–6 h	↑30↓	↑	Herman et al.	1981
	Murine bladder cancer[a]	42	15	– –	?	Tacker and Anderson	1982
	Lewis lung tumor[a]	42	15	– –	?	Weinstein et al.	1979
Fluorouracil	Murine cervix cancer[a]	43, 45	30	↑	?	Fujiwara et al.	1984

[a] Data from animal studies.

apeutics as the main mechanism underlying the enhanced effect observed on combining hyperthermia and drugs. The relation is, however, not so simple, as shown in Table 1. There is no universal increased concentration and increased efficacy of the drugs given with hyperthermia, and there is no positive interaction for all drugs where increased intracellular concentration has been demonstrated. There may be differences among different tumours, and the topic is further complicated by the fact that chemotherapeutics differs in their uptake mechanism (active transport, passive diffusion) or exclusion (passive diffusion, active efflux). The drug charge may also change due to temperature and pH shift, affecting the drug uptake.

It has been shown that hyperthermia increases the intracellular cisplatin concentration in cisplatin-sensitive and -resistant strains of CHO cells (Wallner et al. 1986) and causes preferential accumulation of the drug in P388 leukaemia cells compared to normal murine bone marrow (Alberts et al. 1980). Thus increased drug uptake may contribute to the observed effects in some tumours with some drugs, but it is not the only cause of the increased cell killing.

Increasing the temperature makes the fungicide polyene antibiotics (amphotericin B; Hahn 1982) cytotoxic for mammalian cells, probably by interaction with sterols in the membranes. Similarly, local anaesthetics and alcohols which have effects on membranes turn cytotoxic when combined with hyperthermia, thus suggesting a common membrane target.

DNA Effects

Alkylation is a process where reactive parts of the chemotherapeutic drug are covalently bound to macromolecules. Although there are many binding sites, impaired DNA replication will ultimately lead to cellular reproductive death. From Arrhenius' analysis of survival curves, low activation energies, implying an alkylating process, have been suggested for the increased effect of thio-TEPA (Johnson and Pavelec 1973) and nitrosoureas (Hahn 1978). Direct DNA damage has been shown as *single-strand breakage* of DNA on examination alkaline sucrose sedimentation profiles after combination of hyperthermia and the monofunctional alkylating agent methyl methanesulfonate (Bronk et al. 1973; Ben-Hur and Elkind 1974), bleomycin (Meyn et al. 1979; Kubota et al. 1979; Smith et al. 1986), cisplatin (Meyn et al. 1980) and peptichemio (Djordjevic et al. 1978) as well as structural DNA changes (electron spin resonance, Chapman et al. 1983; neutral nucleoid sedimentation. Smith et al. 1986). Several experiments have demonstrated reduced repair of single-strand breaks at higher temperatures (Bronk et al. 1973; Ben-Hur and Elkind 1974; Djordjevic et al. 1978; Kubota et al. 1979; Meyn et al. 1979; Smith et al. 1979). A direct inhibition of the *DNA repair enzyme polymerase β* has also been demonstrated by hyperthermia alone (Spiro et al. 1982).

The role of alkylation has recently been supported by the evidence that several chemotherapeutics (BCNU, doxorubicin, bleomycin, mitomycin, see Arrik and Nathan 1984; Issels et al. 1984; cisplatin, Hromas et al. 1987; chlorambucil, Wang and Tew 1985) may exert their cytotoxic activity via *oxygen radical production,* as do radiation and even hyperthermia alone, by depleting cellular glutathione, which can function as an oxygen radical detoxifier (Arrik and Nathan 1984; Shrieve et al. 1986). NADH dehydrogenase in mitochondria contributes to production of semiquinone free radical intermediates from anthracyclines (Gervasi et al. 1986), and the new drug lonidamine, which has a particular effect on mitochondria under acidic conditions, may have similar effects interfering with the electron transport (Kim et al. 1984). It is thus possible that hyper-thermia, ionizing radiation and most chemotherapeutic drugs may exert their cell-killing activity through similar cellular effects, leading to a unifying concept of mechanism of action for all modalities.

Effect of Prior Heating on Drug Sensitivity

We have experimentally shown that hyperthermia and chemotherapy (cyclophospha-mide, BCNU and cisplatin) have the best antitumoral effect when they are given simul-taneously or the drug is injected immediately before heating (Dahl and Mella 1982, 1983). Similar results have been reported for most combinations in vitro. If Chinese hamster cells were exposed to simultaneous hyperthermia (43° C) and doxorubicin first an increased cytotoxic activity occurred, then the cells became refractory (Hahn and Strande 1976). Similar heat-induced drug tolerance has been observed in other cells pre-heated at 43° C (Morgan et al. 1979) or 41° C (Dahl 1982) before exposure to doxorubi-cin. Preheating for 3 h at 40° C protected against killing of bleomycin and BCNU (Mor-gan et al. 1979). Chinese hamster cells treated with actinomycin D (Donaldson et al. 1978) were rendered resistant for continuous exposure of the drug after 30 min. Using the concept of step-down heating an initial exposure of cells to temperatures above 45° C increased the effect of adriamycin and bleomycin, but not cisplatin, compared to that achieved by continuous heating at 41.5° C (Herman et al. 1984). Thus prior heating may modify the final effect of the combined modality approach.

Side Effects

We have not seen any substantial increase of local side effects during local heating com-bined with drugs, in contrast to the report by Honess and Bleehen (1980) of skin reaction after local hyperthermia (44° C) and cyclophosphamide and BCNU. We have, however, seen a substantial increase of the renal toxicity due to systemic hyperthermia (41° C) ac-companying our local heating in rats at 44° C (Mella et al. 1987) contrasting with the re-port by Beckley et al. (1982) of no renal toxicity in 10 patients treated with cisplatin (70 mg/m^2) during whole body hyperthermia (41.8° C). Increased liberation of catechol-amines after injection of doxorubicin during whole-body hyperthermia has been related to observed ventricular irritability and cardial dysfunction (Kim et al. 1979). Heat may recall prior bleomycin skin damage (Kukla and McGuire 1982).

A reduced LD_{10} dose have been observed for mice treated with cyclophosphamide, methyl-CCNU and vincristine (Rose et al. 1979). Of all drugs tested, only melphalan showed a therapeutic gain when tumour-bearing mice were treated with whole-body hy-perthermia (41° C for 45 min) and different drugs (Honess and Bleehen 1985). Whole-body hyperthermia must be performed by experienced people with great care and with monitoring for organ-specific toxicity combined with systemic drugs. The tolerance limit for heat alone (42° C) may not be optimal for interaction with chemotherapeutics.

Local hyperthermia can safely be used clinically, and two small randomized studies show increased tumour response (Arcangeli et al. 1980; Kohno et al. 1978). On regional hyperthermic perfusion the increased dose response at high drug doses (Mella 1985) can be of value, as drug concentrations 10 times the systemic levels can be used (Vaglini et al. 1985), but the tolerated temperatures generally are in the lower range (41–42° C).

Modern radical surgery seems to offer similar local tumour control as reported in many studies of combined modality approach underlining the need for controlled clinical studies.

Conclusion

The preclinical experimental data indicate that significant antitumour effects can be achieved by combining hyperthermia with chemotherapeutics. The alkylating agents (melphalan, cyclophosphamide), including mitomycin C, nitrosoureas and cisplatin, seems the best choice for further clinical studies. Bleomycin seems to be activated at higher temperatures, above a threshold at 43° C, but the role of doxorubicin is still ambiguous. Antimetabolites and vinca alkaloids seems not to be very effective combined with hyperthermia.

Some drugs with limited clinical activity have antitumour effects when combined with hyperthermia (electron-affinic drugs, lonidamine etc.). These drugs are of particular interest, as local hyperthermia could direct their actions to specific locations or they could function as sensitizers for hyperthermia.

Properly designed clinical studies seem highly warranted, as combination of chemotherapeutics and hyperthermia would then be founded on a firm experimental basis.

References

Alberts DS, Peng Y-M, Chen GH-S, Moon TE, Cetas TC, Hoescheie JD (1980) Therapeutic synergism of hyperthermia-cis-platinum in a mouse tumor model. JNCI 65: 455–461

Arcangeli G, Cividalli A, Lovisolo G, Mauro F, Creton G, Nervi C, Pavin G (1980) Effectiveness of local hyperthermia in association with radiotherapy or chemotherapy: comparison of multimodality treatments on multiple neck node metastases. In: Arcangeli G, Mauro F (eds) Proceedings of the First Meeting of the European Group of Hyperthermia in Radiation Oncology. Masson Italia, Milano, pp 257–265

Arrick BA, Nathan CF (1984) Glutathione metabolism as a determinant of therapeutic efficacy: a review. Cancer Res 44: 4224–4232

Ballard BE (1974) Pharmacokinetics and temperature. J Pharm Sci 63: 1345–1358

Barlogie B, Corry PM, Drewinco B (1980) In vitro thermochemotherapy of human colon cancer cells with cis-dichlorodiammineplatinum (II) and mitomycin C. Cancer Res 40: 1165–1168

Bates DA, Mackillop WJ (1986) Hyperthermia, adriamycin transport, and cytotoxicity in drug-sensitive and -resistant Chinese hamster ovary cells. Cancer Res 46: 5477–5481

Bates DA, Mackillop WJ (1987) Hyperthermia enhances melphalan transport and cytotoxicity in Chinese hamster ovary cells (Abstr). 35th Annual Meeting of the Radiation Research Society, Atlanta

Beckley SA, Madajewicz S, Highby D, Takita H, Bhargava A, Wajsman Z, Pontes JE (1982) Combination of systemic hyperthermia with chemotherapy. 13th International Cancer Congress, Seattle

Ben-Hur E, Elkind MM (1974) Thermal sensitization of Chinese hamster cells to methyl methanesulfonate: relation of DNA damage and repair to survival response. Cancer Biochem Biophys 1: 23–32

Block JB, Harris PA, Peale A (1975) Preliminary observations on temperature-enhanced drug uptake by leukemic leukocytes in vitro. Cancer Chemother Rep 59: 985–988

Braun J, Hahn GM (1975) Enhanced cell killing by bleomycin and 43° hyperthermia and the inhibition of recovery from potentially lethal damage. Cancer Res 35: 2921–2927

Bronk BV, Wilkins RJ, Regan JD (1973) Thermal enhancement of DNA damage by an alkylating agent in human cells. Biochem Biophys Res Commun 52: 1064–1070

Chapman IV, Leyko W, Gwozdzinski K, Koter M, Grzelinska E, Bartosz G (1983) Hyperthermic modification of bleomycin-DNA interaction detected by electron spin resonance. Radiat Res 96: 518–522

Clawson RE, Egorin MJ, Fox BM, Ross LA, Bachur NR (1981) Hyperthermic modification of cyclophosphamide metabolism in rat hepatic microsomes and liver slices. Life Sci 28: 1133–1137

Dahl O (1982) Interaction of hyperthermia and doxorubicin on a malignant, neurogenic rat cell line (BT$_4$C) in culture. Natl Cancer Inst Monogr 61: 251–253

Dahl O (1983) Hyperthermic potentiation of doxorubicin and 4'-epi-doxorubicin in a transplantable neurogenic rat tumor (BT$_4$A) in BD IX rats. Int J Radiat Oncol Biol Phys 9: 203–207

Dahl O (1986) Hyperthermia and drugs. In: Wathmough DJ, Ross WM (eds) Hyperthermia. Blackie, Glasgow, pp 121–153

Dahl O, Mella O (1982) Enhanced effect of combined hyperthermia and chemotherapy (bleomycin, BCNU) in a neurogenic rat tumour (BT$_4$A) in vivo. Anticancer Res 2: 359–364

Dahl O, Mella O (1983) Effect of timing and sequence of hyperthermia and cyclophosphamide on a neurogenic rat tumour (BT$_4$A) in vivo. Cancer 52: 983–998

Djordjevic O, Kostic L, Brkic G (1978) The combined effects of hyperthermia and a chemotherapeutic agent on DNA in isolated mammalian cells. In: Streffer C, et al. (eds) Cancer therapy by hyperthermia and radiation. Urban and Schwarzenberg, Munich, pp 278–280

Dodion P, Riggs CE, Akman SR, Bachur NR (1986) Effect of hyperthermia on the in vitro metabolism of doxorubicin. Cancer Treat Rep 70: 625–629

Donaldson SS, Gordon LF, Hahn GM (1978) Protective effect of hyperthermia against the cytotoxicity of actinomycin D on Chinese hamster cells. Cancer Treat Rep 62: 1489–1495

Double EB, Strohbehn JW, de Sieyes DC, Alborough DP, Trembly BS (1982) Therapeutic potentiation of cis-dichlorodiammineplatinum (II) and radiation by interstitial microwave hyperthermia in a mouse tumor. Natl Cancer Inst Monogr 61: 259–262

Eksborg S, Ehrsson H (1985) Drug level monitoring: cytostatics. J Chromatogr 340: 31–72

Fisher GA, Hahn GM (1982) Enhancement of cis-platinum (II) diamminedichloride cytotoxicity by hyperthermia. Natl Cancer Inst Monogr 61: 255–257

Fujiwara K, Kohno I, Miyao J, Sekiba K (1984) The effect of heat on cell proliferation and the uptake of anti-cancer drugs into tumour. In: Overgaard J (ed) Hyperthermic oncology, vol 1. Taylor and Francis, London, pp 405–408

Gerad H, Egorin MJ, Whitacre M, van Echo DA, Aisner J (1983) Renal failure and platinum pharmacokinetics in three patients treated with cis-diamminedichloroplatinum (II) and whole-body hyperthermia. Cancer Chemother Pharmacol 11: 162–166

Gervasi PG, Agrillo MR, Citti L, Danesi R, del Tacca M (1986) Superoxide anion production by adriamycinol from cardiac sarcosomes and by mitochondrial NADH dehydrogenase. Anticancer Res 6: 1231–1236

Giovanella BC, Lohman WA, Heidelberger C (1970) Effects of elevated temperatures and drugs on the viability of L1210 leukemia cells. Cancer Res 30: 1623–1631

Hahn GM (1978) Interactions of drugs and hyperthermia in vitro and in vivo. In: Streffer C, et al. (eds) Cancer therapy by hyperthermia and radiation. Urban and Schwarzenberg, Munich, pp 72–79

Hahn GM (1982) Hyperthermia and cancer. Plenum, New York

Hahn GM, Shiu EC (1983) Effect of pH and elevated temperatures on the cytotoxicity of some chemotherapeutic agents on Chinese hamster cells in vitro. Cancer Res 43: 5789–5791

Hahn GM, Shiu E (1986) Adaption to low pH modifies thermal and thermo-chemical responses of mammalian cells. Int J Hyperthermia 4: 379–387

Hahn GM, Strande DP (1976) Cytotoxic effects of hyperthermia and adriamycin on Chinese hamster cells. JNCI 57: 1063–1067

Har-Kedar I (1975) Effect of hyperthermia and chemotherapy in EMT6 cells. In: Wizenberg MJ, Robinson JE (eds) Proceedings of the International Symposium on Cancer Therapy by Hyperthermia and Radiation. American College of Radiology, pp 91–93

Hassanzadeh M, Chapman IV (1982) Thermal enhancement of bleomycin-induced growth delay in a squamous carcinoma of CBA/Ht mouse. Eur J Clin Oncol 18: 795–797

Hecquet B, Meynadler J, Bonneterre J, Adenis L, Demaille A (1985) Time dependency in plasmatic protein binding of cisplatin. Cancer Treat Rep 69: 79–83

Herman TC (1983) Effect of temperature an the cytotoxicity of vindesine, amsacrine, and mitoxantrone. Cancer Treat Rep 67: 1019–1022

Herman TC, Cress AE, Sweets C, Gerner EW (1981) Reversal of resistance to methotrexate by hyperthermia in Chinese hamster ovary cells. Cancer Res 41: 3840–3843

Herman TS, Henle KJ, Nagle WA, Moss AJ, Monson TP (1984) Effect of step-down heating on the cytotoxicity of adriamycin, bleomycin, and cis-diamminedichloroplatinum. Cancer Res 44: 1823–1826

Honess DJ, Bleehen NM (1980) Effects of the combination of hyperthermia and cytotoxic drugs on the skin of the mouse foot. In: Arcangeli G, Mauro F (eds) Proceedings of the Ist Meeting of the European Group of Hyperthermia in Radiation Oncology, Masson Italia, Milano, pp 151–155

Honess DJ, Bleehen NM (1982) Sensitivity of normal mouse marrow and RIF-1 tumour to hyperthermia combined with cyclophosphamide or BCNU: Lack of therapeutic gain. Br J Cancer 46: 236–248

Honess DJ, Bleehen NM (1985) Thermochemotherapy with cisplatinum, CCNU, BCNU, chlorambucil and melphalan on murine marrow and two tumours: therapeutic gain for melphalan only. Br J Cancer 58: 63–72

Honess DJ, Donaldson J, Workman P (1985) The effect of systemic hyperthermia on melphalan pharmacokinetics in mice. Br J Cancer 51: 77–84

Hromas RA, Andrews PA, Murphy MP, Burns P (1987) Glutathione depletion reverses cisplatin resistance in murine L1210 leukemia cells. Cancer Lett 34: 9–13

Issels RD, Bigalow JE, Epstein L, Gerweck LE (1984) Enhancement of cysteamine cytotoxicity by hyperthermia and its modification by catalase and superoxide dismutase in Chinese hamster ovary cells. Cancer Res 44: 3911–3915

Johnson HA, Pavelec M (1973) Thermal enhancement of thio-TEPA cytotoxicity. JNCI 50: 903–908

Joiner MC, Steel GG, Stephens TC (1982) Response of two mouse tumours to hyperthermia with CCNU or melphalan. Br J Cancer 45: 17–26

Kim JH, Kim SH, Alfieri A, Young CW, Silvestrini B (1984) Lonidamine: a hyperthermic sensitizer of HeLa cells in culture and of the Meth-A tumor in vivo. Oncology 41: 30–35

Kim YD, Lees DE, Lake CR, Whang-Peng J, Schuette W, Smith R, Bull J (1979) Hyperthermia potentiates doxorubicin-related cardiotoxic effects. JAMA 241: 1816–1817

Klein ME, Frayer K, Bachur NR (1977) Hyperthermic enhancement of chemotherapeutic agents in L1210 leukemia. Blood [Suppl] 50: 223

Kohno I, Kaneshige E, Fujiwara K, Sekiba K (1978) Thermochemotherapy (TC) for gynecologic malignancies. In: Overgaard J (ed) Hyperthermic oncology, vol 1. Taylor and Francis, London, pp 753–756

Kubota Y, Nishimura R, Takai S, Umeda M (1979) Effect of hyperthermia on DNA single-strand breaks induced by bleomycin in HeLa cells. Gann 70: 681–685

Kukla L, McGuire WP (1982) Heat-induced recall of bleomycin skin changes. Cancer 50: 2283–2284

Lin P-S, Cariani PA, Jones M, Kahn PC (1983) Work in progress: The effect of heat on bleomycin cytotoxicity in vitro and on the accumulation of [57]-bleomycin in heat-treated rat tumors. Radiology 146: 213–217

Magin RL, Sisik BI, Cysyk RL (1979) Enhancement of bleomycin activity against Lewis lung tumors in mice by local hyperthermia. Cancer Res 39: 3792–3795

Magin RL, Cysyk RL, Litterst CL (1980) Distribution of adriamycin in mice under conditions of local hyperthermia which improve systemic drug therapy. Cancer Treat Rep 64: 203–210

Marmor JB, Kozak D, Hahn GM (1979) Effects of systemically administered bleomycin or adriamycin with local hyperthermia on primary tumor and lung metastases. Cancer Treat Rep 63: 1279–1290

Mella O (1985) Combined hyperthermia and cis-diamminedichloroplatinum in BD IX rats with transplanted BT$_4$A tumours. Int J Hyperthermia 1: 171–183

Mella O, Dahl O (1985) Timing of combined hyperthermia and 1,3-bis-(2-chloroethyl)-1-nitrosourea or cis-diamminedichloroplatin in BD IX rats with BT$_4$A tumours. Anticancer Res 5: 259–264

Mella O, Eriksen R, Dahl O, Laerum OD (1987) Acute systemic toxicity of combined cis-diammine-dichloroplatin and hyperthermia in the rat. Eur J Clin Oncol, (in press)

Meyn RE, Corry PM, Fletcher SE, Demetriades M (1979) Thermal enhancement of DNA strand
    breakage in mammalian cells treated with bleomycin. Int J Radiat Oncol Biol Phys 5: 1487–1489
Meyn RE, Corry PM, Fletcher SE, Demetriades M (1980) Thermal enhancement of DNA damage
    in mammalian cells treated with cis-diamminedichloroplatinum (II). Cancer Res 40: 1136–1139
Mimnaugh EG, Waring RW, Sikic BI, Magin RL, Drew R, Litterst CL, Gram TE, Guarino AM
    (1978) Effect of whole-body hyperthermia on the disposition and metabolism of adriamycin in
    rabbits. Cancer Res 38: 1420–1425
Mizuno S, Amagai M, Ishida A (1980) Synergistic cell killing by antitumor agents and hyperther-
    mia in cultured cells. Gann 71: 471–478
Morgan JE, Honess DJ, Bleehen NM (1979) The interaction of thermal tolerance with drug cyto-
    toxicity in vitro. Br J Cancer 39: 422–428
Murthy MS, Khandekar JD, Travis JD, Scanlon EF (1984) Combined effect of hyperthermia (HT)
    and platinum compounds in vivo and in vitro on murine and human tumor cells. In: Overgaard
    J (ed) Hyperthermic oncology, vol 1. Taylor and Francis, London, pp 421–424
O'Donnel JF, McKoy WS, Makuch RW, Bull JM (1979) Increased in vitro toxicity to mouse bone
    marrow with 1,3-bis(2-chloroethyl)-1-nitrosurea and hyperthermia. Cancer Res 39: 2547–2549
Oliverio VT (1973) Toxicology and pharmacology of nitrosourea. Cancer Cemother Rep 4: 13–20
Osieka R, Magin RL, Atkinson ER (1978) The effect of hyperthermia on human colon cancer
    xenografts in nude mice. In: Streffer C, et al. (eds) Cancer therapy by hyperthermia and radia-
    tion. Urban and Schwarzenberg, Munich, pp 287–290
Ostrow S, van Echo D, Egorin M, Witacre M, Grochow L, Aisner J, Colvin M, Bachur N, Wiernik
    PH (1982) Cyclophosphamide pharmacokinetics in patients receiving whole-body hyperthermia.
    Natl Cancer Inst Monogr 61: 401–403
Overgaard J (1976) Combined adriamycin and hyperthermia treatment of a murine mammary car-
    cinoma in vivo. Cancer Res 36: 3077–3081
Reinhold HS, Endrich B (1986) Tumour microcirculation as a target for hyperthermia. Int J Hyper-
    thermia 2: 111–137
Riviere JE, Page RL, Dewhirst MW, Tyczkowska K, Thrall DE (1986) Effect of hyperthermia on
    cisplatin pharmacokinetics in normal dogs. Int J Hyperthermia 2: 351–358
Rose WC, Veras GH, Laster WR Jr, Schabel FM Jr (1979) Evaluation of whole-body hyperthermia
    as an adjunct to chemotherapy in murine tumors. Cancer Treat Rep 63: 1311–1325
Shrieve DC, Li GC, Astromoff A, Harris JW (1986) Cellular glutathione, thermal sensitivity, and
    thermotolerance in Chinese hamster fibroblasts and their heat-resistant variants. Cancer Res 46:
    1684–1687
Skibba JL (1982) Effects of hyperthermia on hepatic drug disposition. 13th International Cancer
    Congress, Seattle
Skibba JL, Jones FE, Condon RE (1982) Altered hepatic disposition of doxorubicin in the per-
    fused rat liver at hyperthermic temperatures. Cancer Treat Rep 66: 1357–1363
Smith PJ, Mircheva J, Bleehen NM (1986) Interaction of bleomycin, hyperthermia and a calmodu-
    lin inhibitor (trifluoperazine) in mouse tumour cells. II. DNA damage, repair and chromatin
    changes. Br J Cancer 53: 105–114
Spiro IJ, Denman DL, Dewey WC (1982) Effect of hyperthermia on CHO DNA polymerase α and
    β. Radiat Res 89: 134–149
Suzuki K (1967) Application of heat to cancer chemotherapy – experimental studies. Nagoya J
    Med Sci 30: 1–21
Tacker JR, Anderson RU (1982) Delivery of antitumor drug to bladder cancer by use of phase tran-
    sition liposomes and hyperthermia. Cancer Res: 1211–1214
Teicher BA, Kowal CD, Kennedy KA, Sartorelli AC (1981) Enhancement by hyperthermia of the
    in vitro cytotoxicity of mitomycin C toward hypoxic tumor cells. Cancer Res 41: 1096–1099
Twentyman PR, Morgan JE, Donaldson J (1978) Enhancement by hyperthermia of the effect of
    BCNU against the EMT6 mouse tumor. Cancer Treat Rep 62: 439–443
Vaglini M, Andreola S, Attili A, Belli F, Marolda R, Nava M, Prada A, Santinami M, Cascinelli N
    (1985) Hyperthermic antiblastic perfusion in the treatment of cancer of the extremities. Tumori
    71: 355–359
Van der Linden, Sapareto SA, Corbett TH, Valeriote FA (1984) Adriamycin and heat treatments in
    vitro and in vivo. In: Overgaard J (ed) Hyperthermic oncology, vol 1. Taylor and Francis, Lon-
    don, pp 449–452

Von Szczepanski L, Trott KR (1981) The combined effect of bleomycin and hyperthermia on the adenocarcinoma 284 of the C3H mouse. Eur J Clin Oncol 17: 997–1000

Wallach DFH (1977) Basic mechanisms in tumor thermotherapy. J Mol Med 2: 381–403

Wallner KE, DeGregorio MW, Li GC (1986) Hyperthermic potentiation of cis-diamminedichloro-platinum (II) cytotoxicity in Chinese hamster ovary cells resistant to the drug. Cancer Res 46: 6242–6245

Wang AL, Tew KD (1985) Increased glutathione-S-transferase activity in a cell line with acquired resistance to nitrogen mustards. Cancer Treat Rep 69: 677–682

Wang BS, Lumanglas AL, Ruszala-Mallon V, Wallace RE, Durr FE (1984) Effect of hyperthermia on the sensitivity of human colon carcinoma cells to mitoxanthrone (MX). Proc Am Assoc Cancer Res 25: 341

Warren RD, Bender RA (1977) Drug interactions with antineoplastic agents. Cancer Treat Rep 61: 1231–1241

Weinstein JN, Magin RL, Yatvin MB, Zaharko DS (1979) Liposomes and local hyperthermia: selective delivery of methotrexate to heated tumors. Science 204: 188–191

Yamada K, Someya T, Shimada S, Ohara K, Kukita A (1984) Thermochemotherapy for malignant melanoma: combination therapy of ACNU and hyperthermia in mice. J Invest Dermatol 82: 180–184

Yamane T, Koga S, Maeta M, Hamazoe R, Karino T, Oda M (1984) Effects of in vitro hyperthermia on concentration of adriamycin in Ehrlich ascites cells. In: Overgaard J (ed) Hyperthermic oncology, vol 1. Taylor and Francis, London, pp 409–412

Yerushalmi A (1978) Combined treatment of a solid tumour by local hyperthermia and actinomy in D. Br J Cancer 37: 827–832

Zupi G, Badaracco G, Cavaliere R, Natali PG, Greco C (1984) Influence of sequence on hyperthermia and drug combination. In: Overgaard J (ed) Hyperthermic oncology, vol 1. Taylor and Francis, London, pp 429–432.

Thermostability of Cytostatic Drugs in Vitro and Thermosensitivity of Cultured Human Lymphoblasts Against Cytostatic Drugs

B. Voth, H. Sauer, and W. Wilmanns

Institut für Strahlenbiologie, Gesellschaft für Strahlen- und Umweltforschung, Universität München, Schillerstraße 42, 8000 München 2, FRG

Introduction

The interaction of hyperthermia and antineoplastic therapy has attracted new interest recently. Experimental in vitro and in vivo studies showed an increased cytotoxic effect on malignant cells when hyperthermia was combined either with radiation therapy or chemotherapy (Hahn 1979; Herman et al. 1982; Kai et al. 1986; Mini et al. 1986; Overgaard 1976; Overgaard and Overgaard 1972). A large variety of cytostatic drugs are effective as antineoplastic agents, but only a few of them have yet been tested under hyperthermic conditions. Therefore 16 different cytostatic drugs were tested to find out whether they are suitable for further clinical studies, especially in combination with the application of regional hyperthermia. At first the thermostability of these drugs was tested in vitro. Then the influence of hyperthermia on the cytotoxic efficiency of the drugs was examined. Furthermore, the lowest temperatures at which an increased cytotoxicity occurred were tested. Since the application of regional hyperthermia can increase whole-body temperature to 39° C, it is important to know whether there are cytotoxic drugs which show an increased cytotoxicity only at temperatures above 39° C. These drugs would not increase systemic side effects such as bone marrow toxicity. Increased side effects could otherwise be limiting factors of the treatment. Drugs ideal for combination with regional hyperthermia develop their increased activity only in the heated tumor region.

All experiments were done in vitro. The cells used were cultured permanently growing human lymphoblasts. The metabolic activity of the cells as a function of heat and of heat combined with cytostatic drugs was studied by the incorporation of radiolabeled nucleosides into the DNA of the cells.

Material and Methods

Cell Line

A permanently growing human lymphoblast cell line (LS2 cell line), originally derived from a patient with acute undifferentiated leukemia, was used for all experiments. The cells were maintained in suspension in RPMI 1640 medium supplemented with 20% fe-

Recent Results in Cancer Research, Vol. 107
© Springer-Verlag Berlin · Heidelberg 1988

tal calf serum, streptomycin (100 µg/ml), penicillin (100 IU/ml), and L-glutamine (3 mM) at 37° C in a humidified atmosphere with 5% CO_2. The cultures were divided every 48 h. The doubling time of logarithmically growing cells was 35 h.

Chemicals and Drugs

The following agents were used: adriamycin (ADM), 4′-epi-adriamycin (4′-EA) daunomycin (Dauno), and 5-fluorouracil (5-FU) from Farmitalia; methotrexate (MTX) and dacarbazine (DTIC) from Medac; 4-OH-cyclophosphamide (4-OH-CYCLO) and ifosfamide (IFO) from Degussa Pharma Gruppe Asta; cytarabine (ARA-C) from Upjohn; mitoxantrone (Mitox) from Cyanamid; vincristine (VCR), vinblastine (VBL), and vindesine (VDS) from Lilly; cisplatin (DDP) and etoposide (VP 16–213) from Bristol; 5-fluorouracildeoxyriboside (5-FUdR) from La Roche.

RPMI 1640 medium, fetal calf serum, L-glutamine, penicillin, and streptomycin were obtained from Seromed; HCl, trichloroacetic acid (TCA), and toluene from Merck; omnifluor from NEN Chemicals; ^{3}H-deoxyuridine (^{3}H-dUR) and ^{3}H-thymidine (^{3}H-dTR) from Amersham-Buchler.

Heat and Drug Treatment

All heat and drug treatments were performed on exponentially growing cells with less than 10% dead cells (trypan blue exclusion method) at an initial concentration of 2000 cells/µl in sterile microtiter plates (Nunc) in a shaking waterbath (Heraeus). The incubation temperatures ranged from 37° C to 43° C. Aliquots of 100 µl pure cell suspension or drug-cell suspension were preincubated for 30 min. The preincubation temperature varied according to the experiment. After the preincubation time the microtiter plates were put on ice and the radioactivity at a final concentration of 2 µmol/l ^{3}H-dTR or ^{3}H-dUR (specific activity 0.5 mCi/µmol) was added and the cells were incubated for another 60 min in the shaking waterbath. The incubation temperature again depended on the experiment. Parallel to all hyperthermia experiments identical controls were incubated at 37° C throughout the whole experiment. Parallel to all drug experiments there was always one control, which was incubated under the same temperature conditions but with 0.9% NaCl instead of the cytostatic drug.

Precipitation of Nucleic Acids

TCA precipitable radioactivity was measured. The cells were harvested (Titertek cell harvester) and precipitated with ice-cold o.1 N HCl and 5% (v/v) TCA on specially designed filters. The filters were dried for 30 min at $+80°$ C, cooled for 20 min at $+4°$ C, and finally an omnifluor-toluene suspension (5 g omnifluor/l toluene) was added. The radioactivity was counted in a liquid scintillation counter (Tri Carb, Packard).

Table 1. Drug concentrations which produce a 30% or 50% decrease in DNA metabolism

Drugs	30% reduction (mol/l)	50% reduction (mol/l)
1. *Antimetabolites*		
5-FU	1.5×10^{-4}	3.6×10^{-4}
5-FUdR	1.2×10^{-9}	2.5×10^{-9}
MTX	2×10^{-7}	4×10^{-7}
ARA-C	2×10^{-9}	6×10^{-9}
2. *Anthracycline and anthrachinone*		
Dauno	8×10^{-7}	1.2×10^{-6}
ADM	8×10^{-6}	1.2×10^{-5}
4'-EA	2×10^{-6}	5×10^{-6}
Mitox	5×10^{-7}	1×10^{-6}
3. *Vinca alkaloids*		
VCR	3×10^{-5}	4×10^{-5}
VBL	2.5×10^{-5}	3.6×10^{-5}
VDS	2×10^{-5}	4×10^{-5}
4. *Alkylating agents*		
IFO	6×10^{-3}	7×10^{-3}
4-OH-CYCLO	1×10^{-4}	2×10^{-4}
DTIC	7×10^{-4}	9×10^{-4}
5. *Other cytostatic agents*		
VP 16-213	2.3×10^{-6}	5×10^{-6}
DDP[a]	1.7×10^{-4}	(1.7×10^{-4})

[a] A 50% reduction of nucleoside incorporation could not be achieved in spite of the high cisplatin concentration. *DDP* was used in a concentration of 1.7×10^{-4}. *M* in all experiments.

Results

Different drug concentrations were defined which had comparable cytostatic effects on DNA metabolism at 37° C and produced a 30% or 50% decrease in nucleoside incorporation. The cytostatic effect was compared to controls with 0.9% NaCl instead of the cytostatic drug. The results are shown in Table 1. For cisplatin, extremly high concentrations were needed to show an effect on the DNA metabolism in LS2 cells. A 50% reduction of nucleoside incorporation could not be achieved at the highest cisplatin concentration (1.7×10^{-4} *M*).

Thermostability of Cytostatic Drugs In Vitro

In concentrations which produce a 50% reduction of nucleoside incorporation (Table 1) drug solutions were incubated in 0.9% NaCl for 90 min at different temperatures (37, 39, 41, 43° C). Next the preheated drugs were added to cell suspensions. These drug-cell suspensions were then preincubated for 30 min at 37° C. Afterwards the radioactivity was added. After another 60 min the TCA precipitable radioactivity was measured. Table 2 shows the results. Except for etoposide, all tested drugs were thermostable at an incubation time of 90 min and temperatures up to 43° C. Etoposide became ineffective after being incubated for 90 min at 43° C.

Table 2. Thermostability of cytostatic drugs in vitro

Drugs	39° C	41° C	43° C
1. *Antimetabolites*			
5-FU	+	+	+
5-FUdR	+	+	+
MTX	+	+	+
ARA-C	+	+	+
2. *Anthracycline and anthrachinone*			
Dauno	+	+	+
ADM	+	+	+
4'-EA	+	+	+
Mitox	+	+	+
3. *Vinca alkaloids*			
VCR	+	+	+
VBL	+	+	+
VDS	+	+	+
4. *Alkylating agents*			
IFO	+	+	+
4-OH-CYCLO	+	+	+
DTIC	+	+	+
5. *Other cytostatic agents*			
DDP	+	+	+
VP 16-213	+	+	−

+ Thermostable after 90 min incubation.
− Thermoinstable after 90 min incubation.

Thermosensitivity of Cultured Human Lymphoblasts to Cytostatic Drugs

Cell suspensions with drug concentrations which produce a relatively weak (30%) inhibition of nucleoside incorporation were preincubated for 30 min at different temperatures (37–42.5° C). Then the radioactivity was added and the drug cell suspensions were either continuously incubated at the preincubation temperature for 60 min (altogether 90 min of hyperthermia) or further incubated at 37° C (30 min hyperthermia and 60 min normothermia).

Figure 1 shows IFO as an example. ^{3}H-dTR or ^{3}H-dUR incorporation is shown as a function of the incubation temperature. The upper curve for the controls with 0.9% NaCl instead of the cytostatic drug represents the pure heat effect on DNA metabolism in LS2 cells. Up to 41° C the DNA metabolism is not much influenced by heat. Above 41° C heat alone has a cytotoxic effect on LS2 cells. The lower curve represents the combined effect of IFO and hyperthermia. According to the drug concentration which results in a weak cytostatic effect on DNA metabolism at 37° C (Table 1), the ^{3}H-dTR and ^{3}H-dUR incorporation is less than in the control group at 37° C. Up to 40° C there is no significant change in DNA metabolism. Between 40° C and 41° C IFO causes a significant decrease in nucleoside incorporation. The duration of hyperthermia (30 or 90 min) seems to make no difference. Compared with the pure heat effect (upper curves) there is a synergistic effect of drug and hyperthermia for IFO at 41° C.

Different drugs showed different increases in efficacy at hyperthermic temperatures. To be able to quantify drug effects, the cytostatic effect of each drug at 41° C was measured compared with the control group. So-called temperature-enhancement factors

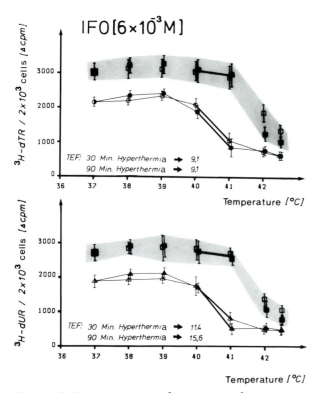

Fig. 1. Ifosfamide *(IFO):* the ³H-dTR resp. ³H-dUR incorporation into the DNA is shown as a function of the incubation temperature. The *upper curve* for the controls with 0.9% NaCl instead of the cytostatic drugs represents the pure heat effect on DNA metabolism in LS2 cells. The *lower curve* represents the combined effect of IFO and hyperthermia. There is a synergistic effect of heat and drug for IFO at 41° C. □, △, ○, 30 min hyperthermia; ■, ▲, ●, 90 min hyperthermia; ◨, ◮, ◑, 37° C throughout the experiment; □, ◫, ■, 0.9% NaCl and ³H-dTR or ³H-dUR; ○, ◑, ●, ³H-dTR incorporation; △, ◮, ▲, ³H-dUR incorporation; *TEF,* temperature-enhancement factor

(TEF) were defined. For each drug the effects were evaluated graphically, as for IFO (Fig. 1), and the slope of the curve between 40° C and 41° C was divided by the slope of the control curve in the same temperature range. The TEFs calculated in such a manner are listed in Table 3. The priority within this table is determined by the TEFs of the shorter hyperthermia influence (30 min). Table 3 shows that the tested drugs could be classified into four different types, depending on their cytostatic efficiency under hyperthermic conditions (41° C). Type 1 (IFO, MTX) showed a very good increase in cytostatic activity. Compared to 37° C there was a more than 50% reduction of nucleoside incorporation. The duration of hyperthermia (30 or 90 min) seemed to make no difference. Type 2 (4-OH-CYCLO, 4'-EA, Mitox, VDS, VCR, VBL, ADM, DTIC) showed a good increase in cytostatic activity. There was a 25%–50% reduction of nucleoside incorporation. The longer duration of hyperthermia (90 min) increased the cytostatic activity of VDS, VCR, VBL and DTIC further. Type 3 (5-FU, Dauno, DDP) showed only a moderate increase in activity. A 15%–25% reduction could be demonstrated. No influence of hyperthermia could be demonstrated for type 4 (5-FUdR, ARA-C, VP 16-213),

Table 3. Thermosensitivity of cultured human lymphoblasts against cytostatic drugs

Drugs	TEF 30 min hyperthermia		TEF 90 min hyperthermia	
	^{3}H-dTR	^{3}H-dUR	^{3}H-dTR	^{3}H-dUR
Type 1: very good increased activity				
IFO	9.1	11.4	9.1	15.6
MTX[a]	(3.3)	12.5	(2.2)	12.5
Type 2: good increased activity				
4-OH-CYCLO	5.7	5.9	4.6	5.6
4'-EA	4.8	5.1	5.1	6.8
Mitox	5.4	4.0	5.4	5.0
VDS	3.3	4.3	4.8	6.6
VCR	3.5	4.0	5.0	6.0
ADM	3.3	3.6	3.7	4.6
VBL	2.3	3.4	4.1	5.8
DTIC	2.2	3.3	4.4	5.5
Type 3: moderate increased activity				
5-FU[a]	(2.2)	3.4	(3.1)	3.4
Dauno	1.6	3.0	2.2	3.0
DDP	2.0	2.5	2.3	3.4
Type 4: no increased activity				
5-FUdR[a], ARA-C, VP 16-213				

TEF, temperature-enhancement factor.
Type 1: > 50% reduction of nucleoside incorporation.
Type 2: 25%–50% reduction of nucleoside incorporation.
Type 3: 15%–25% reduction of nucleoside incorporation.
Type 4: < 15% reduction of nucleoside incorporation.
[a] The ^{3}H-dTR incorporation could not be evaluated as a mean of a cytostatic effect because these drugs cause only a reduction in ^{3}H-dUR incorporation.

with which less than 15% reduction of nucleoside incorporation could be shown for this type.

Analysis of the lowest temperatures at which an increased cytostatic effect occurred showed that up to 39° C, none of the tested drugs was increased in its activity. Above 39° C, different drugs showed an increased reduction of nucleoside incorporation. For those drugs which were obviously well stimulated by hyperthermia (types 1 and 2 in Table 3), the temperatures at which increased activity was first noted were recorded. Group 1 (VCR, ADM, VBL, VDS) showed an increase above 39° C, group 2 (Mitox, 4-OH-CYCLO, IFO, MTX, 4'-EA, DTIC) above 40° C.

Discussion

The observations made in this study show that with the exception of etoposide, all tested drugs are thermostable for 90 min up to 43° C. These drugs are likely to be effective in combination with hyperthermia in vivo. In our test system etoposide became ineffective after 90 min incubation at 43° C. Therefore the use of etoposide in combination with hyperthermia cannot be recommended.

In agreement with other results it could be demonstrated that in vitro some cytostatic drugs develop increased activity under hyperthermic conditions (Hahn et al. 1975; Hahn and Shiu 1983; Herman 1983; Loven et al. 1986). Since nowadays a large variety of cytostatic drugs are available for cancer treatment, those cytostatic drugs which showed a very good or good increased activity in our in vitro system should be tested in further clinical studies in combination with hyperthermia. These drugs are ifosfamide, methotrexate, 4-OH-cyclophosphamide, 4'-epi-adriamycin, mitoxantrone, vindesine, vincristine, adriamycin, vinblastine and dacarbazine. They are likely to show an increased cytostatic effect under hyperthermic conditions in vivo. Since no increased activity could be demonstrated up to 39° C, no increased systemic toxicity is to be expected when using these drugs in combination with regional hyperthermia. Concerning cisplatin, there is a discrepancy between our results and other published data. We needed extremly high cisplatin concentrations to inhibit DNA metabolism. Therefore we believe that our test system, especially the LS2 type of lymphoblastoid culture cells, is not suitable to evaluate the cisplatin effect and our cisplatin data are of only limited value. Other results have demonstrated that cisplatin shows increased cytostatic activity under hyperthermic conditions (Fisher and Hahn 1982; Wallner et al. 1986).

References

Bates DA, Mackillop WJ (1986) Hyperthermia, adriamycin transport and cytotoxicity in drug-sensitive and -resistant Chinese hamster ovary cells. Cancer Res 46: 5477–5481

Fisher G, Hahn GM (1982) Enhancement of cis-platinum (II) diamminedichloride by hyperthermia. Natl Cancer Inst Monogr 61: 255–257

Hahn GM (1979) Potential for therapy of drugs and hyperthermia. Cancer Res 39: 2264–2268

Hahn GM, Shiu EC (1983) Effect of pH and elevated temperatures on the cytotoxicity of some chemotherapeutic agents on Chinese hamster cells in vitro. Cancer Res 43: 5785-5791

Hahn GM, Braun J, Har-Kedar J (1975) Thermotherapy: synergism between hyperthermia (42–43°) and adriamycin (or bleomycin) in mammalian cell inactivation. Proc Natl Acad Sci USA 72: 937–940

Herman TS (1983) Effect of temperature on the cytotoxicity of vindesine, amsacrine and mitoxantrone. Cancer Treat Rep 67: 1019-1022

Herman TS, Sweets CC, White DM, Gerner EW (1982) Effect of heating on lethality due to hyperthermia and selected chemotherapeutic drugs. JNCI 68: 487–491

Kai H, Matsufuji H, Sugimachi K, Okudaira Y, Inokuchi K (1986) Combined effects of hyperthermia, bleomycin and x rays on Ehrlich ascites tumor. J Surg Res 41: 503–509

Loven D, Lurie H, Hazan G (1986) Enhanced effect of systemic cyclophosphamide by local tumor hyperthermia in mice. Cancer Treat Rep 70: 509–512

Mini E, Dombrowski J, Moroson BA, Bertino JR (1986) Cytotoxic effects of hyperthermia, 5-fluorouracil and their combination on a human leukemia T-lymphoblast cell line, CCRF-CEM. Eur J Clin Oncol 22: 927–943

Overgaard J (1976) Combined adriamycin and hyperthermia treatment of a murine mammary carcinoma in vivo. Cancer Res 36: 3077–3081

Overgaard K, Overgaard J (1972) Investigations on the possibility of a thermic tumor therapy I. Eur J Cancer Clin Oncol 8: 65–78

Wallner KE, DeGregorio MW, Li GC (1986) Hyperthermic potentiation of cis-diamminedichloroplatinum (II) cytotoxicity in Chinese hamster ovary cells resistant to the drug. Cancer Res 46: 6242–6245

A Rapid in Vitro Assay for Predicting Thermochemosensitivity of Human Cancer: Comparison with Clonogenic Assay*

K. Yamada[1], H. A. Neumann[1], H. H. Fiebig[1], R. Engelhardt[1], and H. Tokita[2]

[1] Medizinische Klinik, Albert-Ludwigs-Universität, Hugstetter Straße 55, 7800 Freiburg, FRG
[2] Research Institute, Chiba Cancer Center, Chiba, Japan

Introduction

A rapid in vitro assay based on morphological changes in the nucleus has, recently, been reported for predicting human cancer chemosensitivity (Tokita et al. 1986a, b; Takamizawa et al. 1985). This method provides a realistic potential with regard to the following advantages: (1) We can apply it to any kind of tumors, including intestinal tumors contaminated with bacteria. Consequently, the proportion of tumors yielding sensitivity data is high. (2) The results are obtained within 4 h for alkylating agents and *Vinca* alkaloids, within 8 h for antibiotic and antimetabolic agents. (3) The assay can be done in a clinical laboratory on a regular basis because of its simplicity.

Quite recently, we reported the first attempt to predict response to thermochemotherapy using this simple method (Yamada et al. 1988). Then, clinical usefulness as an in vitro predictive assay for thermochemotherapy was strongly suggested.

In the present study, we compare the new assay with clonogenic assay, which has already been established on grounds of clinical response (Salmon et al. 1980; Mann et al. 1982; von Hoff et al. 1983; Neumann et al. 1985), in order to extend our experimental basis for evaluating the clinical potential of this method on thermochemotherapy.

Materials and Methods

Tumors

Eight human tumors, comprising three malignant melanomas, two lung carcinomas, two colon carcinomas and one leukemia, were utilized. Tumor materials passaged in nude mice were taken, except leukemia K-562. Tumor tissues were cut into 0.5 mm with a razor blade on a teflon sheet to obtain the small cell clumps. These were placed in a stainless steel strainer (0.5 mm diameter) to remove fibrous tissues. The medium containing tumor cells was then transferred to a tube (Falcon 2070) and centrifuged at 800 r.p.m. for 30 s three times to avoid contamination of the necrotic part and fat.

* This study was supported by Bundesministerium für Jugend, Familie, Frauen und Gesundheit, Bonn.

Table 1. Drug concentrations in the two test systems

Drug	Clonogenic assay (μg/ml)	Cytotoxic test (μg/ml)
Melphalan	10^{-3}–10^{-1}	0.5
Mitomycin C	10^{-3}–10^{-1}	0.3
Vincristine	10^{-4}–10^{-2}	0.1

Drugs

Melphalan, mitomycin C and vincristine were tested. Drug concentrations are shown in Table 1. Each drug was dissolved with 0.9% NaCl solution and serially diluted with RPMI 1640 medium supplemented with 10% fetal calf serum.

Culture Assay

Tumor cells were incubated in 0.4 ml medium (RPMI 1640) supplemented with 10% fetal calf serum, containing antitumor agent as shown in Table 1, at 37° C for 4 h for the normothermic groups and for the hyperthermic groups at 43° C for 1 h and afterward at 37° C for 3 h. After incubation, tumor cells were pumped down with 23-gauge needle injector to obtain pure single cells. The medium containing tumor cells was centrifuged at 800 rpm for 5 min and then displaced by calf serum. The calf serum containing tumor cells was centrifuged again at 800 rpm for 5 min and the precipitate was placed on glass slides. Subsequently, the cells were fixed in absolute methanol for 3 min and stained with Giemsa.

Data Evaluation

More than 200 tumor cells per glass slide were examined through the high-power field of a microscope. Changes noted in the nucleus, karyorrhexis changes and other degenerative changes with each drug were compared with the control group under the statistical analysis (X^2 test).

The results were expressed in terms of positive, half-positive and negative data. Positive and half-positive data indicate significant difference from the control group on karyorrhexis changes and on other degenerative changes, respectively.

Preparation of Clonogenic Assay

Clonogenic assay was carried out in samples of the same human tumors which were examined concurrently by the cytotoxic test. After obtaining a single cell suspension, viable cells 1×10^5 were plated in the presence of 30% fetal calf serum in Iscove's modified Dulbecco's medium with antitumor agents at concentrations (Table 1) chosen over the 2–3 log concentration range as proposed by Salmon et al. (1981). Methylcellulose (final concentration: 0.9%) was used as a viscous support. Cells were incubated at 37° C in the presence of 7.5% CO_2 for 10–14 days. Hyperthermic dishes were incubated at 43° C for 1 h and afterward stored at 37° C. Aggregates of more than 30 cells with a diameter of 80–100 μm were considered as colonies. Drug effect with and without heat was expressed as percent survival of tumor cell colonies.

Results

Karyorrhexis changes in the nucleus as shown in Fig. 1 were induced after short-term incubation with drug tested in our cytotoxic test. Moreover, gain of karyorrhexis changes with condensed chromatin was observed after 43° C hyperthermia for 1 h and postincubation for 3 h at 37° C.

The results of the comparison with clonogenic assay are shown in Tables 2 and 3. The same human tumors were tested against the same drugs in both assays. Between the two test systems, thermosensitivity to 43° C and response to thermochemotherapy were investigated.

Fig. 1. Karyorrhexis changes in the nucleus developed after incubation for 4 h with Melphalan *(left)*. Gain of karyorrhexis changes was observed after 43° C hyperthermia *(right)*

Table 2. Thermosensitivity to 43° C hyperthermia in the two test systems

Case	1	2	3	4	5	6	7	8
Clonogenic assay	73	49	88	18	54	84	30	70
Cytotoxic test	−	+	−	+	±	−	+	−

At the level of less than 50% survival of colony as compared with normothermic control dishes in clonogenic assay, there was a strong correlation between the two test systems. Results given as percent survival of colony as compared with normothermic control dishes.

+ Significant difference from the normothermic control on karyorrhexis changes in the nucleus ($p<0.05$).

± Significant difference from the normothermic control on other degenerative changes in the nucleus ($p<0.05$).

Table 3. Response to thermochemotherapy in the two test systems

Case	1	2	3	4	5	6	7	8
Clonogenic assay								
Melphalan	−	+	−	−	+	−	−	−
Mitomycin C	−	−	−	−	−	−	−	−
Vincristine	−	−	−	−	−	−	+	−
Cytotoxic test								
Melphalan		+	−	−	+	−	−	−
Mitomycin C	−	±	−	−	−	±	−	−
Vincristine	−	±	−	−	−	−	+	−

In some cases karyorrhexis changes in the nucleus (positive data) only were defined as an activity criteria in our cytotoxic test, but in general parallel data were obtained from the two test systems.
+ Enhanced colony reduction under hyperthermic conditions at each drug concentration.
± Significant difference with the hyperthermic control on karyorrhexis changes in the nucleus ($p < 0.05$).
− Significant difference with the hyperthermic control on the other degenerative changes in the nucleus ($p < 0.05$).

Thermosensitivity to 43° C

The results of thermosensitivity to 43° C are shown in Table 2. The data of the clonogenic assay are expressed in percent survival of colony as compared with normothermic control dishes. In our cytotoxic test, thermosensitivity data are represented as positive, half-positive and negative. More than 70% survival of colony was observed in four tumors (cases 1, 3, 6 and 8) in the clonogenic assay, which corresponded to the results in cytotoxic test. These tumors seem to be resistant to 43° C hyperthermia for 1 h.

Analysis of less than 50% cutoff in the clonogenic assay showed strong correlation with the cytotoxic test for tumors sensitive to 43° C.

Response to Thermochemotherapy

Table 3 shows the thermochemosensitivity in the two test systems. In the clonogenic assay, a drug was defined as positive if it enhanced colony reduction under hyperthermic conditions. Positive and half-positive data in the cytotoxic test are those showing significant differences from hyperthermic controls in karyorrhexis changes and other degenerative changes (X^2 test, $p < 0.05$).

The clonogenic assay predicted thermal enhancement of drugs for three tumors: melphalan in cases 2 and 5, vincristine in case 7. Five tumors showed no difference between normothermic and hyperthermic conditions in all drugs tested. The positive data in our cytotoxic test corresponded to the results of the clonogenic assay. However, some drugs were also predicted as half-positive for two tumors (cases 2 and 6) in the cytotoxic test.

Discussion

Although therapeutic applications of hyperthermia, especially in combination with radiation (Kim et al. 1982; Scott et al. 1983; Arcangeli et al. 1983) or chemotherapy (Stehlin 1980; van der Zee et al. 1983; Storm et al. 1984), have been reported in recent years, it is urgently necessary to develop a rapid and accurate predictive assay for clinical response in order to improve the effectiveness of hyperthermia treatment. Several attempts to predict response to thermochemotherapy (Mann et al. 1983; Andrysek et al. 1985; Neumann et al. 1985) have already been reported, mainly using the clonogenic system developed by Hamburger and Salmon (1977). Mann et al. (1983) particularly have demonstrated a good correlation between clonogenic assay and clinical response in combination therapy involving hyperthermia and drug treatment of human malignant melanomas.

However, it is accepted that the clinical success rate of clonogenic assay may vary considerably (von Hoff 1983). In addition, it takes 2–3 weeks to obtain the results. Therefore, it seems necessary to develop methods which are more rapid, simple and reliable as predictive assays. We previously reported the ability to predict response to thermochemotherapy using in vitro assay based on morphological changes in the nucleus (Yamada et al. 1987). It was strongly suggested that this simple method may be clinically useful as a predictive assay for thermochemosensitivity.

The purpose of the present study was to define the activity criteria of our cytotoxic test by comparing it to the clonogenic assay. The results indicate that correlation between clonogenic assay and our cytotoxic test was dependent upon the criteria for each test system. At the level of less than 50% survival of colony as compared with normothermic control in the clonogenic assay, there was a high relationship with our cytotoxic test for tumors sensitive to 43° C hyperthermia. The tumors which showed more than 70% survival of colony seemed to be resistant to 43° C hyperthermia, which corresponded to the results observed in the cytotoxic test. In respect of thermochemotherapy, when only karyorrhexis changes in the nucleus (positive data) were selected as an activity criterion in the cytotoxic test, parallel data were obtained in the two test systems.

In general, karyorrhexis changes in the nucleus are induced after short-term incubation with alkylating agents or *Vinca* alkaloids (Tokita et al. 1982). Moreover, gain of karyorrhexis changes was found after 43° C hyperthermia in the present study. These morphological changes often occurred in the presence of condensed chromatin. This condensation of chromatin after hyperthermia has been pointed out by some investigators (Overgaard 1976; Fajardo et al. 1980). Although karyopyknosis was also observed after 43° C hyperthermic incubation in some tumors in the present study, karyorrhexis changes were predominant.

It will be necessary to establish new morphological criteria for the combination therapy of hyperthermia and other antitumor agents, such as antibiotic and antimetabolic agents, because these agents develop karyopyknosis change in the nucleus (Tokita et al. 1982).

While the number of tumors tested was small, our findings with the rapid in vitro assay encourage us to continue comparative study with clonogenic assay, and, moreover, to attempt a comparison with clinical response.

Summary

We previously reported a rapid in vitro assay based on morphological changes in the nucleus in order to predict response to thermochemotherapy. It was strongly suggested that this simple method may be clinically useful. In the present study, a comparison with the clonogenic assay was carried out on eight different human tumors (three malignant melanomas, two lung carcinomas, two colon carcinomas and one leukemia). Melphalan, mitomycin C and vincristine were tested.

Correlation between the two test systems was dependent upon the criteria for each test system. At the level of less than 50% survival of colony as compared with normothermic dishes in clonogenic assay, there was a high correlation between the two test systems for sensitive tumor to 43° C. In respect of response to thermochemotherapy, when only karyorrhexis changes in the nucleus were selected as an activity criteria in our cytotoxic test, parallel data between the two test systems were obtained.

References

Andrysek O, Bláhová E, Gregora V, Rezneý Z (1985) An experimental model for predicting the synergism of hyperthermia with cytostatics. Neoplasma 32: 93–101

Arcangeli G, Dividalli A, Nervi C, Creton G (1983) Tumor control and therapeutic gain with different schedules of combined radiotherapy and local external hyperthermia in human cancer. Int J Radiat Oncol Biol Phys 9: 1125–1134

Fajardo LF, Egbert B, Marmor J, Hahn GM (1980) Effects of hyperthermia in a malignant tumor. Cancer 45: 613–623

Hamburger AW, Salmon SE (1977) Primary bioassay of human tumors stem cells. Science 197: 461–463

Kim JH, Hahn EW, Ahmed SA (1982) Combination of hyperthermia and radiation for malignant melanoma. Cancer 50: 478–482

Mann BD, Kern DH, Giuliano AE, Burk MW, Campbell MA, Kaiser LR, Morton DL (1982) Clinical correlation with drug sensitivities in the clonogenic assay. Arch Surg 117: 33–36

Mann BD, Storm FK, Morton DL, Bertelsen CA, Korn EL, Kaiser LR, Kern DH (1983) Predictability of response to clinical thermochemotherapy by the clonogenic assay. Cancer 52: 1389–1394

Neumann HA, Fiebig HH, Löhr GW, Engelhardt R (1985) Effects of cytostatic drugs and 40.5° C hyperthermia on human bone marrow progenitors (CFU-C) and human clonogenic tumor cells implanted into mice. JNCI 75: 1059–1066

Overgaard J (1976) Influence of extracellular pH on the viability and morphology of tumor cells exposed to hyperthermia. JNCI 56: 1243–1250

Salmon SE, Albert DS, Durie BGM, Meysken FL, Soehnlen B, Chen HS, Moon T (1980) Clinical correlation of drug sensitivity in the human tumor stem cell assay. Recent Results Cancer Res 74: 300–305

Salmon SE, Meysken FL, Albert DS (1981) New drug in ovarian cancer and malignant melanoma: in vitro phase II screening with the human tumor stem cell assay. Cancer Treat Rep 65: 1–12

Scott RS, Johnson RJ, Kowal H, Krisnamsetty RM, Story K, Clay L (1983) Hyperthermia in combination with radiotherapy: a review of five years experience in the treatment of superficial tumors. Int J Radiat Oncol Biol Phys 9: 1327–1333

Stehlin JS (1980) Hyperthermic perfusion for melanoma of the extremities: experience with 165 patients, 1967 to 1979. Ann NY Acad Sci 335: 352–355

Storm FK, Sillberman AW, Ramming KR, Kaiser LR, Harrison WH, Elliot RS, Haskel CM, Sarna G, Morton DL (1984) Clinical thermochemotherapy: a control trial in advanced cancer patient. Cancer 53: 863–868

Takamizawa H, Sekiya S, Iwamisawa H, Ishige H, Tokita H, Tanaka N (1985) A new in vitro chemosensitivity test: individualized chemotherapy against ovarian cancer and its clinical effect. Gan To Kagaku Ryoho 12: 2293–2297

Tokita H, Tanaka N, Ueno T (1982) Morphological changes of tumor cells by antitumor drugs – comparison of in vivo and in vitro. 41st Japanese Cancer Congress, Kyoto

Tokita H, Tanaka N, Fujimoto S, Nakao K (1986a) The in vitro cytotoxic test for predicting human colon and rectum cancer chemosensitivity. Med Biol 112: 171–175

Tokita H, Tanaka N, Ueno T, Fujimoto S, Nakao K, Takada N, Takamisawa H (1986b) In vitro cytotoxic test for predicting human cancer chemosensitivity (Abstr 3883). 14th International Cancer Congress, Budapest

Van der Zee J, van Rhoon GC, Wike-Hooley JL, Faithful NS, Rheinhold HS (1983) Whole-body hyperthermia in cancer therapy: a report of a phase I–II study. Eur J Cancer Clin Oncol 19: 1189–1200

Von Hoff DD (1983) Send this patient's tumor for culture and sensitivity. N Engl J Med 308: 154–155

Von Hoff DD, Clark GM, Stogdill BJ, Sarosdy J, O'Brien MT, Casper JT, Mattox DE, Page CP, Cruz AB, Sandbach JF (1983) Prospective clinical trial of a human tumor cloning system. Cancer Res 43: 1926–1931

Yamada K, Neumann HA, Fiebig HH, Engelhardt R, Tokita H (1988) A rapid in vitro assay for predicting human cancer thermochemosensitivity. Recent Results Cancer Res (in press)

The Cytotoxic and Bleomycin-Sensitizing Effect of Hyperthermia on Different Cell Lines

H. Eichholtz-Wirth

Institut für Strahlenbiologie, Gesellschaft für Strahlen- und Umweltforschung, Universität München, Schillerstraße 42, 8000 München 2, FRG

Hyperthermia is known to increase the effect of bleomycin on cell survival, as first demonstrated by Hahn et al. (1975) and confirmed in various cell lines (Mizuno and Ishida 1983; Hermann 1983). The purpose of our experiments was to study the influence of hyperthermic temperatures on sensitivity to bleomycin in three cell lines that show marked differences in drug sensitivity under normothermic exposure conditions: is cell sensitization by heat dependent on the effectiveness of the cytostatic action, and can differences in drug sensitivity of various cell lines be compensated by simultaneous heat exposure?

Clonogenic cell survival was determined for exponentially growing HeLa, CH and HaK cells as a function of bleomycin concentration at a given exposure time of 8 h (Fig. 1 A) or as a function of exposure time at constant drug concentration of 5 µg/ml (Fig. 1 B; for further details of growth conditions, see Eichholtz-Wirth and Hietel 1986). At 37° C survival curves are biphasic with a sensitive initial phase and a resistant final portion of the survival curve. There is a striking difference in sensitivity between the cell lines: after 24 h and 5 µg/ml, survival is 0.65 (HaK), 0.24 (CH), and 0.1 (HeLa). The biphasic course of the survival curve is due not to age response or genetic selection, as demonstrated by fractionation experiments, but rather to induced resistance which clears within 4 h after the end of exposure. These results partly agree with data of Terasima et al. (1972), who studied the survival response of bleomycin on four mammalian cell lines of different origins. These cell lines all exhibited an upward concavity; however, the sensitivity differed only slightly between the cell lines.

Cell survival was then studied after 1 h exposure to hyperthermic temperatures up to 43° C. Figure 2 A describes the cytotoxic effect of heat alone for the three cell lines. HeLa cells are the most sensitive, in that 0.1 survival level is achieved in 1 h at 42° C, whereas CH and HaK cells both require 43° C for the same effect. Heat is then combined simultaneously with bleomycin (1 h, 5 µg/ml, Fig. 2 B). There is an enhanced cytotoxic effect in all three cell lines which is most evident in CH cells. For HeLa cells, the combined heat and drug effect can only be measured up to 42° C, as 43° C alone kills more than 99% of the cells.

To quantitate the enhancing effect of combined modality treatment, the observed surviving fraction was compared to the product of the surviving fractions of heat and drug alone (Bhuyan 1979): no difference indicates additivity of the single effects; a surviving fraction smaller than expected indicates that cells are sensitized by one agent to the action of the other. For both HeLa and HaK cells, the effect is only slightly more

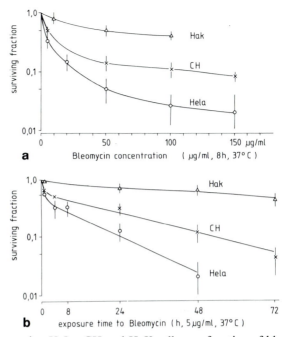

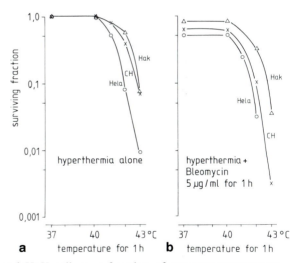

Fig. 1a, b. Cell survival of exponentially growing HeLa, CH and HaK cells as a function of ble-omycin concentration (at 8 h, **a**) or exposure time (at 5 μg/ml, **b**) at 37° C. Each point represents the mean (±SD) of at least 10 dishes analyzed on at least three different occasions

Fig. 2a, b. Cell survival of HeLa, CH and HaK cells as a function of exposure temperature. **a** Hyperthermia alone (1 h); **b** simultaneous heat and drug treatment (5 μg/ml, 1 h)

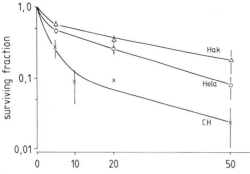

Fig. 3. Cell survival as a function of bleomycin concentration at isoeffective (surviving fraction about 0.5) hyperthermic temperature: simultaneous treatment for 1 h at 41° C *(HeLa)* and 42° C *(CH, HaK)*

than additive, whereas for CH cells there is a big increase in effectiveness, resulting in more than 1 log additional cell kill at 43° C.

To confirm the results of drug sensitization by hyperthermia, 1 h of heat treatment that results in 50% cell kill, i.e. 41° C for HeLa cells and 42° C for CH and HaK cells, was combined simultaneously with exposure to bleomycin concentrations between 5 and 50 µg/ml. Figure 3 shows the data corrected for the effect of heat alone. Although these data cannot be compared to the 37° C survival data of Fig. 1, because exposure time is only 1 h rather than 8 h, there is a reversal in the ranking of sensitivity. At 37° C HeLa cells are much more sensitive to bleomycin than CH cells, whereas in combined exposure CH cells are about 4 times more sensitive than HeLa cells. It is interesting to note that despite the pronounced sensitizing effect of hyperthermia, the biphasic structure of the curves is not altered, nor is the relation between the sensitive and the resistant components significantly changed. These results are in accordance with data of Braun and Hahn (1975), but in contrast to data of Meyn et al. (1979), using CHO cells, and Lin et al. (1983), using V79 cells, who both describe exponential survival curves at 43° C. Since hyperthermia and bleomycin are believed to have different cell cycle phase specificity (Westra and Dewey 1971; Barranco et al. 1973), our data suggest that the biphasic nature of the dose-response curves for bleomycin is not due to cell cycle specificity.

In summary, there are big differences between the three cell lines in their sensitivity to bleomycin exposure at hyperthermic temperatures. In HeLa cells the main effect is obtained by hyperthermia alone, whereas in CH cells the effect of hyperthermia on sensitivity to bleomycin is of equal importance. HaK cells are rather resistant towards both treatment modalities and the resistance cannot be overcome by combining them.

References

Barranco SC, Luce JK, Romsdahl MM, Humphrey RM (1973) Bleomycin as a possible synchronizing agent for human tumors in vivo. Cancer Res 33: 882–887

Bhuyan BK (1979) Kinetics of cell kill by hyperthermia. Cancer Res 39: 2277–2284

Braun J, Hahn GM (1975) Enhanced cell killing by bleomycin and 43° hyperthermia and the inhibition of recovery from potentially lethal damage. Cancer Res 35: 2921–2927

Eichholtz-Wirth H, Hietel B (1986) The relationship between cisplatin sensitivity and drug uptake into mammalian cells in vitro. Br J Cancer 54: 239–243

Hahn GM, Braun J, Har-Kedar J (1975) Thermochemotherapy: synergism between hyperthermia (42–43°) and adriamycin (or bleomycin) in mammalian cell inactivation. Proc Natl Acad Sci USA 72: 937–940

Hermann TC (1983) Temperature dependence of adriamycin, cisdiamminedichloroplatinum, bleomycin, and 1,3-bis(cloroethyl)-1-nitrosourea cytotoxicity in vitro. Cancer Res 43: 517–520

Lin PS, Hefter K, Jones M (1983) Hyperthermia and bleomycin schedules on V79 chinese hamster cell cytotoxicity in vitro. Cancer Res 43: 4557–4561

Meyn RE, Peter PD, Corry M et al. (1979) Thermal enhancement of DNA strand breakage in mammalian cells treated with bleomycin. J Radiat Oncol Biol Phys 5: 1487–1489

Mizuno S, Ishida A (1983) Enhancement of bleomycin cytotoxicity by hyperthermia, ethanol, and local anesthetics. Cancer Treat Symp 1: 1–6

Terasima T, Takabe Y, Katsumata TM et al (1972) Effect of bleomycin on mammalian cell survival. JNCI 49: 1093–1100

Westra A, Dewey WC (1971) Heat shock during cell cycle of Chinese hamster cells in vitro. Int J Radiat Biol 19: 467–477

Pilot Studies of Microwave-Induced Brain Hyperthermia and Systemic BCNU in a Rat Glioblastoma Model[*]

O. Mella, A. Mehus, and O. Dahl

Department of Oncology, Haukeland Hospital, University of Bergen, 5021 Bergen, Norway

Introduction

Glioblastoma is a human tumour with aggressive behaviour but low metastatic capacity, representing a local cerebral problem. However, the prognosis is poor with an ultimate fatal outcome despite current therapy. The nitrosoureas, especially BCNU, may result in temporary tumour regression. Hyperthermia may increase the effect of radiation or chemotherapy. Exploration of hyperthermia in the treatment of aggressive gliomas is in part delayed by the lack of suitable animal models. A significant interaction between BCNU and hyperthermia has been shown in a subcutaneous rat glioma (Dahl and Mella 1982). For a glioblastoma model, the foot tumour has several disadvantages: (1) a tumour bed which gives a more well-defined tumour than in the brain; (2) possibly different vascular drug penetration; (3) skin and subcutaneous tissue more resistant to hyperthermia than the brain; (4) a greater temperature gradient during hyperthermia (normal foot temperature about 30° C).

We recently presented the BT$_4$An rat tumour as an in vitro and in vivo model for glioma invasion (Bjerkvig et al. 1986). The tumour has been developed as an intracerebral glioma therapy model. It has several characteristics of aggressive gliomas, including extensive local invasion of the brain, perivascular infiltration, focal necrosis and CNS characteristic markers (manuscript in preparation). There is 100% tumour take and a predictable symptom-free period after inoculation. In the present experiments, a system for local brain hyperthermia was developed and tested for feasibility of the hyperthermic procedure and animal tolerance. The brain tumour was investigated for sensitivity to BCNU given i.p. and to microwave-induced hyperthermia. Combination of the two modalities was studied to see if enhancement of the tumour effect could also be registered in a tumour with diffuse margins (Overgaard 1978) and to evaluate brain tolerance.

---

[*] Supported by the Norwegian Cancer Society. Technical assistance by Brynhild Haugen, Stig Rasmussen and Svein Magnussen was greatly appreciated.

Materials and Methods

Animals and Tumour

Young rats of both sexes belonging to the inbred strain BD IX, with average weight 278 g, were anaesthetized with fentanyl-fluanisone 0.25–0.5 mg/kg s.c. and diazepam 0.25–0.5 mg/kg i.p. A cranial burr hole was made by a dental drill at the coronal suture 4 mm lateral to the sagittal suture. Exponentially growing BT$_4$An cells were harvested from cell culture, counted in a Burker chamber or Coulter counter, and 10^5 cells were diluted in 1% methylcellulose to a 10-μl volume and manually injected through a 0.07-mm cone-pointed Hamilton syringe with a stopper device 4 mm into the right frontal lobe lateral to the ventricle. Based on results of the first experiment, the tumour cells were injected at 2 mm depth in the second experiment and only male rats were used. Eight rats were inoculated per day and were randomized and treated one by one 14 or 15 days after inoculation with sham procedure, hyperthermia, BCNU or combined hyperthermia and BCNU.

Hyperthermia

The basic equipment was the BSD-1000 clinical hyperthermia system (BSD Medical Corporation, Salt Lake City, Utah, USA). An applicator with a physical aperture of 2×3 cm and length 10 cm was built locally and operated at 701.5 MHz. The aperture was closed with silicone rubber and the applicator was circulated with thermostatically controlled (34.5° C, range 33.8–35.6° C) deionized water for surface cooling. Phantom studies showed a circular heating pattern. A 2-cm scalp skin incision was made in the fentanyl-fluanisone- and diazepam-anesthetized rats. A single sterile, calibrated Bowman probe in a Teflon sheath with a stopper device was inserted at a 45° angle with the horizontal plane 4 mm length from lateral to medial direction through the burr hole made 2 weeks previously. This resulted in temperature measurement at approximately 2.5 mm depth into the brain, representing a point in the medial periphery of most tumours. In the second experiment, the probe was placed at the same angle and depth, but from medial to lateral direction. The applicator was placed in good contact with the scalp, with the field centre straight above the burr hole. BCNU alone rats and control rats were similarly anaesthetized and had brain probes placed and kept in position for 50 min. Hyperthermia was for 45 min at 42.0° C in experiment 1 and 42.5° C in experiment 2. Power output was adjusted manually initially, with computer control after reaching the desired (control) temperature. After treatment, the skin was closed by clips and the rats were given 0.03 mg buprenorphine to counteract respiratory depression and to prolong analgesia.

BCNU

Commercially available drug (Carmustine, Bristol Myers Inc., Syracuse, N.Y., USA) was dissolved in absolute ethanol and diluted in sterile water to a 3.33 mg/ml solution containing 10% (v/v) ethanol. The drug was given i.p. at a dose of 20 mg/kg just prior to hyperthermia or sham treatment. BCNU alone rats and hyperthermia alone rats were given the corresponding volume of 10% ethanol.

Treatment Evaluation

Brain, rectal and cooling water temperatures were registered during treatment and average temperatures were calculated with the BSD-1000 software package, as was average power after achievement of the desired temperature. The rats were observed and weighed daily. Rats with weight loss and symptoms such as passivity, paresis and tear flow, were killed by CO_2 inhalation. Rats dying < 24 h after treatment were counted as toxic deaths and were excluded from survival analysis. Survival was defined as time from tumour inoculation to killing. Weight loss was the lowest weight post-treatment before stabilization or weight regain. Survival and weight loss data in the various groups were compared statistically using the Mann-Whitney test.

Results

In both experiments, the desired temperature was reached in all animals. The applicator proved to be sensitive to the degree of scalp contact and level of water pressure in the applicator, and especially in experiment 1, the induction time to control temperature varied. Due to increased experience, standardization of the hyperthermic procedure was better in experiment 2 and technical data are given in Table 1. In this experiment, mapping of the temperature distribution to the brain surface during the last minutes of treatment was performed in four rats by manual retraction of the probe in 1-mm steps at a fixed applicator output. The temperature was in all mapped rats within 42.0–42.5° C.

Three rats did not complete the 45 min at the desired temperature. Two had evident breathing problems, expressed as a fall in respiratory frequency (both died within hours), and one stopped breathing during treatment. The anaesthesia-induced respiratory depression was more pronounced in hyperthermia-treated rats, but as it was also seen in some fixed rats before microwave output, it in part seemed related to the fixation procedure and pressure from the applicator. None of the six brains from the toxic deaths in experiment 2 showed microscopic brain damage. One of these rats had an exceptionally large tumour, and in all four hyperthermia-treated brains there was vascular dilation and erythrocyte pooling, especially in the tumours, indicating hyperemia.

In experiment 1 there was no difference in survival between the treatment groups (Fig. 1). On macroscopic examination of the brains, several rats had no tumour at the calvarial surface. There was tumour growth in the brain depth and lateral surface, which could mean that these areas were insufficiently heated. Tumour cell inoculation was therefore at 2 mm depth in experiment 2 and the temperature was measured in the lateral tumour periphery. Female rats in experiment 1 had a greater variation in survival than males; thus only male rats were used thereafter. In experiment 2, control and BCNU alone rats had similar survival (Fig. 1, Table 1). Hyperthermia alone did not significantly prolong survival. The combined BCNU and hyperthermia group had the three longest survivors (45, 49 and 66 days), and survival was significantly prolonged above control and BCNU groups ($p < 0.01$). All rats with symptoms had, at section, macroscopic, expansive tumours with 1–3 mm midline shift to the left.

BCNU increased the nadir weight loss after treatment compared to controls or hyperthermia alone ($p < 0.01$). Weight loss was not significantly greater in the combined treatment group than in the BCNU alone group. One hyperthermia alone rat had transient hind leg paresis and possibly urinary bladder paresis. In four hyperthermia or combined modality rats, passivity without paresis was seen 1–2 days following treatment.

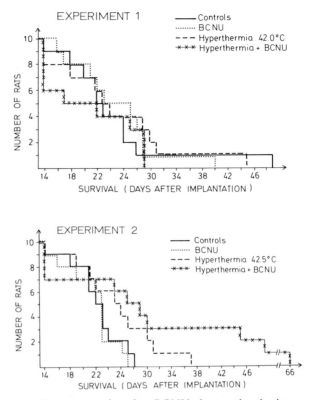

Fig. 1. Survival of BD IX rats with BT$_4$An brain tumours given i.p. BCNU alone or local microwave-induced brain hyperthermia for 45 min with or without BCNU on day 14 or 15 post implantation

Table 1. Parameters registered in experiment 2, with desired intratumoural temperature 42.5° C. Toxic deaths were excluded from survival analysis. Mean values ± SD

Treatment group	Survival (days)	Toxic deaths	Relative weight (nadir)	Control temp. (° C)	Rectal temp. (° C)	Induction time (min:s)	Power at control temp. (W)
Control	23.0 ± 3.1	1/10	0.96 ± 0.01				
BCNU	22.4 ± 3.3	1/10	0.90 ± 0.01				
Hyperthermia	26.4 ± 5.6	1/10	0.95 ± 0.01	42.4 ± 0.1	33.9 ± 0.9	3:12 ± 2:55	59 ± 5
Hyperthermia and BCNU	38.7 ± 15.2	3/10	0.84 ± 0.04	42.3 ± 0.2	34.6 ± 1.1	2:44 ± 1:50	58 ± 6

Discussion

Attempts to explore the potential of hyperthermia in the treatment of human malignant brain tumours have been made previously (Sutton 1971). Progress is delayed by technical problems in brain hyperthermia induction and thermometry, and by insufficient knowledge of brain tolerance. In clinical trials, it may be difficult to estimate the contribution of hyperthermia to tumour effect, as hyperthermia is usually given together with radiation or chemotherapy. Adequate, inexpensive animal models have been lacking.

In the present experiments, both the hyperthermic procedure and the BT$_4$4An brain tumour in inbred rats proved to be suitable for the exploration of hyperthermia as brain tumour therapy. Apart from potentiation of the anaesthesia-induced respiratory depression by the fixation procedure and hyperthermia, treatment was well tolerated when experience was gained (experiment 2). Only one rat showed transient hyperthermia-related paresis and there was no evident short-term brain damage at microscopy in any rat.

The main disadvantage with the model was the small animal size, which greatly restricted probe insertion for thermometry. In these experiments, only one probe was used in order to limit brain trauma. However, in separate studies of hyperthermia in non-tumor-bearing brains with the same technique and desired temperature, the temperature was 40° C 8 mm posterior to the presently measured point in the right hemisphere. Thus the brain hyperthermia in these experiments was fairly localized.

Experiment 2 showed that hyperthermia, when combined with BCNU in a dose without effect on symptom-free survival, significantly extended survival in a brain tumour with poorly limited margins towards surrounding brain. This indicates that the combination should be a candidate for clinical trials in malignant brain tumours, particularly as normal brain damage was not demonstrated to be increased. Although rats treated with BCNU alone were hypothermic (mean brain temperature during sham treatment 32.6° C), weight loss was similar to normothermic rats given the same BCNU dose in our previous experiments. Thus a theoretically attractive protection from the systemic effect of BCNU during hypothermia was not evident.

References

Bjerkvig R, Laerum OD, Mella O (1986) Glioma cell interactions with fetal rat brain aggregates in vitro and with brain tissue in vivo. Cancer Res 46: 4071–4079

Dahl O, Mella O (1982) Enhanced effect of combined hyperthermia and chemotherapy (bleomycin, BCNU) in a neurogenic rat tumour (BT$_4$A) in vivo. Anticancer Res 2: 359–364

Overgaard J (1978) The effect of local hyperthermia alone, and in combination with radiation, on solid tumors. In: Streffer C et al (eds) Cancer therapy by hyperthermia and radiation. Urban & Schwarzenberg, Munich, pp 49–61

Sutton CH (1971) Tumor hyperthermia in the treatment of malignant gliomas of the brain. Trans Am Neurol Assoc 96: 195–199

Pharmacokinetics of Methotrexate During Regional Hyperthermia in a Pig Model[*]

A. Schalhorn, H. Knorr, N. Seichert, P. Schöps, G. Stupp-Poutot, W. Permanetter, and W. Schnizer

Medizinische Klinik III, Klinikum Großhadern, Ludwig-Maximilians-Universität, Marchioninistraße 15, 8000 München 70, FRG

Introduction

Studies of human lymphoblast cell lines have shown that methotrexate (MTX) is one of those substances whose activity is increased under conditions of hyperthermia (Voth et al. 1987). In the recent past, sequential medium-dose MTX/5-fluorouracil (5-FU) therapy, followed by a leucovorin rescue adapted to the kinetics of serum MTX (Bleyer 1978; Jolivet et al. 1983; Sauer and Schalhorn 1980), has become important for the treatment of metastasized colorectal carcinomas (Heim et al. 1986; Herrmann et al. 1984; Leone et al. 1986; Possinger et al. 1984; Schalhorn et al. 1984). On account of the possible potentiating effect on MTX, an examination of these therapy regimens under hyperthermic conditions would appear to be desirable. The primary application in a pig model makes it possible to obtain information on the safety and possible side effects of such therapy under conditions of hyperthermia. In this paper, we report on investigations into the pharmacokinetics of MTX. One of our aims was to establish whether, under constant infusion of 150 mg MTX in the pig, serum levels similar to those seen in man under sequential medium-dose MTX/5-FU therapy could be achieved. A further point we wished to investigate with the aid of this model system was the tissue levels achieved in various organs and the concentrations of MTX in normothermic and hyperthermic muscle tissue. A further question of interest was the the excretion of MTX in the urine.

Methods

A detailed description of the pig model used in this study will be published elsewhere (Seichert et al. in preparation). MTX was infused continuously over a period of 100–120 min. A primary bolus injection of MTX was deliberately dispensed with. MTX was infused at 7.5/mg/kg to an amount of 150 mg. Blood samples were obtained during the heating-up phase (5, 15 and 30 min after initiation), during the plateau phase (30, 45, 60 and 90 min after initiation), on completion of hyperthermia and thereafter at intervals of 30 min and 1 h for a further 5 h. During hyperthermia, urine was collected in 15-min portions and thereafter in 30-min or 1-h portions. Muscle tissue was obtained from both the hyperthermic and the normothermic gluteus. Two samples were taken during the pla-

* Supported by the Deutsche Forschungsgemeinschaft (Scha 299/2-2).

teau phase (30 and 60 min after initiation), and two further samples 2 h and 4–5 h after hyperthermia. Antemortem samples of bile and tissue from the liver, kidneys, spleen and bowel were obtained. In the serum, bile and urine, MTX concentrations were measured by an enzyme immunoassay or enzymatically (Schalhorn et al. 1983). With low tissue concentrations of MTX it was necessary to use a more sensitive radioimmunoassay (Schalhorn et al. 1983).

Results

In Fig. 1, the course of the MTX serum levels during and following continuous MTX infusion is represented. In the cases investigated so far, a rapid increase in serum levels to values of about $10^{-5}M$ is observed. The plateau is achieved after about 40 min and, at the end of treatment, the MTX levels lie between 1 and $2.5 \times 10^{-5}M$. At the end of the infusion, the MTX serum levels decrease biexponentially. Five hours after termination of hyperthermia, they are within the range $0.5–3 \times 10^{-6}M$, and are thus then only about 1/8th–1/20th of the plateau values.

Figure 2 shows – in a single animal – a comparison of the course of the serum MTX levels and the urinary concentration of MTX, together with its accumulative excretion in the urine. The MTX concentration in the urine increases very rapidly to high values, reaching a maximum of $1 \times 10^{-3}M$. The relationship of the concentrations of urine MTX and serum MTX attains a maximum of 66:1. At the end of hyperthermia, the MTX concentration in the urine initially remains at a high level; 5 h after ending hyperthermia it is still in the region of $3.4 \times 10^{-4}M$. In this animal, the cumulative amount of MTX excreted in the urine by the 5th h after termination of hyperthermia (corresponding to about 7 h after initiation of MTX infusion) is 37 mg, i.e., 27% of the infused amount of 138 mg MTX.

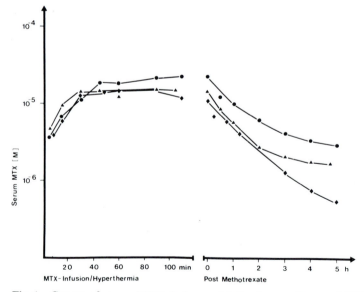

Fig. 1. Course of serum MTX during and following continuous MTX infusion plus simultaneous regional hyperthermia

Figure 3 shows the cumulative MTX excretion in the urine of the animals investigated so far. The results of the above-mentioned individual study are confirmed: the maximum concentrations in the urine increase into the $10^{-3} M$ range (results not shown) and, by the end of the experiment, about 7 h after initiating infusion, approximately 24%–30% of the MTX has been excreted in the urine.

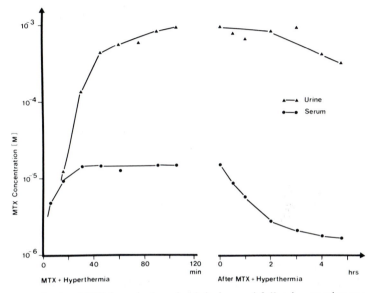

Fig. 2. Comparison of serum and urine MTX in a single animal during and following continuous MTX infusion plus simultaneous hyperthermia

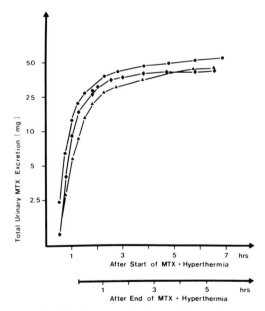

Fig. 3. Cumulative MTX excretion in the urine during and following the combined modality of MTX infusion plus regional hyperthermia

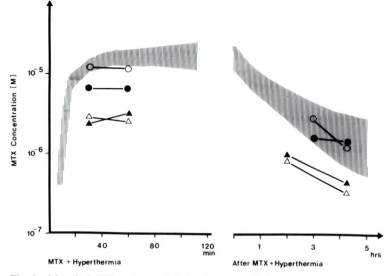

Fig. 4. Muscle MTX under and following MTX infusion plus regional hyperthermia. MTX levels in normothermic and hyperthermic muscle tissue of two animals *(circles, triangles)* are represented by *open* and *closed symbols* respectively. For comparison, the *shaded area* demonstrates the course of serum MTX in the animals investigated so far

In Fig. 4, the MTX concentrations in the muscle tissue during hyperthermia (under continuous MTX infusion) and a number of hours after completion of treatment are represented. Under MTX infusion, tissue levels of the antimetabolite in the region of $10^{-6}M$ are achieved. In the few individual measurements, no differences have so far been determined between the normothermic and hyperthermic muscle regions in this phase. After termination of the MTX infusion and hyperthermia, the MTX concentration in muscle tissue also decreases. The premortem MTX concentrations in the heated muscle region, as also in the normothermic control muscle, are already below $1.45 \times 10^{-7}M$.

Ante mortem, approximately 7 h after the start of infusion and 5 h after termination of hyperthermia, tissue specimens were taken from different organs. Figure 5 shows a comparison of MTX tissue concentrations and the corresponding levels in the urine and bile. At this point in time, the serum levels are between $5 \times 10^{-7}M$ and $3 \times 10^{-6}M$. The MTX levels in the bile ($3-7 \times 10^{-4}M$) and in the urine (up to $3.5 \times 10^{-4}M$) are especially high – higher than the serum levels by a factor of more than 100. With figures in the region of $10^{-5}M$, the levels in the liver and kidney tissue at this time are higher than the corresponding serum levels by a factor of 3–5. The MTX levels in tissue homogenate of bowel and spleen are of the same order of magnitude as the serum levels. The premortem MTX levels in the muscle biopsies are in the range of $3.2 \times 10^{-7}M - 1.43 \times 10^{-6}M$, and are thus only one fifth to a half of the serum concentrations.

Discussion

The results reported here show that the infusion for 100–120 min leads to serum levels of MTX that, in the plateau phase, lie within the region of just over $10^{-5}M$. In patients

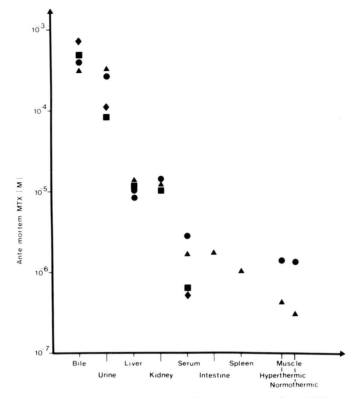

Fig. 5. Comparison of premortem MTX levels in various organs with the corresponding MTX serum and MTX urine levels. The *various symbols* represent individual animals

with metastasized colorectal carcinomas, by administering an MTX bolus injection of 150 mg/m^2 and following this with a 4-h infusion of a further 150 mg/m^2 MTX, in compliance with a therapy protocol proposed by Herrmann et al. (1984), we were able to achieve MTX serum levels in the region of $6.2 \times 10^{-6} M$ to $2.7 \times 10^{-5} M$ (Schalhorn et al. 1984). The measured results reported here thus show that the selected MTX dose and the duration of the infusion are capable, in the pig model, of simulating medium-dose MTX therapy as employed in man within the framework of various therapeutic studies (Heim et al. 1986; Herrmann et al. 1984; Leone et al. 1986; Possinger et al. 1984; Schalhorn et al. 1984). The model is thus highly suitable for studying the influence of medium-dose MTX therapy under normothermic and hyperthermic conditions after transplantation of various human tumours, in particular gastrointestinal tumours, to the pig. In the event of a greater MTX uptake by tumour tissue (Schalhorn et al. 1983, 1986), the metabolism of MTX (Schalhorn et al. 1985, 1986) will be particularly amenable to investigation. Here, the comparative measurement of MTX polyglutamates (Schalhorn et al. 1983, 1985, 1986; Wilmanns and Schalhorn 1985) in normothermic and hyperthermic tumour tissue will be of major importance. Should an increased metabolism to MTX polyglutamates be demonstrable under hyperthermia, this would provide an explanation for any increase in the effectiveness of MTX brought about by hyperthermia (Schalhorn et al. 1986; Wilmanns and Schalhorn 1985). This would justify the further use of MTX within the framework of combined hyperthermia and chemotherapy.

The measurement of the MTX concentration in the urine samples confirmed findings already made in man (Bleyer 1978; Jolivet et al. 1983; Pitman and Frei 1977) and in the monkey (Jacobs et al. 1976), namely that after all the forms of high-dose MTX therapy, very high concentrations of MTX are achieved. With an acidic urine pH, MTX concentrations in excess of $10^{-4}M$ are associated with the danger of precipitation within the renal tubules and consequent nephrotoxicity and delayed MTX clearance, which, in the last resort, can lead to severe damage to mucosa and bone marrow (Jacobs et al. 1976; Pitman and Frei 1977; Sauer and Schalhorn 1980). The investigations thus show that under medium-dose MTX therapy, too, consistent alkalinization of the urine to ensure safe treatment is indispensable.

The relatively high concentrations of MTX in the liver mirror the findings in man and in other mammals (Schalhorn et al. 1983). This preferential storage of MTX compounds in the liver may be explained by the great structural similarity to the physiological folates, for which, of course, the liver is an important storage organ. The high MTX concentrations in the bile confirm the known fact that a portion of the MTX is excreted via the liver. We have detected similar MTX levels, in the region of $10^{-4}M$, in the externally draining bile of a patient treated with medium-dose MTX/5-FU for advanced hepatic metastases of a carcinoma of the colon.

High concentrations of MTX in the kidneys are also known from studies on man (Jacobs et al. 1976; Pitman and Frei 1977). Whether they represent an expression of MTX uptake by the renal tissue itself, or are due to the urine within the renal system, cannot be decided with any degree of certainty.

In the studies carried out so far, in view of the short follow-up period of only 5 h after hyperthermia, we deliberately did not administer leucovorin. In future investigations, however, in particular after regional hyperthermia of transplanted tumours, a leucovorin rescue will be necessary (Sauer and Schalhorn 1980). It will then be possible to establish accurately the long-term effects of medium-dose MTX therapy and regional hyperthermia on the tumour tissue and, in particular, the clinical effectiveness of such an approach.

Summary

In the pig model, regional hyperthermia in the gluteus was combined with the infusion of 150 mg MTX over a period of 100–120 min. The pharmacokinetic data reveal that this approach is capable of simulating the situation that is necessary and achievable in medium-dose MTX therapy of human tumours. Under MTX infusion, serum levels in the region of $10^{-5}M$ are attained. As an expression of a renal and hepatic MTX excretion, high levels of MTX are found both in the urine and in the bile. Especially high concentrations of MTX are found in the liver and kidney tissue. In the normothermic and hyperthermic muscle, very low MTX levels are found. The pharmacokinetic data obtained show that the selected model is suitable for the future investigation of the effect of regional hyperthermia and MTX on transplanted tumours.

References

Bleyer WA (1978) The clinical pharmacology of methotrexate. Cancer 41: 36–51

Heim ME, Flechner H, Edler L, Queißer W (1986) Sequential high-dose methotrexate and 5-fluorouracil in the treatment of advanced colorectal cancer. Tumor Diagn Ther 7: 197–200

Herrmann R, Spehn J, Beyer U, von Franque U, Schmieder A, Holzmann K, Abel U (1984) Sequential methotrexate and 5-fluorouracil: improved response rate in metastastic colorectal cancer. J Clin Oncol 2: 591–594

Jacobs SA, Stoller RG, Chabner BA, Johns DG (1976) 7-Hydroxymethotrexate as a urinary metabolite in human subjects and rhesus monkeys receiving high dose methotrexate. J Clin Invest 57: 534–538

Jolivet J, Cowan KH, Curt GA, Clendennin NS, Chabner BA (1983) The pharmacology and clinical use of methotrexate. N Engl J Med 309: 1094–1104

Leone B, Romero A, Rabinovich MG, Perez JE, Macchiavelli M, Strauss E (1986) Sequential therapy with methotrexate and 5-fluorouracil in the treatment of advanced colorectal carcinoma. J Clin Oncol 4: 23–27

Pitman SW, Frei E III (1977) Weekly methotrexate-calcium leucovorin rescue: effect of alkalinization on nephrotoxicity; pharmacokinetics in the CNS; and use in non-Hodgkin's lymphoma. Cancer Treat Rep 61: 695–701

Possinger K, Schalhorn A, Zellmann K, Wilmanns W (1984) Sequential methotrexate(MTX)/5-fluorouracil(5-FU) therapy with low dose leucovorin rescue of patients with pretreated colorectal tumors. In: Klein HO (ed) Advances in the chemotherapy of gastrointestinal cancer. Perimed, Erlangen, pp 273–277

Sauer H, Schalhorn A (1980) Rationale Grundlagen und Praxis des Citrovorumfaktor(Leucovorin)-Schutzes nach hochdosierter Methotrexat-Therapie. Onkologie 3: 64–71

Schalhorn A, Sauer H, Wilmanns W, Stupp-Poutot G (1983) Methotrexate and methotrexate polyglutamates in human sarcoma metastases after high-dose methotrexate therapy. Klin Wochenschr 61: 1898–1894

Schalhorn A, Possinger K, Sauer H, Wilmanns W, Stupp-Poutot G (1984) Sequential medium-dose methotrexate/5-fluorouracil: low dose leucovorin rescue adapted to methotrexate serum pharmacokinetics. J Cancer Res Clin Oncol [Suppl] 107: 64

Schalhorn A, Wilmanns W, Sauer H, Stupp-Poutot G (1985) The role of methotrexate polyglutamates in methotrexate action. In: Recent advances in chemotherapy – anticancer section. University of Tokyo, Tokyo, pp 229–230

Schalhorn A, Sauer H, Wilmanns W (1986) The importance of polyglutamate formation for the efficacy of methotrexate therapy. J Cancer Res Clin Oncol [Suppl] 111: 12

Voth S, Sauer H, Wilmanns W (1987) Thermostability of cytostatic drugs in vitro and thermosensitivity of cultured human lymphoblasts against cytostatic drugs. Recent Results Cancer Res, (submitted)

Wilmanns W, Schalhorn A (1985) Role of polyglutamates in methotrexate action. Chemioterapia 4: 349–353

Summary of Recent Clinical Experience in Whole-Body Hyperthermia Combined with Chemotherapy[*],[**]

R. Engelhardt

Medizinische Universitätsklinik, Albert-Ludwigs-Universität, Hugstetter Straße 55, 7800 Freiburg, FRG

Introduction

The rationale for clinical application of whole-body thermochemotherapy is based on preclinical data suggesting the possibility of thermal enhancement of selectivity of the cytotoxic action of anticancer drugs (Honess 1983; Neumann et al. 1985b).

The heat sensitivity and the heat + drug sensitivity of human bone marrow progenitor cells seem to be rather uniform (Neumann et al. 1985a). Human tumors, however, are characterized by marked inter- and intraindividual heterogeneity in sensitivity to drug, heat, and drug + heat (Leith et al. 1983; Mella et al. 1984; Runge et al. 1985). Therapeutic gain, therefore, will principally occur in a tumor-individual way or – at the best – in a tumor type-related manner. It thus seems necessary to run clinical trials in a number of tumor types in order to establish their thermochemosensitivity.

In this communication preliminary data are given on three trials concerning the following tumor types:

1. Small cell lung cancer, extensive disease stage (SCLC)
2. Non-small cell lung cancer, inoperable stages (NSCLC)
3. Malignant melanoma, metastasized (MM)

Material and Methods

SCLC Phase III Trial

Patients having histologically proven SCLC, extensive disease stage, were randomized into two treatment arms, one receiving chemotherapy alone and the other receiving chemotherapy + whole-body hyperthermia. Criteria for inclusion were: histologically proven tumor; metastases outside the ipsilateral hemithorax; absence of brain metastases; absence of cardiopulmonary failure; age younger than 65 years; Karnofsky index greater than 50%; no prior chemotherapy; no second tumor; no metallic implants; se-

---

* Dedicated to Prof. Dr. med. G. W. Löhr on the occasion of his 65th birthday.
** Supported by the Bundesministerium für Jugend, Familie und Frauen und Gesundheit, Bonn.

Recent Results in Cancer Research, Vol. 107
© Springer-Verlag Berlin · Heidelberg 1988

rum creatinine level below 1.5 mg/dl and serum bilirubin below 2.0 mg/dl; written informed consent.

Patients randomized to the normothermic treatment received six cycles of chemotherapy according to the ACO protocol: 60 mg/m^2 Adriamycin i.v. on day 1, 200 mg/m^2 cyclophosphamide orally on days 2–5 and 2 mg vincristine i.v. on day 1; cycles were repeated on day 22. Patients being randomized for the hyperthermic treatment received the first three of the six cycles together with whole-body hyperthermia (41 ± 0.5° C/1 h).

For induction of whole-body hyperthermia we used the Siemens unit, consisting of a plexiglass cabin which is preheated by warm air up to 55–60° C. The patients rested on a mattress in which the coil field electrode was connected to a 27-MHz generator. Core temperature was controlled by a gauged rectal probe. During temperature measurement the generator was switched off.

NSCLC Pilot Study

Patients with inoperable NSCLC were entered in a phase I-II trial. The criteria for inclusion were as outlined for SCLC above. All patients received the following chemotherapy: 100 mg/m^2 cisplatin i.v. on day 1 together with forced diuresis; 3 mg/m^2 vindesine i.v. on day 1; 120 mg/m^2 VP-16 i.v. on days 2 and 3. The drugs given on day 1 were applied together with whole-body hyperthermia (41 ± 0.5° C/1 h); drug injection was started, when core target temperature was reached, which was then maintained for 1 h.

MM Pilot Study

In a phase I-II trial patients were entered having metastasized malignant melanoma. The criteria for inclusion were as outlined for SCLC above, with the exception of allowance of cytotoxic pretreatment with DTIC. Chemotherapy was as follows: 80 mg/m^2 cisplatin i.v. on day 1, together with forced diureses and whole-body hyperthermia (41 ± 0.5° C/1 h); 50 mg/m^2 Adriamycin i.v. on day 2 with an interval of more than 24 h between hyperthermia and the Adriamycin injection.

Response

Responses were categorized according to Miller et al. (1981): Complete remission (CR) is disappearance of all signs of measurable disease for a period of at least 1 month. Partial remission (PR) is shrinkage of all measurable tumors by at least 50% (2 diameters) for at least 1 month without new tumor growth. Minor response (MR) is shrinkage of all measurable tumor by less than 50% but at least 25% (2 diameters) for at least 1 month. No change (NC) indicates that measurable tumor has neither increased nor decreased in size for at least 1 month. Progessive disease (PD) is any increase in size of measurable tumor or the appearance of new disease during therapy.

Results

SCLC Phase III Trial

Interim results have so far been calculated on 55 patients in the SCLC trial. Forty-four of these are evaluable, 22 in each arm. Reasons for exclusion were: revised histology (2), revised stage (1), second tumor (1), patient's refusal to continue the therapy (3), death before completion of cycles 1–3 (4).

The median Karnofsky index values in the two groups are identical at 80%. Responses (CR + PR) are 8/22 in the normothermic treatment arm and 15/22 in the hyperthermic treatment arm. Mean duration of response was 105 days (range 42–193 days) in the normothermic arm and 130 days (range 53–392 days) in the hyperthermic arm. Toxicity regarding bone marrow depression were calculated from the nadir values of white blood cells and platelets during the first three cycles. White blood cell mean nadir values were $3170/mm^3$ without hyperthermia and $2440/mm^3$ in the hyperthermic group. This difference is statistically significant ($p = 0.044$; Wilcoxon test). The platelet mean nadir values were $231\,520/mm^3$ for the normothermic arm and $169\,030/mm^3$ for the hyperthermic arm. This difference is not statistically significant.

No impairment of renal, cardiac or liver function occurred. Neurotoxicity was not enhanced. Three patients presented with grade I burns (erythema) in the sacral region.

NSCLC Pilot Study

Fourteen patients were entered in the phase I–II trial of NSCLC. Evaluable are seven patients who have completed two or more treatment cycles. Reasons for exclusions were: rise in creatinine serum level to 2.3 mg/dl (1); target temperature not reached (1); patient's refusal to continue treatment after first cycle (2); PD before starting second cycle (3).

The histological subtypes of the seven evaluable patients are: squamous cell carcinoma (4); large cell carcinoma (2); adenocarcinoma (1). The response rates among the seven patients are as follows, with durations of response, in days, given in parentheses: CR 1/7 (150); PR 2/7 (180); MR 2/7 (30, 80). Toxicity was calculated on the basis of the 18 hyperthermic treatment cycles performed in the seven evaluable patients plus the patients excluded for reasons of toxicity. One of the 18 cycles was followed by an increase of creatinine up to 2.3 mg/dl. This patient was excluded. The white blood cell nadir values were below $3000/mm^2$ in nine of 18 cycles and in two of these nine patients below $2000/mm^3$. Platelet nadir values below $100\,000/mm^3$ occurred in two of 18 treatment cycles.

MM Pilot Study

Fifteen patients have so far been entered in a phase I–II trial of MM. Eleven are evaluable, having completed two or more hyperthermic treatment cycles. Reasons for exclusion were: grade II skin burn (in sacral region) after first cycle (1); protocol violation (1); target temperature not reached (1); PD before start of second cycle (1).

The 11 evaluable patients comprise four females and seven males.

Response rates are as follows, with the durations of response, in days, given in parentheses: CR 0/11; PR 1/11 (235); MR 2/11 (60+, 92); NC 2/11 (60, 90). Toxicity was calculated on the basis of 27 hyperthermic treatment cycles performed in the 11 evaluable patients plus those patients excluded for reasons of toxicity. In these 27 cycles we observed an increase of creatinine to 1.4 mg/dl in one patient; this patient was not excluded, and the creatinine serum level became normal under continued therapy. Skin burn grade II occurred in one of 27 cycles. Bone marrow toxicity was calculated from the white blood cell nadir values, which were below $3000/mm^3$ in twelve of 27 cycles, including four cases below $2000/mm^3$. One patient developed leukocytopenia below $1000/mm^3$. No septic complication developed. Platelet nadir below $100000/mm^3$ occurred in one of 27 cycles.

Discussion

Preclinical data in vitro as well as in vivo showed contradictory results regarding the question whether adriamycin and/or vincristine could be enhanced by 41° C hyperthermia (in vitro: Hahn et al. 1975; Klein et al. 1977; Mizuno and Ishida 1982; Chlebowski et al. 1982; Neumann et al. 1985; in vivo: Kamura et al. 1979; Overgaard 1976; Rotstein et al. 1983; Hinkelbein et al. 1984).

The data we obtained from our phase III trial in SCLC patients rather clearly show such an enhancement in respect to the therapeutic effect as well as to the hematological side effects. The other side effects of the drugs used, i.e., cardiac or neurological effects, were not enhanced.

Preclinical data demonstrated the possibility of a thermal enhancement of cisplatin at 41° C in vitro (Klein et al. 1977; Hahn 1979; Barlogie et al. 1980) and in vivo (Alberts et al. 1980; Honess 1983; Mella 1985).

It is too early to reach any conclusions regarding the response rates in the two pilot studies we are running using cisplatin together with whole-body hyperthermia (41 ± 0.5° C/1 h). Regarding the side effects, however, no life-threatening increase of either the unspecific (i.e., bone marrow) toxicity, or the specific (i.e., renal) toxicity of cisplatin seems to be induced by the hyperthermia. The *lege artis* administration of the drug is, of course, a precondition. The question of whether therapeutic gain is induced by the additional application of hyperthermia in the setting we have used remains open: The increase in quality and duration of response in our phase III trial is obvious, but not yet statistically significant. But even if the increase should attain significance by means of greater patient numbers, the degree of this increase of therapeutic parameters must be set against the stress to the patient, the increase of side effects, and the financial outlay. Although the side effects are not enhanced to a life-threatening degree, it is necessary to look for ways of increasing the selectivity of the thermal enhancement. Preclinical data, as well as preliminary results from clinical trials, seem to indicate that this might be achieved by the addition of nontoxic sensitizing agents, e.g., local anesthetics (Mizuno and Ishida 1982; Chlebowski et al. 1982) or lonidamine (Robins et al. 1986; Cavaliere et al. 1984).

References

Alberts S, Peng YM, Chen HSG, Moon TE, Cetas TC, Höschele ID (1980) Therapeutic synergism of hyperthermia-cis-platinum in a mouse tumor model. JNCI 65: 455–461
Barlogie B, Corry PM, Drewinko B (1980) In vitro thermochemotherapy of human colon cancer cells with cis-dichlorodiammineplatinum(II) and mitomycin-C. Cancer Res 40: 1165–1168

Cavaliere R, Difilippo F, Varanese A, Carlini S, Calabro A, Aloe L, Piarulli L (1984) Lonidamine and hyperthermia: clinical experience in melanoma. Preliminary results. Oncology [Suppl 1] 41: 116–120

Chlebowski RT, Block JB, Cundiff D, Dietrich MF (1982) Doxorubicin cytotoxicity enhanced by local anesthetics in a human melanoma cell line. Cancer Treat Rep 66: 121–125

Hahn GM (1979) Potential for therapy of drugs and hyperthermia. Cancer Res 39: 2264–2268

Hahn GM, Braun J, Har-Kedar I (1975) Thermochemotherapy: Synergism between hyperthermia (42–43° C) and Adriamycin (or bleomycin) in mammalian cell inactivation (cancer chemotherapy/cell membranes). Proc Natl Acad Sci USA 72: 937–940

Hinkelbein W, Menger D, Birmelin M, Engelhardt R (1984) The influence of whole-body hyperthermia on myelotoxicity of doxorubicin and irradiation in rats. In: Overgaard J (ed) Proceedings of the 4th International Symposium on Hyperthermic Oncology, vol 1. Taylor and Francis, London, pp 281–283

Honess DJ (1983) Animal models in the evaluation of therapeutic gain of thermo-chemotherapy. In: Spitzy KH, et al (eds) Proceedings of the 13th International Congress of Chemotherapy. Egermann, Vienna, pp 273/15–273/18

Kamura T, Aoki K, Nishikawa K, Baba T (1979) Antitumor effect of thermodifferential chemotherapy with carboquone on Ehrlich carcinoma. Gann 70: 783–790

Klein ME, Frayer K, Bachur NR (1977) Hyperthermic enhancement of chemotherapeutic agents in L1210 leukemia. Blood [Suppl 1] 50: 223

Leith JT, Heyman P, de Wyngaert JK, Dexter DL, Calabresi P, Glicksman AS (1983) Thermal survival characteristics of cell subpopulations isolated from a heterogeneous colon tumor. Cancer Res 43: 3240–3246

Mella O (1985) Combined hyperthermia and cis-diamminedichloroplatinum in BD IX rats with transplanted BT4A tumors. Int J Hyperthermia 1: 171–183

Mella O, Eriksen R, Dahl O, Laerum OD (1984) Therapy effects and toxicity of hyperthermia and cis-platin in BD IX rats with a transplantable neurogenic tumour (BT4A). In: Overgaard J (ed) Proceedings of the 4th International Symposium on Hyperthermic Oncology, vol 1. Taylor and Francis, London, pp 413–416

Miller AB, Hoogstraten B, Staquet M, Winkler A (1981) Reporting results of cancer treatment. Cancer 47: 207–214

Mizuno S, Ishida A (1982) Selective enhancement of the cytotoxicity of the bleomycin derivate peplomycin by local anesthetics alone and combined with hyperthermia. Cancer Res 42: 4726–4729

Neumann HA, Fiebig HH, Engelhardt R, Löhr GW (1985a) Cytostatic drug effects on human clonogenic tumor cells and human bone marrow progenitor cells (CFU-C) in vitro. Res Exp Med 185: 51–56

Neumann HA, Fiebig HH, Löhr GW, Engelhardt R (1985b) Effects of cytostatic drugs and 40.5° C hyperthermia on human bone marrow progenitors (CFU-C) and human clonogenic tumor cells implated into mice. JNCI 75: 1059–1066

Neumann HA, Fiebig HH, Löhr GW, Engelhardt R (1985c) Effects of cytostatic drugs and 40.5° C hyperthermia on human clonogenic tumor cells. Eur J Cancer Clin Oncol 21: 515–523

Overgaard J (1976) Combined Adriamycin and hyperthermia treatment of a murine carcinoma in vivo. Cancer Res 36: 3077–3081

Robins HI, Longo WL, Gillis W, O'Keefe S, Shecterle LM, Hugander AH, Martin PA, Schmidt G, Dennis WH (1986) Preliminary results of a phase I study combining 41.8° C whole body hyperthermia (WBH) and lonidamine (Lon) (Abstr 138). Proc Am Soc Clin Oncol 5: 36

Rotstein LE, Daly J, Rozsa P (1983) Systemic thermochemotherapy in a rat model. Can J Surg 26: 113–116

Runge HM, Neumann HA, Buecke W, Pfleiderer A (1985) Cloning ovarian carcinoma cells in an agar double layer versus a methylcellulose monolayer system. A comparison of two methods. J Cancer Res Clin Oncol 110: 51–55

Coagulation and Fibrinogenolysis During Whole-Body Hyperthermia

W. Klaubert, K. Eisler, J. Lange, and W. Wilmanns

Klinikum Großhadern, Ludwig-Maximilians-Universität, Marchioninistraße 15, 8000 München 70, FRG

Introduction

It has been shown that hyperthermia alone or in combination with radiation or some cytotoxic drugs causes greater damage in tumor tissue than in normal tissue. This effect is attributed to a higher sensitivity of tumor cells to heat and to changes in the microenvironment of the malignant tissue. Alterations were found in blood flow, oxygen tension and tissue pH. To some extent these alterations may be related to changes in hemostasis, for instance fibrinogen deposition in the tumor favoring tumor necrosis.

Material and Methods

Hyperthermia was performed in six cancer patients (two lung neoplasms, two intestinal tumors, two malignant melanomas). Hyperthermia was induced with the help of the heat exchanger of an extracorporal circuit (ECC). All patients received high doses of heparin (3000 USP/h) and proteinase inhibitor aprotinin (500000 kIU aprotinin/h).

Blood samples were drawn with an anticoagulant (1 part anticoagulant to 9 parts blood) before and after instillation of the ECC, after heating up, after 2 h and 4 h during heating plateau, and after return to normal body temperatures. As an anticoagulant we took 0.13 M tri-Na-citrate-2-hydrate for functional tests and a solution containing tri-Na-citrate-2-hydrate, 0.06 M TES, 0.05 M EDTA, and 1000 kIU/ml aprotinin adjusted to pH 7.4 for the other tests.

Clottable fibrinogen was measured according to Hörmann and Gollwitzer (1966). Serum samples obtained by this procedure were used for F-CB3 estimation. F-CB3 was measured by radioimmunoassay (RIA). The initially described inhibition assay (Gollwitzer et al. 1977) was replaced by a binding assay and calculation performed by use of a computerized spline function program. Normal levels in plasma were below 40 pmol/ml.

PMN-elastase-α-1-proteinase inhibitor complex was measured using enzyme-linked immunosorbent assay (ELISA; Merck).

Fibrinopeptide A was measured by RIA (Nossel et al. 1971) in plasma after fibrinogen was removed by adsorption to bentonite (Byk Mallinkrodt). In our laboratory procedure normal values were ≤ 6 ng/ml.

Recent Results in Cancer Research, Vol. 107
© Springer-Verlag Berlin · Heidelberg 1988

Antithrombin III and heparin were measured by functional tests (Boehringer-Mann-heim), protein C by ELISA (Boeringer-Mannheim), and fibronectin, prothrombin and plasminogen by laser nephelometry (Behring-Werke).

Results

After the implementation of the ECC all plasma protein concentrations showed a con-siderable drop except for F-CB3 (Figs. 1, 2).

During the following heating period (2–4 h), maximal blood temperatures of 44° C and maximal rectal temperatures of 41.8° C were reached (Figs. 1, 2). During the entire

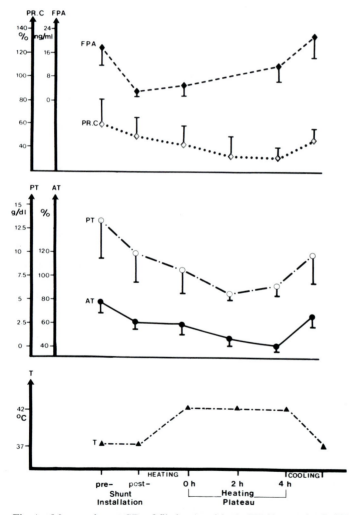

Fig. 1. Mean values ± SD of fibrinopeptide A *(FPA)*, protein C *(PR.C)*, prothrombin *(PT)*, and an-tithrombin III *(AT)* in six cancer patients before, during and one day after WBH. *T*, rectal tempera-ture. Normal values: FPA <5 pmol/ml; PC 80%–120%; PT 5–10 g/dl; AT 80%–120%

course of hyperthermia the plasma levels of heparin were >0.8 USP/ml and PTT was > 180 s.

The instillation of the ECC led to a drop in protein concentration of all proteins except for F-CB3. During the following course of hyperthermia (heating up and heating plateau phase) fibrinopeptide A showed only a slight increase, from 3.0±0.7 to 11.3± 5.2 ng/ml (Fig. 1). Antithrombin III activity decreased from 62.5%±9% to 4%–2.3%± 5.2% (Fig. 2), protein C concentration from 59%±19% to 39%±9% (Fig. 1), prothrombin concentration from 10±3 to 6±1 mg/dl, and plasminogen concentration from 9.1±2.3 to 6.9±1.7 mg/dl (Fig. 2). F-CB3 showed no significant changes, ranging from 25±16 to 28±19 pmol/ml (Fig. 2).

Plasma levels of fibrinogen decreased from 480±61 to 37±84 mg/dl (Fig. 2) and of fibronectin from 28±10 to 19±9 mg/dl (Fig. 2).

After the end of the hyperthermia and return to normal temperatures, those proteins which had decreased during hyperthermia increased.

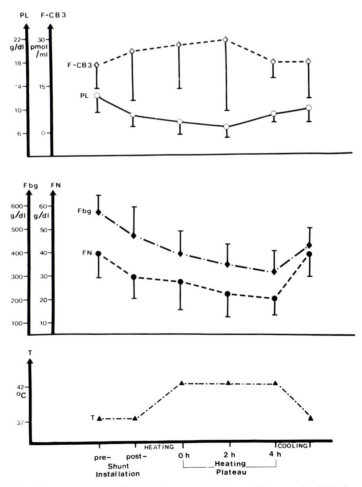

Fig. 2. Mean values ± SD of plasminogen *(PG)*, F-CB3, fibrinogen *(Fbg)*, and fibronectin *(FN)*, in six cancer patients before, during and 1 day after whole-body hyperthermia. *T*, rectal temperature. Normal values: PG 6–25 mg/dl; F-CB3 <40 pmol/ml; Fbg 200–400 mg/dl; FN 25–40 mg/dl

Discussion

Even before starting the actual heating up process, the implementation of the ECC led to a considerable drop in plasma protein concentration.

The low fibrinopeptide A values during the heating period show that thrombin activity is not elevated during whole-body hyperthermia. Obviously the high heparin levels prevent coagulation disorders despite the low levels of antithrombin III.

The F-CB3 test measures a split product of the proteolytically most susceptible α-chain of the fibrinogen molecule. Therefore F-CB3 is a sensitive parameter for proteolysis mediated by different fibrinolytic active proteases, mainly plasmin. Normal F-CB3 values and the only slight decrease in plasminogen levels show that fibrinolysis is not stimulated. This may be attributed to the high doses of the proteinase inhibitor aprotinin which were administered to all patients during hyperthermia.

Although the hemostatic system is not essentially altered, fibrinogen levels decrease during whole-body hyperthermia. The reason is not yet clear, but it has been shown in animal models by electron microscopy studies and by measurement of ^{131}I-fibrinogen that fibrinogen is more abundantly deposited in heated tumors than in control tumors (Lee et al. 1985). Copley (1980) describes a theory postulating a thrombin-independent fibrin clotting in the microcirculation of the tumor. Fibrinogen is adsorbed at damaged sites in the vascular wall of the blood vessels. On adsorbtion of further fibrinogen and other proteins, protein layers develop (Copley 1983), leading to occlusion of the tumor vessels. This process will be favored by low pH (Copley 1980) such as has been demonstrated in tumors especially during hyperthermia, which thus leads to tumor necrosis.

References

Copley AL (1980) Fibrinogen gel clotting, pH and cancer therapy. Throm Res 18: 1–6
Copley AL (1983) The physiological significance of the endothelial fibrin lining as the critical interface in the vessel-blood-organ and the importance of in vivo fibrinogenin formation in health and disease. Throm Res [Suppl] 5: 105–146
Gollwitzer R, Hafter R, Timpl H, Graeff H (1977) Immunological assay for a carboxyterminal peptide of the fibrinogen α-chain in normal and pathological human sera. Thromb Res 11: 859–873
Hörmann H, Gollwitzer R (1966) Analytische Untersuchungen über Rinderfibrinogen und -fibrin. Hoppe Seylers Z Physiol Chem 346: 21–41
Lee SY, Song CW, Levitt SH (1985) Change in fibrinogen turnover in tumors by hyperthermia. Eur J Cancer Clin Oncol 21: 1507–1513
Nossel HL, Younger L, Wilner GD, Procupez T, Canfield RE, Butler VP (1971) Radioimmunoassay of human fibrinopeptide A. Proc Natl Acad Sci USA 68: 2350–2353

Treatment of Superficial Neoplastic Lesions Using Hyperthermia, Radiotherapy, and Chemotherapy

H. Kolbabek, W. Nespor, and G. Koderhold

Sonderabteilung für Strahlentherapie, Ludwig-Boltzmann-Institut für Klinische Onkologie, Allgemeines Krankenhaus Wien-Lainz, Wolkersbergenstraße 1, 1130 Wien, Austria

Introduction

At the Special Department for Radiotherapy of the Community Hospital of Vienna-Lainz, a device for the treatment of superficial neoplastic lesions (down to 3 cm) was installed in October 1985. Since then, over 60 patients have been subjected to a combined modality treatment of radiotherapy (RT) and/or chemotherapy (ChT) with local hyperthermia (HT) in addition (Streffer et al. 1978; Hahn 1979).

Apparatus and Method

The frequency emitted from our device is 915 MHz. The temperatures (T) are measured using probes which are inserted subcutaneously via venflon catheters. The treatment program is adjusted by feedback using these probes, which are multiprobes giving data on three points of temperature measurement. We try to place the probes in such a way as

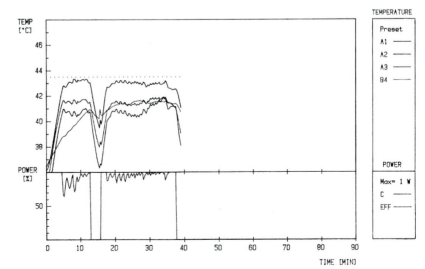

Fig. 1. Treatment plan: fibrosarcoma after surgery

to get information about the center of the tumor, the tumor margin and some normal tissue in the neighborhood.

A water bolus is included in the system for cooling and a more homogeneous energy output (Bolmsjö 1986).

The local HT is performed twice a week, according to the literature (Overgaard 1981; Lindholm et al. 1982; Arcangeli et al. 1983), in order to avoid thermotolerance. HT was given 4 h after RT for the first 6 months of our experience with the combination RT + HT. We now give HT in the first hour after RT or ChT (Overgaard 1980).

We succeed in most cases in achieving and maintaining a temperature in the tumor of 43.5° C or more. Noteworthy is a lowering of the temperature after 13–23 min in se-

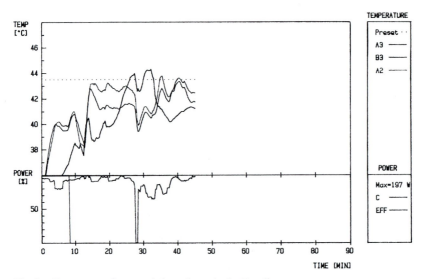

Fig. 2. Treatment plan: neck lymph node, bulky disease

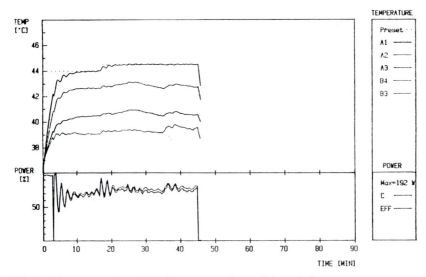

Fig. 3. Treatment plan: Merkel cell tumor in partial remission

lected cases (Fig. 1) and a crossing over of the temperature lines measured at different points in the tumor (Fig. 2). In a case of a large Merkel cell tumor on the forearm, the temperature distribution was rather inhomogeneous in bulky disease, but more homogeneous when a remission set in (Fig. 3).

Patients and Results

Of the more than 60 patients treated with HT, 50 are evaluable. Four groups of patients emerge from our work so far (Table 1).

The first group comprises chest wall recurrences after breast resection and some breast carcinomas which could be treated primarily showing at least a partial remission (PR). Among, the chest wall recurrences, six of 19 are still in complete remission (CR), 13 in PR. Of the five breast carcinomas treated primarily, three achieved CR: a patient admitted for an ulcerating tumor initially showed necrosis of the tumor and has now been free of tumor for 1 year; a second one who had ChT after RT and HT – doxorubicin, vincristine and cyclophosphamide, then methotrexate and fluorouracil – was tumor-free for 5 months but is now lost to follow-up; and a third 86-year-old patient is still in CR after 6 months.

The second group includes patients with tumors of the head and neck (ENT tumors). Here, superficial lesions and lymph nodes are easy to control, but deeper-seated (> 3 cm) tumors, such as tonsil lesions, can respond only partially after a treatment with our

Table 1. Results in 50 patients treated with a regimen including local hyperthermia

Diagnosis	Complete remission/NED	Partial remission	No remission
Breast cancer	3	2	
Breast cancer (op.)	6	13	
Laryngeal cancer (op.)		1	
Tonsillar cancer		2	
Cancer of oral cavity		1	
Parotid cancer (op.)	2		
Struma maligna (op.)		1	
Cancer of sebaceous gland		1	
Rodent ulcer		1	
Neck lymph nodes	1	4	
Merkel cell tumor	1		
Fibrosarcoma	1		
Fibrosarcoma (op.)	1		
Liposarcoma (op.)	1		
Malignant schwannoma (op.)		1	
Malignant melanoma (op.)	1		
Vulvar cancer		1	1
Rectal cancer (op.)	2		
Anal cancer (op.)			1
Inguinal lymph nodes		1	
Total	19	29	2

op., surgery performed.

system. Of the two parotis tumors – a classical indication for local HT – one has been in CR for 8 months; a second parotis tumor relapsed 11 months after RT and HT and was given a second try with RT and HT, resulting again in PR.

A third group of patients treated at our institution was admitted for soft tissue tumors. Of these, a Merkel cell tumor is in CR after 15 months, and a fibrosarcoma on the foot achieved CR only after HT was added to the RT already carried out.

Another fibrosarcoma, in an 81-year-old female, presented in two locations. The first was on the inner thigh: as no HT was available at that time, she was treated with RT alone (60 Gy) after surgery. A second lesion on the outer side of the same thigh was treated with RT (30 Gy only) and HT after surgery. Both sides show no evidence of disease after 18 months; the site treated with RT and HT is less pigmented than the site treated with RT only.

After more experience we did not hesitate to give a full course of RT (60 Gy), adding HT in the later part of the treatment schedule. No damage to the skin was observed; for example, in an 81-year-old male treated after repeated surgery for a recurrent malignant melanoma on the forearm, there is no evidence of disease (NED) after 8 months of follow-up. A liposarcoma on the buttock of a 50-year-old female was treated after surgery with RT and HT: again, NED after 11 months.

A malignant schwannoma in the axilla of a 35-year-old female who also had Recklinghausen's disease displayed PR after pallative surgery, RT and HT, but the patient died after 9 months, lost to follow-up.

Considering these cases of soft tissue tumors, we believe that RT alone should not be given to such patients, even after surgery – which may not be complete. ChT and/or HT seems to be indicated. The now nearly forgotten RT of malignant melanoma is seen in a new perspective (Kim et al. 1982).

The last group of our patients comprises those suffering from tumors of the anogenital region. An 81-year-old female presenting with a vulvar carcinoma achieved PR and was able to walk again; two patients with recurrent superficial manifestations after cancer of the rectum are free of disease; and the lymph nodes in a 83-year-old female with a recurrent rectal carcinoma exhibited PR after HT was added to RT.

A last patient showed no response after RT and HT for superficial recurrences of an anal carcinoma. Combining ChT with HT still produced no response, so palliative surgery was performed.

Thus, our limited experience with the combination of ChT and HT has yielded two PR and two cases of NED so far (Table 2).

Table 2. Results of chemotherapy *(ChT)* and hyperthermia *(HT)*

	ChT	HT	Effect
Breast cancer (op., irrad.)	MTX, FU, ACO	$5\times$	PR
Breast cancer (op., irrad.)	Tam, Doxo	$6\times$	PR
Rectal cancer (op., irrad.)	CisPt, Doxo	$4\times$	NED
Rectal cancer (op.)	FU	$4\times$	NED
Anal cancer (op., rec.)	FU	$6\times$	NR

op., surgery performed; *irrad.*, radiotherapy given; *rec.*, recurrences; *MTX,* methotrexate; *FU,* fluorouracil; *ACO,* doxorubicin + cyclophosphamide + vincristine; *CisPt,* cisplatin; *Tam,* tamoxifen; *Doxo,* doxorubicin; *PR,* partial remission; *NED,* no evidence of disease; *NR,* no remission.

Summary

Local hyperthermia has great promise in the search for a better policy of combined modality treatment of superficial neoplastic lesions. Interstitial HT or whole-body HT should be made available in every tumor treatment center in the near future.

References

Arcangeli G, Cevidalli A, Nervi C, Creton G (1983) Tumor control and therapeutic gain with different schedules of combined radiotherapy and local external hyperthermia in human cancer. Int J Radiat Oncol Biol Phys 9: 1125–1134

Bolmsjö M (1986) Commercially available equipment for hyperthermia. In: Watmough DJ, Ross WM (eds) Hyperthermia. Blackie, Glasgow

Hahn GM (1979) Potential for therapy of drugs and hyperthermia. Cancer Res 39: 2264–2268

Kim JH, Hahn EW, Ahmed SA (1982) Combined hyperthermia and radiation therapy for malignant melanoma. Cancer 50: 478–482

Lindholm CE, Kjellen E, Landberg T, Nilsson P, Persson B (1982) Microwave-induced hyperthermia and ionizing radiation. Preliminary clinical results. Acta Radiol [Oncol] 21 (4): 243–249

Overgaard J (1980) Simultaneous and sequential hyperthermia and radiation treatment of an experimental tumor and its surrounding normal tissue in vivo. Int J Radiol Oncol Biol Phys 6: 1507–151

Overgaard J (1981) Fractionated radiation and hyperthermia: experimental and clinical studies. Cancer 48: 1117–1123

Streffer C, van Bueningen D, Dietzel F, et al. (eds) (1978) Cancer therapy by hyperthermia and radiation. Urban and Schwarzenberg, Munich

Clinical Experience in the Combination of Hyperthermia with Chemotherapy or Radiotherapy

G. C. W. Howard and N. M. Bleehen

Western General Hospital, Crewe Road, Edinburgh EH4 2XU, Great Britain

Introduction

Clinical studies into the effect of hyperthermia alone or in combination with other treatment modalities are now being pursued at many centres. Many different techniques of heat delivery are being investigated. The majority of clinical studies have concentrated on the treatment of superficial tumours with localised hyperthermia. The reason for this is that in general there are still technical problems in heating deep-seated tumours and whole-body hyperthermia is both difficult and potentially dangerous. There are now many controlled studies comparing radiation alone with radiation and hyperthermia. There are many differences between these series, notably in radiation dose and temperatures achieved, but an increase in the response rate of tumours treated with the combined modality is almost universal (Overgaard 1984). The majority of such studies have demonstrated an increase in both partial and complete response rates, in most cases not at the expense of increased toxicity. There have been some reports of an increase in severe skin reactions which may be related to the radiation dose used and the interval between the two modalities (Arcangeli et al. 1983; Howard et al. 1987). It has also been suggested that toxicity may be related to the temperature gradient across the tumour (Dewhirst et al. 1984).

Compared to the extensive clinical studies into the combination of hyperthermia and radiotherapy there have been relatively few investigating the joint action of heat and drugs. Laboratory studies showing enhancement of the cytotoxic activity of many drugs at increased temperatures have led to an interest in such combinations in the clinic. These encouraging laboratory results have yet to be confirmed in clinical studies. The reason for the bias toward investigating heat and radiotherapy is probably the greater ease of study design. Response rates to chemotherapeutic agents are often poorly documented, and it is essential to have controls treated by drug alone. It is rarely possible to obtain heat alone controls in the same patient, and thus assessment of results is difficult. In addition, patients who are considered suitable for hyperthermia are often heavily pretreated and tolerate cytotoxic therapy poorly.

The most encouraging results from combining hyperthermia and drugs are probably still those of Cavaliere and Stehlin using hyperthermic limb perfusion (Cavaliere et al. 1967; Stehlin et al. 1979). Hyperthermic perfusion with melphalan has been shown in one series to significantly increase survival (Rege et al. 1983). Other results are mainly anecdotal, although a significant increase in response of head and neck nodes treated

Recent Results in Cancer Research, Vol. 107
© Springer-Verlag Berlin · Heidelberg 1988

with adriamycin or bleomycin when combined with hyperthermia has been shown by Arcangeli and co-workers (Arcangeli et al. 1980).

Clinical studies undertaken at Cambridge were initially performed using a locally built microwave hyperthermia system (Sathiaseelan et al. 1985). This utilized 915-MHz and 433-MHz microwaves and contact applicators. Multijunction copper constantan thermocouples were used for thermometry. This system was used to treat lesions in combination with radiotherapy and chemotherapy in an early feasibility study. Based on the experience thus gained, further studies were undertaken with both superficial and regional hyperthermia using the BSD 1000 clinical hyperthermia system.

Hyperthermia and Chemotherapy

Seven patients with multiple superficial malignant lesions were treated with chemotherapy and hyperthermia in a pilot study. All histological types were considered suitable as long as there were two or more superficial assessible lesions. Details of these patients and their treatments are tabulated in Table 1. The majority of patients had multiple deposits of either melanoma or squamous cell carcinoma. Patients were treated with CCNU, bleomycin or melphalan. Hyperthermia treatments were timed to coincide with estimated peak serum values of the drug. Patients receiving melphalan had serum drug levels measured on the first day of treatment and subsequent hyperthermia treatments adjusted to coincide with peak values.

The quality of heating in this series was poor, with only 14% of the hyperthermia treatments achieving a minimum intratumour dose equivalent to 60 min at 43° C (min.

Table 1. Details of patients treated with hyperthermia and chemotherapy

Patient no.	Histology	Site	Number of lesions	Chemotherapy	Hyperthermia		Results
					No. of lesions heated	No. of treatments	
1	Melanoma	Left leg	7	CCNU 200 mg oral × 1	5	4	No response in any lesion; unfit for second course
2	Melanoma	Right leg	3	CCNU 200 mg oral × 1	2	4	No response in either lesion; unfit for second course
3	Squamous cell carcinoma	Abdominal wall	4	Bleomycin 30 mg IM × 6 twice weekly	2	6	No response in either lesion
4	Melanoma	Head and neck	2	Melphalan 6 mg/m² daily × 5, three courses	1	6	No response in heated lesion
5	Squamous cell carcinoma	Chest wall	2	Bleomycin 30 mg IM × 6 twice weekly	1	5	Growth delay in heated lesion
6	Squamous cell carcinoma	Scalp	4	Bleomycin 30 mg IM × 6 twice weekly	2	6	Complete response in one heated and one control lesion
7	Adenocarci-noma (breast)	Scalp	3	Melphalan 6 mg/m² daily × 5, three courses	2	6	No response in either lesion

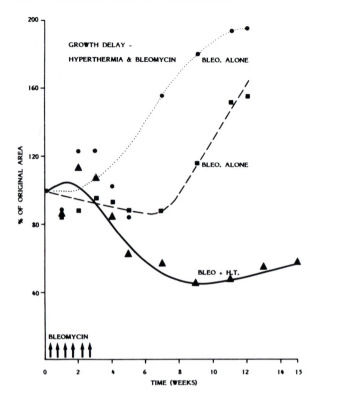

Fig. 1. Growth delay in a lesion treated with bleomycin and hyperthermia (patient 5)

eq. 43° C). The majority of treatments (66%) amounted to a minimum tumour dose of less than 15 min. eq. 43° C.

It can be seen from Table 1 that only two lesions achieved a documented response. One of these was treated with hyperthermia and one was a control. A further heated lesion (patient 5) demonstrated a degree of growth delay compared with drug alone (Fig. 1). The low response rate is probably explained by the poor quality of heat treatments in this series. This experience did, however, highlight the major problems inherent in this sort of investigation. There are a limited number of suitable patients with multiple tumour deposits, such patients often have a large tumour burden and their survival is usually short. Most will be heavily pretreated and may not tolerate further cytotoxic therapy. Two of seven patients did not tolerate a second course. Most chemotherapy regimes are given intermittently and thus courses of hyperthermia may also be prolonged and poorly tolerated.

Superficial Hyperthermia and Radiotherapy

A simple protocol was designed for the treatment of superficial malignancy combining radiotherapy with hyperthermia. In a small series of patients with a total of 41 lesions we have monitored tumour response and local toxicity (Howard et al. 1987). Patients with two or more tumours had radiation alone controls assessed. Where possible a radiation dose of 24 Gy was given in six twice weekly fractions. Hyperthermia treatments were performed immediately after each radiation fraction (within 30 min). In this way we

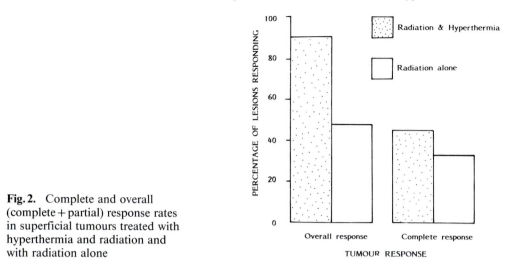

Fig. 2. Complete and overall (complete + partial) response rates in superficial tumours treated with hyperthermia and radiation and with radiation alone

hoped to avoid the effects of thermotolerance whilst gaining a maximum thermal enhancement. Radiation doses ranged from 8 to 54 Gy, with the majority of lesions (36) receiving the protocol dose of 24 Gy.

Hyperthermia treatments were performed using contact microwave applicators (BSD), usually using a frequency of 650 MHz. Some courses of hyperthermia were terminated before the six fractions had been delivered, either because of marked toxicity or owing to complete disappearance of the tumour. Bowman thermistor probes were used to monitor intratumour and normal tissue temperatures. These probes were tracked along blind-ended teflon catheters to monitor the minimum recorded intratumour temperature.

Twenty-one lesions were treated with radiation and 20 with radiation and hyperthermia. The quality of heating in this series of treatments was much improved, with 57% of the 80 treatments exceeding a minimum tumour dose of 30 min. eq. 43° C. Thirty per cent of treatments were still unsatisfactory (< 15 min. eq. 43° C). These mainly concerned large lesions (average area 36 cm²), smaller lesions proving easier to heat.

The overall and complete response rates (90% and 45% respectively) in the heated group were greater than those for radiation alone (48% and 33%) (Fig. 2). When analysed separately, a similar result was obtained for those lesions with internal controls. In the heated group, smaller lesions were more likely to achieve a complete response. In addition, as might be expected, completely responding lesions received better quality heat treatments. This increase in response was at the expense of an increase in skin reaction. Graded on a 0–7 scale, nearly all lesions treated with radiation alone had low-grade reactions (less than grade 4, the first sign of breakdown in the treated field). Forty per cent of lesions treated with the combined modality however had high-grade skin reactions (≥ grade 4) (Table 2). There was also a suggestion that skin reactions developed more quickly in the heated group.

This small series demonstrates many of the features seen in other studies. The difficulty in heating large lesions remains a technical problem resulting in fewer responses. The enhancement of skin toxicity can probably be avoided by suitable timing of the two modalities and more careful skin cooling. Based on these experiences our present protocols are aimed at confirming these results in a randomised fashion and comparing the findings of different numbers of heat treatments.

Table 2. Incidence of high- and low-grade skin reactions in lesions treated with radiation alone and with radiation and hyperthermia

Treatment	Reaction (no. of lesions)			
	Low grade (grades 0–3)		High grade (grades 4–7)	
Radiation alone	20	(95%)	1	(5%)
Radiation and hyperthermia	12	(60%)	8	(40%)

Table 3. Tumour histology of patients treated with pelvic regional hyperthermia

Tumour Type	No. of patients	
Colorectal	13	(43%)
Bladder	6	(20%)
Cervix	5	(17%)
Ovary	2	(7%)
Endometrium	1	
Sarcoma	1	(13%)
Prostate	1	
Vagina	1	

Regional Hyperthermia and Radiotherapy

Thirty patients with extensive pelvic tumours were treated with a combination of regional hyperthermia and radiotherapy in a feasibility study (Howard et al. 1986). An annular phased array applicator was used to induce hyperthermia in patients with a variety of pelvic tumours. This site was chosen as it has been shown to be the region most successfully heated with this equipment (Sapozink et al. 1984; Emami et al. 1984). Patients with extensive pelvic disease where local control was considered unlikely by conventional methods were assessed for treatment. The majority had colorectal or bladder tumours (Table 3). The general medical condition of each patient was carefully assessed prior to treatment, as regional hyperthermia may result in cardiovascular stress. Patients also need to be well motivated to tolerate treatment and overall only 50% of patients assessed were considered suitable.

Patients undergoing regional hyperthermia all had concurrent radiotherapy. Most patients had received prior radiotherapy and some chemotherapy as well. Radiation doses were therefore variable, depending on remaining tolerance. Doses ranged from 12 to 65 Gy (mean 34 Gy). The number of hyperthermia treatments was also designed to suit each individual case. For a prolonged course of radiotherapy, hyperthermia was given once a week. For shorter courses of 2–3 weeks hyperthermia was given twice weekly. The number of treatments varied from one to seven, with a mean of 3.1 per patient.

Bowman thermistor probes were used to monitor temperatures. Sites commonly used were intravesical, rectal and vaginal, with an oral probe to monitor core temperature. Where possible, intratumour temperatures were also measured. All patients had a pretreatment computerised tomographic (CT) scan. This was necessary to monitor later response to treatment and to assess whether an intratumour catheter could be inserted under CT control.

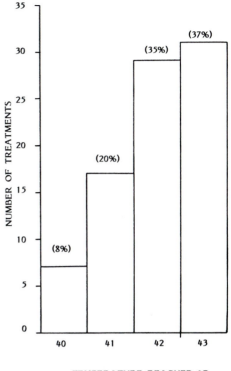

Fig. 3. Pelvic regional hyperthermia: maximum intrapelvic temperatures reached in 84 assessable treatments

NUMBER OF TREATMENTS

TEMPERATURE REACHED °C

Ninety-two pelvic regional hyperthermia treatments have now been performed on 30 patients aged from 36 to 76 years (mean 60 years). We have found that although the equipment is capable of inducing hyperthermia deep within the pelvis, the treatments are poorly tolerated and recorded maximum pelvic heat doses are low (Figs. 3, 4). Intrapelvic temperatures of 42° C or more were recorded in 72% of treatments, but such temperatures are difficult to maintain, with the result that only 20% of treatments achieve a heat dose in excess of 10 min·eq·43° C. The reason for this is poor patient tolerance. The majority of treatments are limited by either local pelvic or generalised discomfort. Less commonly, a rise in systemic temperature or cardiovascular stress has necessitated power reduction or termination of treatment. Although many of these patients were receiving regular analgesics, no specific premedication or analgesia was routinely given. It is likely that treatments would be better tolerated if this were done.

The acute toxicity of the procedure, though limiting the success of the treatment in virtually every case, was short-lived. We have not noted any excessive late toxicity, although this is difficult to assess in such a heavily treated group of patients. Serum liver enzymes and coagulation screening tests have remained unchanged after treatments. On occasions a significant rise in the non-cardiac serum creatine kinase isoenzyme and lactate dehydrogenase was noted. These appear to be non-specific changes and not related to temperatures achieved during treatment. Both response and toxicity are difficult to assess in this group of patients. Twenty-five of the 30 patients treated have been followed up for a minimum of 4 months. Seventeen patients have died, and in all cases death was thought to be a result of their disease. Eight patients remain alive, all of whom have residual disease. An objective response has been seen in eight of 16 assessable patients.

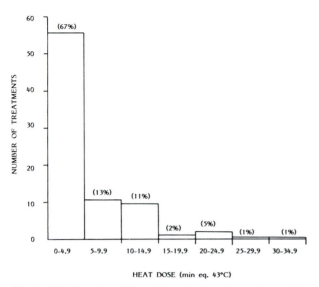

HEAT DOSE (min eq. 43°C)

Fig. 4. Pelvic regional hyperthermia: maximum intrapelvic heat doses recorded during 84 assessable treatments

The dose of radiation was higher in those who responded (range 20–65 Gy, mean 44 Gy) than in those who did not (range 12–30 Gy, 19 Gy). In this group of patients where symptom control was the primary aim of treatment, it is of note that palliation was achieved in 16 of 18 patients who complained of symptoms. This may, however, have occurred with radiotherapy alone.

This experience demonstrates that the annular phased array applicator is technically capable of deep heating in the pelvis. To achieve significant hyperthermia treatments, however, patient tolerance must be improved. Routine analgesia may be one method of doing this, but the dangers are considerable while thermometry remains inadequate. Another approach is to try to steer the area of heating within the pelvis, to spare normal structures and preferentially heat the tumour. There are several methods of doing this, two of which we are investigating. Successful steering of power deposition, resulting in improved treatments, has been achieved using phasing cables (Sathiaseelan et al. 1986). Frequency changes in the range of 55–60 MHz have also resulted in improved heating patterns on occasions.

Conclusions

Clinical hyperthermia investigations performed in this unit to date have been either phase I or phase II studies. Initial experience with the combination of hyperthermia and chemotherapy demonstrated the problems of this type of investigation. Important questions regarding the interaction of heat and drugs in the clinic need to be answered, but the design of such studies is difficult. Superficial lesions treated with the combination of hyperthermia and radiotherapy have shown many of the features of other similar studies, with an increase in response rate but at the expense of increased skin toxicity. Further studies are designed to assess how many heat treatments are necessary to achieve this increased response. Such information would be of great value when considering more toxic procedures such as regional hyperthermia. The pilot study into pelvic regional hyperthermia has highlighted the problems of this procedure. The equipment appears technically capable of inducing therapeutic temperatures within the pelvis; recorded heat doses, however, are very low. If an increase in response of the order of that seen in superficial tumours is to be seen in deep pelvic malignancy, patient tolerance must be improved. This may be achieved by analgesics or even anaesthesia, but until thermometry has improved this is potentially dangerous. Other methods aimed at improving heating patterns, such as steering of the heated volume within the patient, are of promise and warrant further investigation.

References

Arcangeli G, Cividalli A, Lovisolo G, et al. (1980) Effectiveness of local hyperthermia in association with radiotherapy or chemotherapy: comparison of multimodality treatments on multiple neck node metastases. In: Arcangeli G, Mauro F (eds) Proceedings of the 1st Meeting of the European Group of Hyperthermia in Radiation Oncology. pp 257–265

Arcangeli G, Cividalli A, Lovisolo G, Nervi C (1983) Clinical results after different protocols of combined local heat and radiation. Strahlentherapie 159: 82–89

Cavaliere R, Crocatto E, Giovanella B, et al. (1967) Selective heat sensitivity of cancer cells. Cancer 20: 1351–1381

Dewhirst MW, Sim D, Grochowski J (1984) Thermal influence of radiation induced complications vs. tumour response in a phase III randomised trial. In: Overgaard J (ed) Proceedings of the 4th International Symposium of Hyperthermic Oncology, vol 1. Taylor and Francis, London, p 313

Emami B, Perez C, Nussbaum G, Leybovich L (1984) Regional hyperthermia in the treatment of recurrent deep-seated tumours. Preliminary report. In: Overgaard J (ed) Proceedings of the 4th International Symposium on Hyperthermic Oncology, vol 1. Taylor and Francis, London, p 605

Howard GCW, Sathiaseelan V, King GA, Dixon AK, Anderson A, Bleehen NM (1986) Regional hyperthermia for extensive pelvic tumours using an annular phased array applicator: a feasibility study. Br J Radiol 59: 1195–1201

Howard GCW, Sathiaseelan V, Freedman L, Bleehen NM (1987) Hyperthermia and radiation in the treatment of superficial malignancy: an analysis of treatment parameters, response and toxicity. Int J Hyperthermia 3 (1): 1–8

Overgaard J (1984) Rationale and problems in the design of clinical studies. In: Overgaard J (ed) Proceedings of the 4th International Symposium on Hyperthermic Oncology, vol 2. Taylor and Francis, London, p 325

Rege VB, Leone LA, Soderberg CH, et al. (1983) Hyperthermic adjuvant perfusion chemotherapy for stage I malignant melanoma of the extremity with literature review. Cancer 52: 2033–2039

Sapozink MD, Gibbs FA, Gates K, Stewart R (1984) Regional hyperthermia in the treatment of clinically advanced deep-seated malignancy: results of a pilot study employing an annular phased array applicator. Int J Radiat Oncol Biol Phys 10: 775–786

Sathiaseelan V, Howard GCW, Har-Kedar I, Bleehen NM (1985) A clinical microwave hyperthermia system with multipoint real-time thermal dosimetry. Br J Radiol 58: 1187–1195

Sathiaseelan V, Iskander MF, Howard GCW, Bleehen NM (1986) Theoretical analysis and clinical demonstration of the effect of power pattern control using the annular phased array hyperthermia system. IEEE Trans Microwave Theory Tech 34: 514–519

Stehlin JS, Giovanelia B, de Ipolyi PD, Anderson RT (1979) Eleven years experience with hyperthermic perfusion for melanoma of the extremities. Lond J Surg 3: 305–307

Combined Treatment of Advanced Bladder Carcinoma

K.-H. Bichler, S. H. Flüchter, W. D. Erdmann, W. L. Strohmaier, and J. Steimann

Abteilung für Urologie, Universität Tübingen, Calwerstraße 7, 7400 Tübingen, FRG

Introduction

Extended carcinomas of the bladder mostly run an unfavorable course. As a rule these tumors are resistent to radiation therapy. Intravenous administration of cytostatics promises success as described recently (Cant et al. 1985; Jakse et al. 1985). A concomitant disadvantage, however, is that the drugs in the high concentrations needed for treatment give rise to systemic toxic effects. Attempts to increase tissue concentration of cytostatics by performing a transarterial perfusion of carcinoma gave encouraging results (Aigner 1985). A further rise of cytostatic tissue concentration seems possible if a cytostatic microspheres (Spherex; Pharmacia, Freiburg) carcinoma infusion (CMCI) is performed synchronously. Injected intraarterially, the 45-μm microspheres of starch occlude tumor vessels and cause (a) an accumulation of cytostatics in the center of the tumor, but with low concentrations in the periphery, and (b) transient ischemia in the tumor, the starch molecules being degraded by the body's amylase (Davis et al. 1984; Flüchter et al. 1986, 1987). An additional activation of cytostatic effectivity must be assumed if adjunctive transurethral high-frequency hyperthermia (TUHH) of the bladder is employed (Fig. 1). This technique when used alone has a damaging effect on transitional cell carcinoma of the bladder (Bichler et al. 1982; Harzmann 1980). This paper describes first experiences of this combined treatment schedule in the field of urology.

Patients and Methods

Twelve patients with advanced carcinoma (11 with transitional cell carcinoma, one with adenocarcinoma) received CMCI (10 mg mitomycin C – except for the patient with adenocarcinoma, who received 50 mg cisplatin – plus 900 mg Spherex). TUHH followed 24 h later. In *six patients with regional carcinoma ($T_3/4N_0M_0$)* the treatment with mitomycin C was practised for curative purposes in one series of three courses within 2 weeks. One week later the tumor was removed surgically (cystectomy, $n=3$) or transurethrally ($n=3$) (Bichler et al. 1982). In five of *six patients with metastasising tumor ($T_3/4N_1M_1$)* treatment consisted of two series of three courses each using mitomycin C with a nontherapy interval of 3 weeks. Subsequently a diagnostic transurethral resection was performed. The aim of treatment was palliation. In the case of adenocarcinoma only two courses of cisplatin were given. Response to therapy was evaluated according to ob-

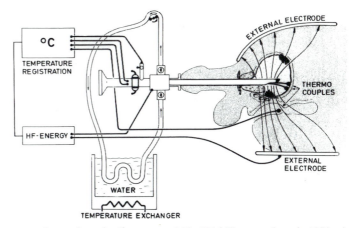

Fig. 1. Transurethral high-frequency hyperthermia (frequency 300–500 kH, wave length 1000 m) using an inner active electrode (24 F endoscope with irradiation zone and teflonised shaft and tip) and an external inactive electrode (belt-like squamous electrode of copper-tin-lead-nickel alloy tied around the hypogastrium). For thermometry and thermoregulation thermocouples were used. A 0.17% NaCl solution (which, like the bladder tissue, has a specific electrical conductivity of 3.5 mS/cm) served as bladder filling and cooling medium (39 °C). A continuous temperature of 43 °C can be achieved in the bladder wall for the duration (1 h) of treatment

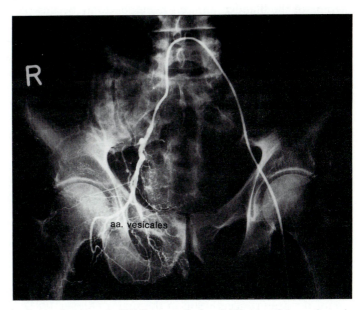

Fig. 2. Intraarterial CMCI: transfemoral approach (Seldinger technique). The tip of the catheter is to be inserted into the internal iliac artery downstream from the origin of the superior gluteal artery, so that most of the drug and the microspheres perfuse the bladder via the vesical arteries

jective criteria (CT, bone scan, ivp, X-ray tomography) before and 3 and 6 months after treatment.

Results

T3/4N0M0 Tumors. In one preparation after cystectomy and two after transurethral removal only tumor necrosis was found. In some others the tumor remained unaffected, and in others it was widespread regressively and necrotically altered. The perivesical fatty tissue showed constricted arteries, edema, fibrosis, and macrophages loaded with hemosiderin as evidence of effects of therapy.

T3/4N1M1 Tumors. Mostly after the first CMCI, the patients' general state of health improved and pain was alleviated or disappeared. In the patients with bone metastases and restriction of movement the ability to walk and move was improved. A stabilisation which could be measured objectively as a reduced tumor volume persisted for 12–34 weeks. The evident rapid progression of the tumor stopped immediately after the start of therapy. Five patients died from their tumor disease. Section demonstrated vast areas of extended necroses of the tumor treated.

Case Report. A 41-year-old man with urachal carcinoma, who had previously undergone partial bladder resection, extirpation of perivesical lymph node metastasis, radiation therapy of the bladder, and left hemilobectomy because of metastases, presented with extended local recurrence with infiltration of the sigmoid colon as well as metastases of the lung on both sides. Before the third cisplatin CMCI of the bladder was performed, a fistula of the sigmoid colon had developed. After cystectomy with an en bloc resection of the sigmoid, remains of the carcinoma could be found in a 3-mm area only. Systemic cisplatin therapy was followed by left lobectomy. In the right lobe of the lung 5 lentil-sized metastases were removed out of the unaffected tissue. Apart from vital carcinoma tissue, histologic study also revealed tumor necroses, but not to the same extent as in the bladder-sigmoid preparation which had been treated intraarterially as well.

Side Effects. In one patient peripheral ischialgic neuritis occurred and 6 weeks later a transitory suppression of erythropoiesis was seen. In two cases a maculate marmoration of the left gluteal area was seen after the third ipsilateral infusion. These findings have not yet completely receded.

Discussion

CMCI allows a direct cytostatic effect upon the tumor. The microspheres cause a further rise of the drug concentration in the tumor (Aigner 1985). There may – as seen in the treatment of liver tumors (Aigner 1985) – be an additive or a synergistic carcinoprive activity caused by the tumor ischemia induced by microspheres. In bladder carcinoma the TUHH causes group cell necrosis, later a subtotal tumor necrosis, and 3 months later stroma hyalinosis (Bichler et al. 1982; Harzmann 1980). The *destructive effect of the combination of CMCI and TUHH* is additive. Synergistic procedures, especially a further activation of cytostatic effects by adjunctive hyperthermia, have recently been under discussion. The results demonstrate that a curative treatment with CMCI and TUHH is possible for the nonmetastasising bladder carcinoma. Histological study of preparations after removal show that one may expect to find tumor necrosis or at least reduction of the tumor mass in all cases. Effects can even be proven in the perivesical fatty tissue. Before radical cystectomy, in case of carcinoma penetrating deeply into the bladder wall, CMCI and TUHH should always be applied with the aim of debulking or destroying the primary tumor and any perivesical micrometastases. It remains our hope that with this method of treatment the currently unsatisfactory survival rates for advanced bladder carcinoma can be improved. Patients with metastasising bladder carcinoma may expect palliation of symptoms.

References

Aigner KR (1985) Regionale Chemotherapie der Leber: Isolierte Perfusion, intraarterielle Infusion und Resektion. In: Eckhardt S, Holzner JK, Nagel GA (eds) Contributions to oncology. Karger, Basel

Bichler KH, Harzmann R, Flüchter SH, Erdmann WD (1982) Fortschritte der transurethralen Elektroresektion des Harnblasenkarzinoms. Urologe [A] 21: 3–8

Bichler KH, Harzmann R, Fastenmaier K, Flachenecker G, Altenähr R, Gericke D, Flüchter SH (1982) Ergebnisse der lokalen transurethralen Hochfrequenzhyperthermie beim Blasenkarzinom. Urologe [A] 21: 12–19

Cant JD, Brausi M, Soloway MS (1985) Adjuvant chemotherapy for locally advanced bladder cancer. World J Urol 3: 115–119

Davis SS, Illum L, McVie JG, Tomlinson E (1984) Microspheres and drug therapy: pharmaceutical, immunological and medical aspects. Elsevier, Amsterdam

Flüchter SH, Bichler KH, Walter E, Laberke HG, Müller-Schauenburg W, Nelde HJ, Rothe KF (1986) Intraarterielle synchrone Mikrosphären-Zytostatikainfusion urologischer Tumoren. In: Nagel GA, Sauer R, Schreiber HW (eds) Aktuelle Onkologie 28, Mitomycin '85. Zuckschwerdt, Munich, pp 172–184

Flüchter SH, Bichler KH, Laberke HG, Walter E, Müller-Schauenburg W, Schulz DJ (1987) Z Regionale Tumortherapie (in press)

Harzmann R (1980) Lokale Hochfrequenzhyperthermie beim Harnblasen-Karzinom. Urban und Schwarzenberg, Munich

Jakse G, Fritsch E, Frommhold H (1985) Combination of chemotherapy and irradiation for nonresectable bladder carcinoma. World J Urol 3: 121–129

Experience with an Annular Phased Array Hyperthermia System in the Treatment of Advanced Recurrences of the Pelvis

P. Steindorfer, R. Germann, and M. Klimpfinger

Klinik für Chirurgie, Medizinische Hochschule, Universität Graz, Auenbruggerplatz, 8036 Graz, Austria

Introduction

Within the last decade hyperthermia (HT) has come more and more into the domain of oncologists. The biological principles of action of heat in the treatment of different malignancies have been presented by many authors, mainly radiobiologists (Bichel et al. 1979; Jorritsma et al. 1985; Overgaard 1978). Not only the synergism of heat and ionizing radiation (Bichel et al. 1979; Jorritsma et al. 1985; Overgaard 1978; Short and Turner 1980; Steindorfer 1985), but also the possible enhancement of the cell-killing effect of different cytotoxic drugs by HT, have been demonstrated within many experimental studies (Dahl 1983; Dahl and Mella 1983; Hahn and Strande 1976; Hahn et al. 1975; Marmor 1969; Mella 1985; Overgaard 1976; Steindorfer 1985). The biologists investigated the effect of so-called moderate HT on normal and tumour tissue in vitro and in vivo. They found that HT between 41.5° C and 45° C had a lot of effects on the tumour cell, depending on exposure time and temperature level (Bichel et al. 1979; Short and Turner 1980; Steindorfer 1985). Metabolism, synthesis of nuclear acids and proteins, cell membranes, cytoskeleton and microsomes are all affected by HT (Steindorfer 1985). The sensitivity of different cells against heat depends on many factors, e.g. pH level of the tissue, oxygen radicals, blood flow rate, cell cycle and previous heat exposure (Short and Turner 1980; Steindorfer 1985). Cells which have been heated before, however long the interval, are able to resist heat for at least 120 h. This particular effect is called "thermotolerance" (TT). The so-called "heat shock proteins" (HSPs) are thought to be responsible for TT (Jorritsma et al. 1985; Li et al. 1982; Muller et al. 1985; Steindorfer 1985). These proteins would better be termed "stress proteins", because they are released not only by heat stress, but also by various other agents such as alcohol, different cytotoxics, and heavy metals (Steindorfer 1985). While the mechanism of TT protects the heated cells against a repeated heat stress, the HSPs are detectable within the cell culture by immunofluorescence. The more rapid the increase of temperature, the higher the temperature, and the longer the time of exposure, the lower are the numbers of thermotolerant cells found within the cultured tumour cell lines. These biological investigations provided the rationale for initiating several clinical pilot studies of the use of HT in combination with radiotherapy (RTX) and/or chemotherapy (CTX) in treatment of advanced malignant disease.

Recent Results in Cancer Research, Vol. 107
© Springer-Verlag Berlin · Heidelberg 1988

Technical Aspects of Regional Hyperthermia

Let us start by attempting to define HT for clinical practice, as in our experience there is a difference between the in vitro and the in vivo situation: HT is an externally induced temperature of more than 41.5° C, within a living organism or a part of it, where the physiological temperature regulation of the body is partially overcome with the aim of achieving a therapeutic effect.

Using this definition, one should plan a transfer of energy to the tumour as localized as possible. However, one will encounter a lot of parameters influencing the temperature level within the desired region. To give some slight idea of how complex and difficult it is to carry out a proper therapy within a tumour area in human beings, we make a few mathematical side-steps. To describe the thermal situation of the individual or of parts of it, one should try to consider all the biological parameters influencing the temperature distribution (Short and Turner 1980). The following "bio-heat equation" stands for one-dimensional transfer of energy to a homogeneous tissue:

$$\frac{d(\Delta T)}{dT} = \frac{1}{C}(Wa + Wm - Wc - Wb)$$

where T is tissue temperature in $°$ K, C is specific heat of the tissue in J/kg/K, Wa is energy absorbed by the tissue, Wm is heat energy produced by metabolism, Wc is loss of energy by convection and Wb is loss of energy by blood flow. For a lot of mathematical models of temperature distribution of living tissue under HT conditions, the part $Wm - Wc - Wb = 0$, so the equation is then

$$\frac{d(\Delta T)}{dT} = \frac{Wa}{C} \quad \text{or} \quad \frac{d(\Delta T)}{dT} = \frac{SAR}{C}$$

where SAR is the "specific absorption rate" of energy within the different tissues. Most of the phantom studies were based on this very simplified equation, which permits a very precise model of the source of energy and the phantom material, but phantoms cannot be compared with patients (Short and Turner 1980; Turner 1984). The main problem in getting enough information about the temperature distribution in the heated area within the patient is the difficulty of temperature measurement in deep-seated tumours. The "thermal mapping procedure" is a very proper and adequate method for superficial tumours, e.g. those within the head and neck region (Gibbs 1983). In the Annular Phased Array System (described below) the mapping procedure is not very easy to handle and depends on the number of probes inserted within the tumour area. The frequency of probes being held in the desired location during the whole therapy is limited, although in a certain number of cases we inserted the probes during surgery.

Design of the Study

Only pretreated patients with advanced recurrences of the pelvis after surgery, RTX and/or CTX were entered in the study. All patients had a rectal cancer recurrence – we did not want to mix different tumour types because of differences in biological reaction to different treatment modalities. This study was also designed as a pilot study, so we selected two different patient groups:

Group A: All patients had been irradiated previously and had a minimal residual RTX dose of at least 2000 R, which is an inadequate dose in such large tumours. All patients had previous radical or palliative surgery, and a few of them had also received CTX, mainly systemic 5-fluorouracil (5-FU).

Group B: Patients with a residual dose lower than 2000 R, a few of them with previous CTX. All of them had previous surgery, as in group A.

Before starting the therapy, all patients were investigated by CT scan, ECG, BCC, renal function test, X-ray of the lungs, and ultrasound of the liver. The performance status of the patients was more than 50% according to Karnofsky's index scale. Distant metastases were excluded before starting therapy. Tumour volume was estimated from CT scan files of the pelvic region.

The treatment plan in group A was a sequential combination of regional HT (RHT) with high-voltage RTX at a daily fraction of 200 R. RHT was given twice a week for at least 40 min at a temperature level of about 42° C within the tumour area. For the RHT we are using the BSD 1000 hyperthermia device with 16 microwave antennas, so-called applicators, in a phased array. These antennas are connected to water boluses filled with deionized water to achieve better coupling to the surface of the patient's body. The input energy is controlled by a computer which is directed by the inserted Bowman thermistor probes (Bowman 1976). This system, the Annular Phased Array System (APAS), is widely used for RHT (Oleson et al. 1986; Short and Turner 1980; Turner 1984). The average frequency we used during the RHT treatment ranged between 60 and 70 MHz, the average energy used being about 1000 W.

The treatment protocol of group B was quite different regarding the systemic therapy. We combined the RHT with 5-FU, the cytotoxic drug used most commonly in colorectal cancer. RHT was given twice a week as in group A, but during each HT treatment the patient received 1000 mg 5-FU intravenously. We started the infusion about 1 h before the HT procedure and continued during the whole therapy. On the days without HT the patients received 750 mg 5-FU orally.

During the whole RHT procedure the patients were supervised by intensive care nurses and physicians. Blood pressure, heart rate, core temperature and subjective sensations of the patients were monitored continuously.

After the end of treatment another CT scan was done to look for necrosis and reduction of tumour volume. Nevertheless it is very difficult to evaluate the response in non-measurable disease like recurrences of the pelvic area. Thus we decided to perform second- or third-look operations after finishing HT. Unfortunately we can present only four cases of second-look surgery, but the histopathological findings and the local situation were of interest.

Results

When we started using the APAS in HT treatment in 1983, we emphasized CT scan as a remission control. We have since done second-look operations in four patients and have to view the results more critically. Thus we cannot evaluate the remission rates in our series objectively, except by surgery and histopathology. Even CT scan in combination with serological markers (CEA, AP) and needle biopsies is not able to produce any objective remission criteria, because of the difficulties in discriminating between necrotic tissue and vital tumour. To evaluate our response, we used the guidelines of the UICC for non-superficial tumours.

Table 1. Remission rates

	CR	PR	NC	PD	NE
Group A HT+RTX (20–32 Gy)	0	4	3	2	1
Group B HT+CTX (1000 mg/750 mg 5-FU)	1 (4)[a]	6	2	2	2

CR, complete remission; *PR*, partial remission; *NC*, no change; *PD*, progressive disease; *NE*, not examined.
[a] Four patients are in CR after surgery, but if only the HT treatment is taken into consideration there was only one case of CR because a few viable tumour cells were found within the resection material after histological investigation in three of the four patients.

Remission Rates

Remission rates are shown in Table 1.

Duration of Response

Due to the advanced nature of the disease in the patients we treated, a significant increase of survival cannot be expected. The follow-up time is also too short and the number of patients too small to evaluate the influence of RHT on survival, but there is an improvement of survival if one takes into account that in these patients, salvage regimens and symptomatic pain relief constitute the only therapy. The mean follow-up time of our patients is at present 23 ± 14 months.

Technical Aspects of Remission

The tumour temperature values reproduced here represent not the centre of the tumour, but the average temperature of all inserted probes. Analysing the average energy used (Fig. 1) and comparing it with the remission rates, we found a significant correlation, but

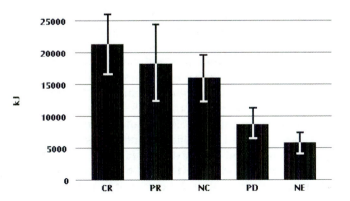

Fig. 1. APAS treatment 1983–1986: energy

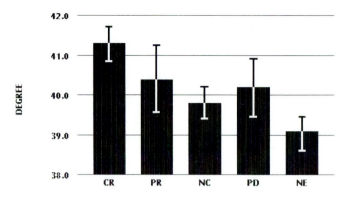

Fig. 2. APAS treatment
1983–1986: temperature

this was not the case for the temperature (Fig. 2). We reached the desired temperature in nearly 80% of the treated cases, but the temperature changes from treatment to treatment, so we could measure the temperature very precisely only at one spot, and this may explain the results. Thus, objective treatment planning before starting RHT is nearly impossible.

Surgical Aspects of Remission and Case Reports

Four patients underwent second- or multiple-look operation after finishing the HT treatment.

The first patient had a 7×8 cm recurrent tumour mass within the presacral region with invasion of the apex of the urinary bladder. His pretreatment consisted of surgery, RTX 57 Gy and one course of 5-FU into the internal iliac artery. He was treated in group B with 15 single HT treatments. Four weeks after the end of HT treatment we resected his pelvic mass and performed histological investigation of the resected specimen. He is at this time still in remission after 38 months follow-up. The histological investigation showed a subtotal necrosis or necrobiosis of the tumour but nevertheless, within the resected area, viable tumour cells.

The second patient, a 46-year-old male, was referred to our clinic with a persistent vesicointestinal fistula due to tumour infiltration with recurrent infections of the urinary tract. He had already received external high-voltage RTX and interstitial radioimplants. He was also treated in group B and received 12 single HT treatments in combination with 5-FU. Five weeks after the end of HT treatment, we resected the tumour-bearing organs: urinary bladder, prostate, rectosigmoid, part of ileum and ascending colon. Again, histomorphological investigation of the resected tumour showed subtotal necrobiosis. Only a few vital tumour cells within the treated area could be demonstrated. There was no evidence of ineffective temperature distribution within the measured areas, but we did not have the possibility of investigating the resected specimen histologically regarding temperature distribution.

The third patient was a 58-year-old female with a presacral recurrent disease of a rectal cancer. She had a cutaneous tumour fistula within the perineal region and had already been treated with external RTX of the pelvic region. She was also treated in group B and had 14 single HT treatments in combination with 5-FU. After finishing the HT treatment the patient did not want to undergo surgery immediately, and we did the second-look operation half a year after the end of HT. There was no evidence of tumour cells within the resected specimen, but fibrosis of the tissue close to the previous tumour-bearing site. This patient is still in complete remission.

The fourth patient had undergone resection in another hospital 3 months after RHT. The histological investigation was performed there and their pathologists, and could like our own, found only a few tumour cells within a necrobiotic area.

Histological Aspects of Response

After second-look operation we performed histopathological investigation of the resected specimen. The tumour was embedded in paraffin after fixation with formaldehyde, then stained with haematoxylin-eosin. Within the investigated sections the tumour showed mainly necrotic or necrobiotic areas, but we also found intact tumour formations. The question of whether the tumour cells have any biological activity can be answered only by further immunohistochemical and immunohistological investigations. Thus nobody can definitely assess the tumour response objectively without histological investigation of the pretreated area. There is currently no non-invasive technique of checking tumour response objectively. Not even NMR yields information on response criteria like in measurable disease. That is why we could find only one patient with a complete response in our treated series. According to the guidelines of the UICC, the patients could be evaluated only as partial remissions. Nevertheless, RHT is effective for deep-seated tumours; one can achieve tumour regression according to the effectiveness of temperature distribution within the treated area. There is no possibility of correlation between temperature and histopathological findings in deep-seated human tumours after RHT. The objective criterion of effectiveness of therapy should be the aim of our future investigation of clinical use of RHT. Heavily pretreated patients with large tumours should not be the best marker of effectiveness of a new treatment modality.

Conclusions and Questions

Why Was Tumour Regression No Better? Was It Due to the Treatment Setup?

These questions should be set to physicists with the aim of improving the quality of deep-heating and temperature-monitoring devices. The thermometry is very difficult, and the new multipoint fiberoptic technique should give us more and detailed information on the treated area. The results should be able to be checked much better than now.

The mapping procedure within the APAS is very time-consuming and difficult without the mapping device. We tried to do the procedure at the end of each single HT treatment, but it was very difficult to reset the Bowman probe into its previous position, so we decided to do no mapping without the necessary equipment. Thus it is possible that the results are related to ineffective temperature distribution within parts of the treated tumours, and the aim of future treatment setups is to get more detailed information of the treated area.

Are the Results Related to the Negative Selection of Patients?

Far advanced and heavily pretreated patients are without any doubt a difficult marker for the effectiveness of RHT, but so far nobody has been able to present a proper clinical deep-heating protocol which takes in account all the experimental considerations of RHT. Thus more clinical studies are needed to get better information about the treatment parameters. We correlated our results with temperature distribution and average energy consumption and found a significant relationship only with the energy used. The temperature distribution showed from one single treatment to another a discrepancy of some degrees between the centre and the periphery of the tumours. It is well known that large tumours respond better to HT than small ones, due to the poor vascularization and low pH levels of the necrotic centre. The tumour bed in the close vincinity to normal tissue does not show this sensitivity, due to increased blood flow during RHT. Despite these known facts, we found vital tumour cells within the necrotic area and not close to the tumour bed. This raises more questions: Are these cells resistant to heat because of thermotolerance? Or were they heated inadequately? Or does the synergism of the cytotoxic drug with RHT not apply to poorly vascularized parts of the tumour parts due to low drug concentrations? Or are these cells resistant to chemotherapy too?

Is There a Synergistic Effect of RHT with RTX?

This question has already been answered for superficial tumours by many clinical studies, but the evaluation of response in deep tumours is much more difficult. As mentioned before, the tumour response at deep-seated locations such as the pelvic region is very hard to assess. It is very difficult to talk about response without having any objective data about the tumour control. To check the synergistic effect of HT with RTX objectively without using invasive methods like surgery is only possible for superficial tumours. In our small series there was no significant difference between group A and group B, if we take into account the histopathological findings in the resected tumours of group B (see Fig. 3). Nevertheless, it should be possible to improve the results of the RTX group combining RHT and RTX in primary treatment, with higher doses of radiation, and resecting the tumour after HT and RTX, as a kind of neo-adjuvant protocol of HT and radiation.

Is There Any Evidence of Synergism in Clinical Use of Thermochemotherapy?

We have to take into account that there is little knowledge about the synergistic effects of different cytotoxics and HT in clinical use. That is why we suggest using only well-known regimens in combination with HT, in order to prevent errors and pitfalls in response interpretation. The objective assessment of tumour response by means of multiple-look operations after finishing HT is a major and risky procedure and needs an experienced surgical oncologist. However, without doubt, the histological evaluation gives the only objective information on the tumour response. We performed all four second-look operations in group B and found a fairly good response in these heavily pretreated tumours. The figures are far too small to allow any conclusions, but nevertheless there are some signs that synergistic effects may be taken into account when using cytotoxics in combination with RHT. In future, we should consider using thermochemother-

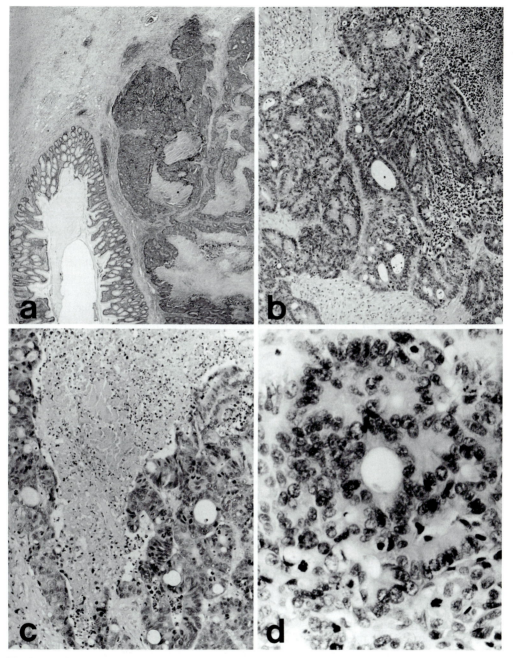

Fig. 3a–d. Tumour recurrence of an adenocarcinoma of the rectum: microscopic appearance after pretreatment, hyperthermia and resection. **a** Invasive tumour formations in the neighbourhood of non-neoplastic mucosa, ×16. **b** Moderately well differentiated adenocarcinoma with large necrotic parts, ×80. **c** As **b,** ×128. **d** Intact tumour formations out of the tumour bed, ×640. Formol fixation, paraffin embedding, haematoxylin-eosin staining

apy perioperatively, in so-called neo-adjuvant protocols in primary treatment, to get more and more detailed information on tumour response. Only single agents or established cytotoxic regimens should be used in combination with HT. Randomized trials should be started only after careful consideration and under strict supervision, and should be carried out with the same HT device.

Is the RHT Procedure Tolerable for the Patient?

In our very early series of single treatments we saw a necrosis of the mucosa of the urinary bladder in one patient, although we drained the bladder by catheter during the whole treatment period. After this experience we decided to cool the bladder during the RHT procedure by means of iced Ringer's solution. We had no more side effects which caused an interruption of the treatment, but there is strong evidence of uncomfortable sensations and sometimes hot spots within the treated areas without any burns or damage to the normal tissue. Follow-up of our patients has not revealed any evidence for damage to normal tissue due to RHT treatment, except the one case of mucosal necrosis of the urinary bladder mentioned above. None of the patients had such severe side effects during RHT therapy that the treatment had to be interrupted. We gave no RHT to the upper abdomen at that time because we wanted to collect more data on the pelvic area before we start with the more difficult area of the liver and pancreas.

What Are the Main Technical Problems During the RHT Therapy Within the APAS?

Obtaining information about the temperature distribution within the tumour during RHT is one of the main problems using the APAS. We did not have the temperature-mapping device and so our information about temperature is related to the number of probes inserted in the tumour sites. The thermal mapping procedure is nearly impossible to perform without the mapping device, including the software. Even with the device, thermal mapping is a complicated procedure and prone to many pitfalls. Better monitoring could be achieved using a multisensor temperature probe, which gives more detailed information of the different compartments of the tumour without using complicated stepper motors to move a single-point probe (Gibbs 1983; van der Zee et al. 1986).

The next and most time-consuming problem in using the APAS is the repositioning of the patient, if the E-field measurements make a readjustment necessary. The three-bolus system makes the readjustment of patients within the APAS very lengthy and complicated for the patient. It would be useful to develop a better water bolus system, perhaps one similar to that employed in the MAPAS (mini annular phased array for limb treatments; BSD Medical Corp., Salt Lake City).

The complication mentioned above, the case of mucosal necrosis of the urinary bladder, was due to the high thermal sensitivity of the urothelial epithelium. That is why we decided to cool the bladder with iced Ringer's solution according to the temperature measured in the bladder by our "core temperature" probe, which breaks off energy at a bladder temperature of more than 39.5° C. After that we had no more complications within the bladder.

To sum up the technical problems, we can say that the RHT treatment within the APAS is a time-consuming and complicated procedure and should be considered very carefully before being offered to tumour patients. RHT is effective, but care should be taken with the setup and with patient selection. More effort should be invested in development of better temperature-measurement devices. More serious data should be gathered in clinical use of RHT in order to prevent charges that the procedures lacks scientific foundation.

References

Bichel P, Overgaard J, Nielsen OS (1979) Synergistic cell cycle kinetic effect of low doses of hyperthermia and radiation on tumor cells. Eur J Cancer 15: 1191

Bowmann RR (1976) A probe for measuring temperature in radio frequency-heated material. IEEE Trans Microwave Theory Tech 24: 43

Dahl O (1983) Hyperthermic potentiation of doxorubicin in a transplantable neurogenic rat tumor (BT$_4$A) in BDIX rats. Int J Radiat Oncol Biol Phys 9: 203

Dahl O, Mella O (1983) Effect of timing and sequence of hyperthermia and cyclosphosphamide on a neurogenic rat tumor BT4a in vivo. Cancer 52: 983-998

Gibbs FA (1983) Thermal mapping in experimental cancer treatment with hyperthermia: description and use of a semiautomatic system. Int J Radiat Oncol Biol Phys

Hahn GM, Strande DP (1976) Cytotoxic effects of hyperthermia and adriamycin on Chinese hamster cells. INCI 57: 1063

Hahn GM, Braun J, Har-Kedar I (1975) Thermochemotherapy: synergism between hyperthermia (42-43°) and adriamycin (or bleomycin) in mammalian cell inactivation. Proc Natl Acad Sci USA 72: 937

Jorritsma JBM, Kampinga HH, Scaf AHJ, Konings AWT (1985) Strand break repair, DNA polymerase activity and heat radiosensitization in thermotolerant cells. Int J Hyperthermia 1: 131-147

Li GC, Petersen NS, Mitchel HK (1982) Induced thermal tolerance and heat shock protein synthesis in Chinese hamster ovary cells. Int J Radiat Oncol Biol Phys 8: 63

Marmor JB (1969) Interactions of hyperthermia and chemotherapy in animals. Cancer Res 39: 2269

Mella A (1985) Combined hyperthermia and cis-diamminedichloroplatinum in BDIX rats with transplanted BT4A tumors. Int J Hyperthermia 1: 171-185

Muller WU, Li GC, Goldstein LS (1985) Heat does not induce synthesis of heat shock proteins or thermotolerance in the earliest stage of mouse embryo development. Int J Hyperthermia 1 (1): 97-103

Oleson JR, Sim DA, Conrad J, Fletscher AM, Gross EJ (1986) Results of a phase I regional hyperthermia device evaluation: microwave annular array versus radiofrequency induction coil. Int J Hyperthermia 2 (4): 327-336

Overgaard J (1976) Combined adriamycin and hyperthermia treatment of a murine mammary carcinoma in vivo. Cancer Res 36: 3077

Overgaard J (1978) The effect of local hyperthermia alone, and in combination with radiation, on solid tumors. In: Streffer C, et al. (eds) Cancer therapy by hyperthermia and radiation. Urban and Schwarzenberg, Munich, p 49

Short GJ, Turner PF (1980) Physical hyperthermia and cancer therapy. IEEE Proc 68 (1): 133

Steindorfer P (1985) Hyperthermie und Chemotherapie solider Tumore (Experimentelle Doppelblindstudie an transplantierten 5123-Morris-Hepatomen). Acta Chir Austriaca [Suppl] 63

Turner PF (1984) Hyperthermia and inhomogenous tissue effects using an annular phased array. IEEE Trans Microwave Theory Tech 32 (8): 874

Van der Zee J, van Putten WLJ, van den Berg AP, van Thoon GC, et al. (1986) Retrospective analysis of the response of tumours in patients treated with a combination of radiotherapy and hyperthermia. Int J Hyperthermia 2 (4): 337-350

Regional Hyperthermia Combined with Systemic Chemotherapy in Advanced Abdominal and Pelvic Tumors: First Results of a Pilot Study Employing an Annular Phased Array Applicator*

R. D. Issels[1, 2], M. Wadepohl[1], K. Tiling[1], M. Müller[1], H. Sauer[1], and W. Wilmanns[1, 2]

[1] Medizinische Klinik III, Klinikum Großhadern, Ludwig-Maximilians-Universität, Marchioninistraße 15, 8000 München 70, FRG
[2] Institut für klinische Hämatologie, Gesellschaft für Strahlen- und Umweltforschung, Landwehrstraße 61, 8000 München, FRG

Introduction

The clinical use of an annular phased array system of electromagnetic wave applicators for regional heating of deep-seated tumors has been previously reported (Sapozink et al. 1984, 1986; Emami et al. 1984; Gibbs et al. 1984; Howard et al. 1986). In these studies the patients received regional hyperthermia combined with radiotherapy. The rationale for this combined treatment modality is based on biological data in vitro and in vivo indicating a synergistic interaction between heat and radiation (Dewey 1984). In clinical studies on superficial human tumors, combined hyperthermia and radiotherapy has been shown to be superior to either modality alone in achieving complete responses (Overgaard 1985).

There is also a strong potential for the combination of heat and chemotherapy (Hahn 1982). Most clinical reports published on thermochemotherapy deal with whole-body hyperthermia (Bull 1984) or hyperthermic limb perfusion (Stehlin et al. 1979). However, very few clinical data are available on regional hyperthermia of deep-seated tumors combined with systemic chemotherapy. The current report details our first experience using such an approach to treat malignancy in the abdomen and pelvis with respect to the feasibility of heating and the observed toxicity of the combined heat and drug treatment.

Materials and Methods

Patient Selection. From July to December 1986, 12 patients with advanced deep-seated tumors of the abdomen (4/12) and pelvis (8/12) were treated (37 treatments in total) with combined regional hyperthermia and chemotherapy. Before entering the study, the patients showed progressive disease and conventional treatment protocols had failed. The median age of the population (6 male, 6 female) was 47 years (range 14–69). The results of histological examinations of the tumors are given in Table 1. The majority of pat-

* This work was supported by grants M 12/85 Wi 1, M 41/85 Wi 2, and M 10/86 Wi 3 from the Deutsche Krebshilfe. Part of this work was presented at the 8th International Congress of Radiation Research, Edinburgh, United Kingdom, July 1987.

Table 1. Histological diagnosis of the tumors

Diagnosis	No.
Soft-tissue sarcoma	7
Adenocarcinoma	4
Teratocarcinoma	1
Total	12

ients had extensive recurrences of pelvic or abdominal tumors with metastatic disease. All patients had undergone at least one previous laparatomy and had been heavily pretreated with radiation and/or chemotherapy. The patients were advised of the experimental nature of the treatment procedure and informed consent was obtained in all cases. For patients considered for this protocol the following were found to be essential: clinical examination, laboratory tests, ECG, chest radiography, CT scan, central venous catheter, epidural catheter, implantation of catheters for thermal mapping, and control by X-ray and CT scan.

Patients with cardiac arrhythmias, congestive heart failure, ischemic heart disease, or Karnofsky status <40% were excluded in view of the cardiovascular stress observed during treatment. For the application of chemotherapy, fluid regimens, and monitoring of central venous pressure a central venous catheter was inserted. In some cases, an epidural blockade was used for deep implantations of thermometry catheters.

Hyperthermia. All patients were treated with the BSD-1000 system and the annular phased array (APA) applicator (BSD Medical Corp., Salt Lake City, USA). The technical characteristics of the system have been described by Turner (1984). The APA consists of 16 radiating apertures arranged in two adjacent rings of eight applicators. The operating frequency used was in the range of 60–80 MHz to obtain deep penetration. The operational details of this system, including patient positioning, surface and systemic cooling, and surface electric fields, have been extensively described (Gibbs et al. 1984).

As an illustration, the APA set-up in the shielded treatment room is shown in Fig. 1. In the APA, patients were in a supine position with their body axis parallel to the long axis of the APA. The APA is powered by four coaxial inputs, each activating a quadrant of the array with power fed from a radio frequency generator through a 2 kW amplifier. All applicators are fed in phase and with the same amplitude. The dimensions of the octagonal aperture are such that the electric field at the center is predominantly polarized along the cylindrical axis of the body. Therefore, the sum of the electric fields from each applicator leads to preferential heating in this central region. For eccentrically located tumors, one to three quadrants of the applicator were disconnected from the power source to achieve a shift of the heating pattern. The treatment goal was to achieve a tumor temperature higher than 42.5° C for a period of 30–60 min. The applied power was reduced if normal tissue exceeded 43° C or tumor tissue exceeded 45° C.

Thermometry. Temperatures were measured with nonperturbing high-resistance thermistors (Bowman 1976) with an accuracy of ±0.1° C. At least 1 day before treatment, Teflon thermometry catheters (Pflugbeil, Munich, FRG) were placed into normal and tu-

Fig. 1. The annular phased array (APA) applicator of the BSD-1000 system for regional heating of deep-seated tumors (Klinikum Großhadern, University of Munich)

mor tissues with X-ray guidance under local and/or epidural anesthesia, followed by CT check-up scan. In some cases the thermometry catheters (Braun, Melsungen, FRG) were inserted during a laparatomy. These catheters were kept in place over the whole treatment period. At the time hyperthermia was started, additional catheters for temperature measurement were inserted in the rectum, vagina, and urinary bladder, and thermistor probes were introduced in the lumen of all catheters. The temperature measurements in normal and tumor tissues were mapped at fixed linear intervals (0.5–1.0 cm) along the thermometry catheters using the thermal mapping system as described by Gibbs et al. (1984). This set of measurements was repeated every 5 min during hyperthermia in order to maximize the amount of information about temperature gradients and their changes in the heated region. Systemic temperatures were intermittently measured outside the directly heated volume using an oral probe. The temperature (T) data were stored on a floppy disk and displayed as $T(°C)$ versus time (min) or $T(°C)$ versus depth (cm) plots. The temperature data were related to the localization of catheter and tumor taken from CT scans and were used for calculations of the temperature distribution in the heated tissues.

Chemotherapy. All patients treated with regional hyperthermia received systemic chemotherapy simultaneously. The major drugs used in combination with hyperthermia are listed in Table 2. Multidrug cycles of chemotherapy were given, with treatments separated by a rest period of 3–4 weeks depending on the recovery from bone marrow toxicity. Patients with soft-tissue sarcoma ($n=7$) received ifosfamide 1500 mg/m^2 (days 1–5) and etoposide 80 mg/m^2 (days 1, 3, 5) with hyperthermia given only on days 1 and 5. Alternatively, cycles of 4′-epi-adriamycin 50 mg/m^2 (days 1 and 2) and vindesine 2 mg/m^2 (days 1 and 2) or cisplatin 50 mg/m^2 (day 1) combined with hyperthermia were used for some of these patients. Patients with colorectal cancer ($n=3$) received 5-fluorouracil 700 mg/m^2 combined with hyperthermia on one day every 3 weeks.

Table 2. Drugs used in combination with hyperthermia and number of times each was used

Ifosfamide (IFO)	19
Etoposide (VP-16)	12
Cisplatin (Cis-Pt)	13
Vindesine (VDS)	10
4′-Epi-adriamycin (4′-EPI-ADR)	5
5-Fluorouracil (5-FU)	4

In general, the patients received infusions (e.g., 0.9% sodium chloride, 5% glucose) through a central venous catheter (6 l/24 h) at least 12 h before heat and drug treatment. The following day, chemotherapy was given as a 1 h i.v. infusion which was started as soon as the tumor target had reached the desired temperature of 42.5° C. Vital signs (e.g., blood pressure, pulse rate, CVP) were monitored carefully during each hyperthermia session and the patients were kept under intensive care for at least 2 h after the end of the treatment.

Results

Hyperthermia. Using the APA of the BSD-1000 system, 37 treatments were given to 12 patients, representing an average of 3.1 treatments per patient. A total of 12 thermochemotherapy sessions were carried out in four patients with abdominal tumors and a total of 25 in eight patients with pelvic tumors. The details of patient and treatment data are given in Table 3 for abdominal tumors and in Table 4 for pelvic tumors.

The heat treatment parameters of both groups are shown in Table 5. In order to evaluate the feasibility of heating during each treatment, the maximum temperatures achieved within the tumor are listed in four different categories. This approach primarily assesses the technical ability of the APA to heat at depth. As shown in Table 5, tumor temperatures higher than 42.5° C were only reached in 2 out of 12 of the treatments of abdominal tumors. However, in 68% of the treatments of pelvic tumors (17/25), the intratumor temperature exceeded 42.5° C. We further analyzed the duration of time for which the tumor temperature was kept above this level. In contrast to the rather disappointing data from the treatments of abdominal tumors, in 48% of the treatments of pelvic tumors (12/25) temperatures $>42.5°$ C were achievable for 31–60 min. With regard to the temperature distribution in pelvic tumors the minimum temperatures during the heat treatment were at least higher than 40° C (see Table 5).

Toxicity. Acute toxicity was defined as occurring within the hyperthermia treatment period. The symptoms of acute toxicity observed in 12 treatments of abdominal tumors and 25 treatments of pelvic tumors are listed in Table 6. Pain within the applicator field (58%) was the major limiting factor for carrying out hyperthermia sessions in the treatment of abdominal tumors. Although pain was commonly observed, it mainly occurred during the heating period ($=0$–30 min) and was controlled by changing the patient's position or by shifting the frequency. However, local pain in the upper thigh and abdominal regions could often only be eliminated by reducing the applied power. Mild tachycardia (>100/min) and extreme anxiety were frequently observed at the same time with no significant increase of systemic temperature in these patients. In treatment of pelvic

Table 3. Patient and treatment data for abdominal tumors

Case	Age (years)	Sex (M/F)	Karnofsky status	Tumor volume (cm³)	No. of heat treatments/ No. of days	Histology	Chemotherapy during heat treatment Drugs	Amount (mg)	Type of pretreatment
1	53	M	40	3690	2/71	Neurofibro-sarcoma	4'-EPI-ADR	210	S/R/C
							VDS	8	
2	58	M	40	1220	2/ 5	Liposarcoma	IFO/VP-16	5500/240	S
3	47	F	50	420	4/43	Liposarcoma	IFO/VP-16	8000/420	S/R/C
4	36	F	50	210	4/31	Leiomyo-sarcoma	IFO/VP-16	10000/800	S/R/C

S, surgery; R, radiation; C, chemotherapy.
Drugs: IFO, ifosfamide; VP-16, etoposide; Cis-Pt, cisplatin; VDS, vindesine; 4'-EPI-ADR, 4'-epi-adriamycin; 5-FU, 5-fluorouracil; DTIC, dacarbazine; BCNU, carmustine; MTX, methotrexate.

Table 4. Patient and treatment data for pelvic tumors

Case	Age (years)	Sex (M/F)	Karnofsky status	Tumor volume (cm³)	No. of heat treatments/ No. of days	Histology	Chemotherapy during heat treatment Drugs	Amount (mg)	Type of pretreatment
5	51	M	50	840	1/ 1	Fibrosarcoma	4'-EPI-ADR	90	C
							VDS	4	
6	64	M	40	280	1/ 1	Adeno-carcinoma (rectum)	5-FU/ DTIC VCR/ BCNU	800/260 2/100	S/R/C
7	26	F	40	230	2/ 4	Adeno-carcinoma (cervix)	Cis-Pt	80	R/C
8	46	F	50	4930	7/72	Terato-carcinoma	IFO/ Cis-Pt	9500/300	C
9	17	M	40	230	6/91	Rhabdomyo-sarcoma	IFO/ VP-16	5000/320	S/R/C
10	14	F	70	80	5/68	Rhabdomyo-sarcoma	4'-EPI-ADR/VDS MTX/Cis-Pt	150/ 12 50/100	S/R/C
11	64	F	50	110	2/22	Adeno-carcinoma (rectum)	5-FU	2000	S/R/C
12	68	M	50	2160	1/ 1	Colorectal carcinoma	5-FU	1100	S

For abbreviations, see footnote to Table 3.

Table 5. Heat treatment parameters

Temperature (° C)	Abdominal		Pelvic		1–30 min		31–60 min	
	No.	%	No.	%	Abdominal	Pelvic	Abdominal	Pelvic
37–40	2/12	17	–	–	1/12	–	1/12	–
40–41.5	6/12	50	5/25	20	4/12	2/25	2/12	3/25
41.5–42.5	2/12	17	3/25	12	1/12	3/25	1/12	–
> 42.5	2/12	17	17/25	68	1/12	5/25	1/12	12/25

No. of abdominal treatments 12, of pelvic treatments 25, total no. of treatments 37.

Table 6. Acute toxicity, defined as occurring within the treatment period

	Abdominal		Pelvic	
	No.	%	No.	%
Pain within the field of the applicator (power dependent)	7/12	58	11/25	44
Systemic temperature > 39° C	1/12	8	2/25	8
Tachycardia > 120/min	–	–	5/25	20
Increase of central venous pressure > 3 cmH$_2$O	3/12	25	8/25	32
Vomiting	–		4/25	16
Arrhythmia	–		1/25	4
Stenocardia	1/12	8	–	

Total no. of treatments 37.

Table 7. Subacute toxicity, defined as beginning within 0–12 h after treatment and persisting for < 24 h

	Abdominal		Pelvic	
	No.	%	No.	%
Vomiting	5/12	42	8/25	32
"Chills"	3/12	25	7/25	28
Systemic temperature > 38° C	4/12	33	7/25	28
Tachycardia > 100/min	5/12	42	12/25	48

Total no. of treatments 37.

tumors, which has a much higher power level, pain was less often observed (44%). Severe tachycardia (> 120/min) occurred in 20% of these patients and, in general, the cardiovascular stress was well tolerated.

Subacute toxicity was defined as beginning within 0–12 h after the end of hyperthermia treatment and persisting for < 24 h thereafter (Table 7). In about one-third of our patients we observed a sudden rise of systemic temperature and pulse rate, and chills within a short time of hyperthermia. Although these symptoms abated with supportive therapy, the etiology of this syndrome (posthyperthermia stress syndrome; PHSS) remains unclear. Screening of serum creatine kinase, lactate dehydrogenase, and serum electrolytes in order to correlate PHSS with thermal injury to normal and/or tumor tissue was further investigated (Issels, Tiling, and Mueller, unpublished results).

In all 37 treatments, the concurrent application of chemotherapy during regional hyperthermia was well tolerated. There was no evidence that hyperthermia increased the onset, duration, or frequency of common acute side effects of the drugs used at the doses and time intervals described. In the four treatments of pelvic tumors in which vomiting was observed (16%), this symptom occurred during the heating-up period and was not drug related. Complications were uncommon, the most serious being a superficial second-degree burn in a patient with rectal carcinoma. In three cases we observed local skin infections at the site of insertion of the thermometry catheters, and in another three cases urinary tract infection was noted as a result of repeated urinary catheterism. All of these complications were resolved with medical management.

Discussion

Noninvasive electromagnetic heating of deep-seated tumors is currently limited to unfocused devices. Several investigators using the APA of the BSD-1000 system have shown that differential heating of tumor and normal tissue was achievable and that toxicity was tolerable (National Cancer Institute 1987).

Our clinical observations of the first 37 treatments on 12 patients in combination with chemotherapy confirm that the APA/applicator is a regional heating device with potential for heating deep-seated tumors of the pelvic region. Of the 25 treatments of pelvic tumors, temperatures in excess of 42.5° C were reached in 68% and maintained for a prolonged period in 48%. Although a high frequency of acute toxicity was observed, the complication rate was quite low, and the complications observed were usually reversible.

The simultaneous application of systemic chemotherapy was well tolerated by the patients, who showed no further increase of common systemic side effects. Due to the preliminary nature of this study and the briefness of the follow-up, no conclusions can be drawn concerning the response of the tumors to this combined treatment. From these preliminary results, however, it appears that thermochemotherapy for deep-seated pelvic tumors is feasible and that further trials are warranted.

References

Bull IMC (1984) Whole body hyperthermia – summary of discussion. Cancer Res (suppl) 44: 4884–4885

Dewey WC (1984) Interaction of heat with radiation and chemotherapy. Cancer Res (suppl) 44: 4714–4720

Emami B, Perez C, Nussbaum G, Leybovich L (1984) Regional hyperthermia in treatment of recurrent deep-seated tumors: preliminary report. In: Overgaard J (ed) Hyperthermic Oncology. Taylor and Francis, London, pp 605–608

Gibbs FA, Sapozink MD, Gates KS, Stewart JR (1984) Regional hyperthermia with an annular phased array in the experimental treatment of cancer: report of work in progress with a technical emphasis. IEEE Trans Biomed Eng 31: 115–119

Hahn GM (1982) Hyperthermia and Cancer. Plenum, New York

Howard GCW, Sathiaseelan V, King GA, Dixon AK, Anderson A, Bleehen NM (1986) Regional hyperthermia for extensive pelvic tumors using an annular phased array applicator: a feasibility study. Br J Rad 59: 1195–1201

National Cancer Institute (1987) NCI Report of HT Equipment Evaluation Contractors Group, 21 February 1987, Atlanta, Georgia, USA

Overgaard J (1985) Hyperthermic Oncology. Taylor and Francis, London

Sapozink MD, Gibbs FA, Gates KS, Stewart JR (1984) Regional hyperthermia in the treatment of clinically advanced, deep seated malignancy: results of a pilot study employing an annular array applicator. Int J Radiat Oncol Biol Phys 10: 775-786

Sapozink MD, Gibbs FA, Egger MJ, Stewart JR (1986) Abdominal regional hyperthermia with an annular phased array. J Clin Oncol 4: 775-783

Stehlin JS, Giovanella B, Ipolyi PD, Anderson RF (1979) Result of eleven years experience with heated perfusion for melanoma of the extremities. Cancer Res 39: 2255-2257

Turner PE (1984) Regional hyperthermia with an annular phased array. IEEE Trans Biomed Eng 31: 106-114

Deep Microwave Hyperthermia for Metastatic Tumors of the Liver

Z. Petrovich, B. Langholz, M. Astrahan, and B. Emami

Department of Radiation Oncology, University of Southern California School of Medicine, 1441 Eastlake Avenue, Room 34, Los Angeles, CA 90033, USA

Introduction

A large number of human tumors have liver as a common site of metastatic disease. The prognosis of patients with this involvement is very poor. In a study of 390 patients with untreated metastatic liver disease the median survival was 75 days, with less than 7% surviving 1 year (Jaffe et al. 1968). In the same study, the proportion of the liver involved or the tumor volume was found to be an important factor influencing the duration of survival. In addition to the poor survival, patients with liver metastases frequently have poor quality of life due to distressing signs and symptoms, such as severe upper abdominal pain, nausea, and jaundice requiring therapeutic intervention. Unfortunately, therapy of hepatic metastases is not very effective. Hepatic artery ligation with or without infusion chemotherapy was found to be effective in selected patients. This treatment, however, has an unacceptable morbidity and mortality (Fortner et al. 1972). The use of an implanted pump for the administration of chemotherapeutic agents directly into the hepatic artery (HAI) has been investigated. In studies using 5-fluorodeoxyuridine (FUDR) and mitocycin C or FUDR and dichloromethotrexate via HAI, excellent response rate and good palliation were noted (Niederhuber et al. 1984; Shepard et al. 1985). This route of administration of chemotherapy (CT) may result in significant complications, including hepatitis in nearly half of the patients (Shepard et al. 1987). A combination of HAI with FUDR and liver irradiation was reported by Byfield et al. (1984). This combination was found to be effective for selected patients with liver metastases; however, there was again treatment toxicity, which resulted in one mortality in 28 patients so treated. Radiation therapy (RT) alone was reported in the early 1950s to be an effective and well-tolerated treatment for hepatic metastases (Phillips et al. 1954). In a more recent report by Sherman et al. (1978) 55 patients with symptomatic liver metastases were treated with eight daily radiation treatments of 3 Gy each, with 56% receiving simultaneous CT. Excellent palliation was noted with a low incidence of toxicity. Results of a study using a rapid course of RT without CT in 109 patients was reported by the Radiation Therapy Oncology Group (Borgelt et al. 1981), confirming Sherman et al.'s 1978 findings. Liver irradiation is a safe and effective treatment of hepatic metastases, provided that radiation tolerance of this organ is not exceeded (Kaplan and Bagshaw 1968).

The present paper reports the experience of several medical centers using deep microwave hyperthermia (HT) in the management of metastatic liver disease.

Recent Results in Cancer Research, Vol. 107
© Springer-Verlag Berlin · Heidelberg 1988

Materials and Methods

From 1981 to 1985, 44 patients with tumors metastatic to the liver, were treated with deep microwave HT in five medical centers in the USA. Adenocarcinoma (79%) was the most frequent diagnosis, with colon (73%) being the most frequent site of origin. Sarcoma (14%) was next in frequency. There were 30 (68%) males and 14 (32%) females with ages ranging from 25 to 79 years (median 54 years). Follow-up extended from 1 to 251 weeks, with an average of 43 weeks. Initial performance status (IPS) was <50 on the Karnofsky scale (Karnofsky and Burchenal 1949) in 18 (41%), >50 in 20 (45%), and in six cases (14%) it was not recorded. Pain was a dominant symptom at presentation in 22 patients (50%). Previous treatment consisted of CT in 12 (27%) and RT in 10 (23%). Multiple agents and different schedules were used in the 12 patients with previous CT (Table 1). In the 10 patients who received RT, the dose ranged from 20 to 50 Gy with an average of 40 Gy.

All 44 patients received deep microwave HT with a BSD-1000 Annular Phased Array (BSD Medical Corporation, Salt Lake City, Utah) (Turner 1983). The total number of HT treatments administered to these 44 patients was 150, with a range from 1 to 8 and an average of 3.4 treatments. In addition to HT, 19 (43%) patients were given RT, among whom four also received CT. HT-CT combination was given to 17 (39%) patients, including four who also received RT. HAI with 5-fluorouracil (5-FU), 1000 mg/day for 10 days, was given to nine of these patients, doxorubicin (ADR) to two, while three received cisplatin intravenously, and one streptozocin (Table 2). The remaining 12 patients received HT alone. Thermal dose (TD) was defined as the number of minutes at 42.5° C. Of the 44 patients treated, four (9%) had TD >100 and 24 (54%) a lower dose. Therapeutic temperature was not reached in 16 patients (36%). All patients had multipoint thermometry, including at least one probe being placed into the tumor. Patients were carefully monitored during each HT session. Vital signs were recorded every few mi-

Table 1. Prior chemotherapy

Agent(s)	Route	n	%
5-FU	IV	4	33
5-FU	Hepatic artery	2	17
5-FU-Mito C-MeCCNU	IV	3	25
ADR-CTX-VCR	IV	2	17
Cisplatin	IV	1	8
Total		12	100

Table 2. Current chemotherapy

Agent	Route	Dose	n	%
5-FU	Hepatic artery	1000 mg/day × 10	9	60
Cisplatin	IV	100 mg/day	3	20
ADR	Hepatic artery	50 mg/day	2	13
Streptozocin	IV	750 mg × 4	1	7
Total			15[a]	100

[a] In two of the 17 patients, CT schedule was not available.

nutes. Response was defined as complete (CR) when there was a total tumor regression as seen with CAT scan. Partial response (PR) was >50% regression, and nominal response (NR) was regression <50%.

Results

Objective tumor response was observed in 16 (36%) patients. One had CR, four had PR, and 11 had a lesser degree of regression (Table 3). Among the 19 patients who received RT with HT, 10 (53%) had an objective response. Of the 12 patients receiving HT alone, four (33%) had NR, while of the 13 HT-CT patients, two (15%) had a response. Patients who received therapeutic temperature had more frequent (46%) tumor regression than those where TD was 0 (18%) (Table 4). Patients with IPS <60 had less frequent therapeutic temperature than patients with IPS >70 (44% vs 80%; $p=0.04$, two-sided Fisher's exact test).

Table 3. Response by treatment administered

Treatment	Progression	No response	NR	PR	CR	Not evaluated	Total
HT	0	5	4	0	0	3	12
HT-RT	1	3	5	2	0	4	15
HT-CT	4	0	1	1	0	7	13
HT-CT-RT	0	1	1	1	1	0	4
Total	5	0	11	4	1	10	44

NR/PR/CR, nominal/partial/complete response.

Table 4. Tumor response by thermal dose

Thermal dose	Progression	No response	NR	PR	CR	Not evaluated	Total
0	0	4	3	0	0	9	16
>0	5	5	8	4	1	5	28
Total	5	9	11	4	1	14	44

NR/PR/CR, nominal/partial/complete response.

Table 5. Survival by response

Survival (months)	Progression	No response	NR	PR	CR	Not evaluated	Total
<6	2	7	5	0	0	6	20
6–12	0	0	4	3	0	3	10
>12	2	1	1	1	1	1	7
Total	4	8	10	4	1	10	37[a]

NR/PR/CR, nominal/partial/complete response.
[a] The seven remaining patients are known to have died, but no dates of death are available.

Among the seven patients surviving > 12 months, three (43%) had an objective response. Of the 10 patients surviving $> 6 < 12$ months, seven (70%) had a response. Patients who survived < 6 months had experienced 25% response rate (Table 5).

Of the 22 patients presenting with pain, five (23%) had complete relief, three (14%) had partial relief, and the majority (64%) had no relief. Treatments were well tolerated by 18 (41%) of the 44 patients in the study. Fair tolerance was noted in 11 (25%), while 15 (34%) tolerated the treatment poorly. Pain during HT was seen in 23%; it responded well to power reduction.

Discussion

This study has demonstrated that HT in combination with RT and/or CT can safely be used in the management of patients with hepatic metastases. It is well tolerated by patients, provided that appropriate monitoring support is available. The objective response rate of 36% is reasonably good considering the poor general condition in one-third of the patients and prior unsuccessful treatment in half of the patients. The response rate was higher (53%) among the patients who were treated with HT-RT combination than in others who did not receive RT. It was also higher among patients who received therapeutic temperature, irrespective of TD. In patients with nonambulatory status, therapeutic temperature was reached less frequently when compared with patients who had a better performance status (29% vs 57%).

At the present time, there is no effective therapy for patients with metastatic liver disease and large tumor volume. HT in combination with RT and/or CT may be a useful treatment in this difficult clinical situation (Sapozink et al. 1986a, b; Stewart et al. 1984; Petrovich et al. 1987). Randomized studies are needed to objectively evaluate the value of hyperthermia in the treatment of metastatic tumor to the liver. It may be of interest to study deep microwave HT with RT in the treatment of previously untreated primary tumors of the liver.

Summary

Between 1981 and 1985, 44 patients with advanced metastatic carcinoma of the liver were treated with deep microwave hyperthermia (HT) in five medical centers in the US. This HT was given with a BSD-1000 Annular Phased Array (BSD Medical Corporation, Salt Lake City, Utah). Of the 44 patients treated, 18 (41%) were in poor general condition and scored < 60 on the Karnofsky scale. In 50% upper abdominal pain was a major presenting symptom. Prior chemotherapy (CT) had been given in 12 (27%) patients, while 10 (23%) had received prior radiotherapy (RT). Colon (73%) was the most frequent site of the primary tumor, and adenocarcinoma (79%) was the most frequent histological diagnosis. A total of 150 HT treatments were given, with an average of 3.4. HT alone was administered to 12 (27%), HT-RT to 15 (34%), HT-CT to 13 (30%) and HT-RT-CT to four (9%). Therapeutic temperature was reached in 28 (64%) patients. The majority (66%) tolerated treatment well. Due to the poor general condition of over one-third of the patients, prior therapy in 50% and the presence of advanced tumor in all, it is not surprising to see a response rate of only 36%. The response rate was 53% among patients receiving RT in addition to HT and 46% in patients who had therapeutic temperature. Survival ranged from < 1 to 63 months, with an average of 11 months. Relief of pain was observed in 8 of 22 patients who presented with this symptom.

HT can be safely delivered to patients with metastatic tumor to the liver. There is a need for evaluation of this treatment in combination with RT and/or CT in previously untreated patients with primary and metastatic liver disease.

References

Borgelt BB, Gelber R, Brady LW, Griffin T, Hendricson FR (1981) The palliation of hepatic metastases: results of the Radiation Therapy Oncology Group Pilot Study. Int J Radiat Oncol Biol Phys 7: 587–591

Byfield JE, Barone RM, Frankel SS, Sharp TI (1984) Treatment with combined intraarterial 5-FUdR infusion and whole-liver radiation for colon carcinoma metastatic to the liver. Preliminary results. Am J Clin Oncol 7: 319–325

Fortner JG, Mulcare RJ, Solis A, Watson RC, Golbey RB (1972) Treatment of primary and secondary liver cancer by hepatic artery ligation and infusion chemotherapy. Ann Surg 127: 162–173

Jaffe BM, Danegan WL, Watson F, Stratt JS (1968) Factors influencing survival in patients with untreated hepatic metastases. Surg Gynecol Obstet 127: 1–11

Kaplan HS, Bagshaw MA (1968) Radiation hepatitis: possible prevention by combined isotopic and external beam radiation therapy. Radiology 91: 1214–1220

Karnofsky DA, Burchenal JH (1949) The clinical evaluation of chemotherapeutic agents in cancer. In: MacLeod CM (ed) Evaluation of chemotherapeutic agents. Columbia University Press, New York, pp 191–205

Niederhuber JE, Ensminger W, Gyres J, Thrall J, Walker S, Cozzi E (1984) Regional chemotherapy of colorectal cancer metastatic to liver. Cancer 53: 1336–1343

Petrovich Z, Emami B, Astrahan M, Langholz B, Luxton G (1987) Regional hyperthermia with BSD-1000 Annular Phased Array in the management of recurrent deep seated malignant tumors. Strahlentherapie, 163: 430–433.

Phillips R, Karnofsky DA, Hamilton LD, Nickson JJ (1954) Roentgen therapy of hepatic metastases. AJR 71: 824–834

Sapozink MD, Gibbs FA, Egger MG, Stewart RJ (1986a) Regional hyperthermia for clinically advanced deep-seated pelvic malignancy. Am J Clin Oncol 9: 162–169

Sapozink MD, Gibbs FAJ, Eggar M, Stewart JR (1986b) Abdominal regional hyperthermia with an annular phased array. Am J Clin Oncol 4: 775–783

Shepard KV, Levin B, Faintuch Y, Doria MI, DuBrow R, Riddell RH (1987) Hepatitis in patients receiving intraarterial chemotherapy for metastatic colorectal carcinoma. Am J Clin Oncol 10: 36–40

Shepard KV, Levin B, Karl RC, Faintuch J, DuBrow RA, Hagle M, Cooper RM, Beschorner J, Stablein D (1985) Therapy for metastatic colorectal cancer with hepatic artery infusion chemotherapy using a subcutaneous implanted pump. Am J Clin Oncol 3: 161–169

Sherman DM, Weichselbaum R, Order SE, Cloud L, Trey C, Piro AJ (1978) Palliation of hepatic metastases. Cancer 41: 2013–2017

Stewart RJ, Bagshaw MA, Corry PM, Gerner EW, Gibbs FA, Hahn GM, Pademaker LP, Oleson JR (1984) Hyperthermia as a treatment of cancer. Cancer Treat Symp 1: 135–145

Turner PF (1983) Regional hyperthermia with Annular Phased Array. IEEE Trans Biomed Eng 31: 106–114

Future Trends in Heating Technology of Deep-Seated Tumors

P. F. Turner, T. Schaefermeyer, and T. Saxton

BSD Medical Corporation, 420 Chipeta Way, Salt Lake City, UT 84108, USA

Introduction

Earlier predictions of the future trends in thermometry have been published and these provide additional insights into this issue (Oleson 1984; Lin 1984; Cetas and Roemer 1984). A primary benefit desired from hyperthermia is to provide an effective adjunctive therapy for deep-seated tumors which have responded poorly to conventional therapies. Optimism regarding this benefit is encouraged by the results of combined treatments of superficial tumors (Overgaard 1984). Researchers in hyperthermia feel that its major benefit may someday be found in the deep-heating of tumors (Oleson 1984). Many patient deaths occur from failure to control the primary tumor in deep tumor sites. Mortality in adult males is highest when the primary sites of disease occur in the lungs, colorectal area, prostate, or pancreas. In adult females the highest rate occurs when the primary sites are the breasts, lungs, colorectal area, or ovaries. A common problem with the conventional therapies of surgery, radiation, and chemotherapy is the failure to control these primary disease sites. These disease sites are often located many centimeters below the skin surface, which creates a difficult problem for hyperthermia because the physics of penetrating deeply does not lend itself to increased precision. Improving the control of these primary sites is expected to improve the cure rates for patients without metastatic disease.

So far, a substantial degree of normal tissue heating in the area of the tumor has been inevitable if the whole tumor is to be heated to the desired temperature. Larger tumor volumes cannot be heated to uniform or even to therapeutic temperatures within the whole tumor (Gibbs 1984; Oleson et al. 1986). It has been encouraging to observe positive clinical responses even though tumor temperatures have been lower and adjunctive therapy less successful than hoped for. In earlier studies a much lower radiation dose was necessitated by previous radiation treatment, resulting in a substantial likelihood of failure to control the primary tumor site. With improvements, hyperthermia is expected to become a more acceptable treatment in the earlier stages of these deep tumor sites.

The present hyperthermia equipment will only partly heat deep tumors to the desired temperatures. This is due to the lack of adequate heating techniques as well as the thermodynamic effects of blood flow within the treatment site. Many tumors have centrally suppressed blood flow relative to the surrounding normal tissue; this can in part compensate for a lack of selective energy targeting. Unless a heating technique is used which can adapt the energy absorption pattern to compensate for unpredictable blood flow effects. Portions of the tumor will be at subtherapeutic temperatures. Presently, the current systems use the temperature feedback of the invasive tumor sensors (representing the tumor temperature) to control the heating power to maintain an effective temperature, while keeping normal tissues below excessive temperatures. It remains with the clinician to decide the permissible heating of normal tissues and prescribe temperature and time of treatment for the disease site.

Thermometry

Thermometry is necessary in hyperthermic treatment to quantify the tumor and normal tissue heating. The more thermometry is used, the more temperature variation in tumor and normal tissue is seen. Often, the center of a large mass has little blood flow, and thus temperatures will rise substantially in that zone compared with other tumor regions. A single temperature sensor in the center of the deep tumors can cause a false impression of the temperature in other tumor locations.

The use of the closed-tip catheter developed by BSD Medical in conjunction with Deseret Medical has facilitated the thermal mapping of a tissue area, enabling many temperature measurements to be made by a single sensor. The Bowman and optical probes have low longitudinal thermal smearing, enabling accurate spatial measurement through the catheter wall. The use of thermal mapping becomes more complicated during ultrasound treatments because of the self-heating of the plastic catheter and the longitudinal thermal smearing of metallic probes.

Dunscombe and McLellan (1986) and Kuhn and Christensen (1986) have reported that the deep heating modalities of electromagnetics (EM) and ultrasound (US) create artifact effects on many temperature sensors. EM fields create self-heating artifacts in metallic sensors, but this can be overcome by using very high resistance non-metallic leads (Bowman 1976) or non-conductive fiberoptic sensors. However, US energy causes self-heating of these non-metallic sensors and even the metallic sensors are reported to have measurable artifact unless the diameter is very small and the US frequency is below 3 MHz.

Multisensor junctions along the temperature probe have been developed for both the optical and the metallic sensors. Although these multisensors may remove the need for thermal mapping in some clinical applications, it has become apparent that even these probes are frequently mapped along the catheters for additional temperature data. These added temperature measurements have been displayed by computer color graphics displaying temperatures as:

- Time versus temperature multicolor plots
- Temperature versus distance along the catheter
- Color bar temperature charts
- Treatment site pictures with local temperatures indicated by numbers, colors, or symbols

The rate of heating or cooling is being monitored to provide information about the deposition of heat as well as local cooling effects. Within the next few years, microcomputers will provide improved treatment guidance as the methods for display of the information are improved and interactive computer treatment guidance become better developed.

A future improvement in thermometry will be the development of non-invasive techniques. This step will certainly make hyperthermia more acceptable to both clinicians and patients. The methods being proposed for non-invasive thermometry include microwave radiometry, US radiometry, active microwave, active US (both time of propagation and reflective), CT scanning of the effects of temperature changes on tissue densities, and MRI thermal image changes (Christensen 1982; Bolomey et al. 1983; Hill and Goldner 1985; N'Guyen et al. 1979). Presently microwave radiometry is being clinically utilized, but provides poor spatial resolution and temperature accuracy. It is expected that as non-invasive thermometry is developed, improved treatments may be possible if compatible with the deep heating devices which can focus and steer the deep heating pattern. Substantial research and development costs will be incurred and considerable time required before a feasible method for non-invasive thermometry can be fully developed which will approach current clinical expectations of temperature accuracy and spatial resolution. Even when developed the cost of implementation of such a method may become a major barrier to clinical use. It is reasonable to expect, however, that initial non-invasive thermometry achievements will be capable of moderately acceptable thermal and spatial accuracy.

Deep Heating Techniques

Interstitial Techniques

The future deep heating techniques will be both externally and internally applied and will include both EM and US methods. There is a current trend toward the use of interstitial techniques because of the ability to selectively place the heat in the tumor tissues with minimal heating of the normal tissues. This does not insure, however, that a better treatment will be accomplished by interstitial methods. The success of interstitial techniques is still dependent upon the implant adequately filling the entire tumor and heating that whole volume. The variability of the thermodynamics within the tumor presents the same variation to the interstitial array as seen in non-invasive techniques. Heating of the tumor will be improved over external heating methods if the interstitial applicator's heating patterns can be dynamically adjusted for the temperatures in the treatment zone. The most promising technique is that using phase-coherent microwave interstitial arrays (Mechling and Strohbehn 1986). Proper selection of implant applicator locations, amplitude, and radiating phase can greatly improve the resultant temperature pattern (Turner 1984a; Trembly et al. 1986). Computer-aided interstitial pretreatment planning is now available with the BSD-500 hyperthermia system. This makes it possible to begin optimization of the interstitial array parameters to create more desirable initial heating conditions. Even these attempts to optimize treatment setup must still rely on real-time measurement of the tissue temperature and modification of the necessary parameters. The unpredictable nature of blood flow has a dramatic effect on the temperature pattern, and must be observed to be known. The development of thermometry built into an interstitial applicator has provided a valuable source of temperature information from which to adjust the control parameters (patent pending). It is still too soon to know all the ways in which these pretreatment methods may integrate into the treatment control. It may be possible to combine numerical predictions of specific absorption rate (SAR) and measurements of the tissue cooling rates when power is temporarily turned off, to estimate the thermal pattern (Clegg et al. 1985). Significant effort has been expended to compare the use of phase-coherent and non-synchronous interstitial arrays. It has been generally observed that the phase-coherent technique requires fewer implanted applicators to achieve the same heat uniformity as the non-synchronous arrays.

Localized current field (LCF) methods have also attracted interest. The performance of this technique is similar to that of the non-synchronous microwave interstitial technique, requiring more inserted applicators than the synchronous array (typically 1 cm spacing). These applicators operate as electrode pairs and the technique is considered basically equivalent to the microwave method (Strohbehn 1984 b).

The ferromagnetic seed method has not been well accepted or applied at this time because there is no direct absorption of power by the tissues. This method has the least uniform temperatures for the same implant spacing of all of the interstitial methods. It has been suggested that by using biocompatible materials, these implanted seeds or rods could be left in place indefinitely. This would allow repeated treatments, possibly for many years, without modifying the implant. The interesting property of the ferromagnetic seeds is that the power absorption is self-regulating by the seed temperature. This means that as the properly designed seed reaches therapeutic temperatures it reduces it own absorption of power to limit further temperature rise. It has been reported that a seed temperature of about 50° C may be necessary to maintain a minimum of 42° C in the implanted zone with moderate or low blood flow (Paliwal 1987). With high blood flow regions, it is estimated that portions of the tumor will still be below 42° C. The heating from these seeds is the same as that achieved by placing hot resistive wires down catheters to heat the tumor region. Hot resistive wires can also be self-regulating. If, however, it is not practical for the catheter to remain in place, it would be preferable to surgically implant biocompatible seeds. It would still be unclear whether the temperature distribution desired was achieved without measurement of the thermal pattern. For these reasons, little use of the ferromagnetic seed method is expected. Acceptance will relate more to compatibility with established surgical procedures, which may result in a preference for the interstitial methods based on surgical convenience rather than on improved thermal expectations.

External Techniques

Electromagnetic Non-invasive Methods

Capacitive Method

The EM method of heating deep-seated tumors has been utilized since the early 1900s. This continues even today as the primary method of inducing deep non-invasive hyperthermia. The capacitive method was utilized far more than any other method in the early days of hyperthermia development (Geyser 1916), largely due to the very simple physics and devices required. One of the early problems noted with this method was that superficial fat layers tend to heat 10–17 times more than other tissues. This is because these tissues have a much higher resistance to the series current generated within the tissues than do those tissues of the body with a higher water content. In recent years, significant investigations of the capacitive method have demonstrated that deep hyperthermia is possible with this technique primarily in regions with little superficial fat. The majority of this work has been in countries where the general population is noted to have low superficial body fat. In Japan, clinical treatments have been reported which demonstrate this capability (Hiraoka et al. 1985).

It has been found helpful to pre-cool fatty tissue surfaces for 20–30 min with a 5° C bolus to avoid excessive temperatures during treatments with fatty layers over 1.5 cm thick. If the deep tumor has substantially reduced blood flow the power absorbed during the 30-min typical induction period may be sufficient to achieve therapeutic tumor temperatures. The mechanisms of this dramatic precooling is still not fully determined. It is possible that precooling the fat layer reduces its conductivity, thereby reducing its power absorption. It has been observed that with this pre-cooling, there is an increase in blood flow within the skin layers. Whether this effect occurs in the fat layer has not yet been determined. Some researchers have demonstrated that limited steering capabilities may be possible with capacitive arrays of three of more electrodes (Morand and Bolomey 1987). This work demonstrates that heating could become more localized in the tissue neighboring one or more of the electrodes. However, since the capacitive field lines are divergent within the tissue a non-invasive central power focusing is not a possibility. Capacitive heating fields will also heat more superficially adjacent to a smaller electrode in paired electrodes. The selective heating of the fat layers is unlikely to be overcome completely by these methods. It is, therefore, unlikely that the capacitive method will be the future non-invasive method of choice; it will, however, offer some practical solutions for natural body orifice and cavity heating. For the present this technique will continue to be used clinically because of the simplicity of the operation and design. There are many specific body sites where superficial tumors may be heated well using this method, such as the head and neck, where superficial fat is less common.

Inductive Method

Another method of inducing tissue heating with EM is that of magnetic coupling. An example of a magnetic heating device is the concentric coil. This device induces cylindrical currents to flow within the conductive tissues of the body, thereby producing heat. There is no current flow in the central region of the body with this device, so no heating is directly applied in the body central zone. However, there is little heating within the surface fat layers. This technique has the limitation that the heating field is not steerable or alterable, and thus it is not expected to be utilized in future systems as a general rule (Gibbs 1984; Oleson et al. 1986). The device is simple to use and construct, with no contacting devices to the patient other than a dielectric bed platform.

Another magnetic field device is the coil electrodes pair. This device provides two deeper heating zones, which can be moved somewhat by changing the surface coil's position or angle. The presence of a second hot zone, as well as the superficial tissue heating, is considered a fundamental problem with this technique. The lack of heat pattern control is also sufficient to classify this method as unsuitable for future deep heating requirements. Some attempts to moderate the second hot zone by superimposing a capacitive field which is synchronous with the inductive field have been reported. This can alter the heating pattern to make it more like a capacitive heating field, but this method is not expected to provide noteworthy improvements.

Helical Coil

The helical coil was reported as having the property of producing axially aligned E-fields and central heating (Chute and Vermeulen 1981; El-Sayed and Abdel-Hamid 1981). Others successfully applied this axial E-field mode of operation to develop hyperthermia applicators (Ruggera et al. 1985; Hagmann and Levin 1985). The helical coil has been incorrectly classified as an inductive applicator because it appears like a coil inductor. Its preferred operation mode is actually quite similar to that of the Annular Phased Array (APA), which also forces the dominant E-field to be aligned with the body (Hagmann et al. 1985). The two devices even operate at the same frequency and at about the same radiator size. The helical coil must be operated at a narrow frequency range where it becomes resonant. The resulting electric field gradients along the length of the coil induce an axial current to flow lengthwise along the conductive body tissues placed inside the coil. The axial magnetic fields induced by the coil currents cause circumferential cylindrical currents to flow in the outer body region similar to the Magnetrode cylindrical coil. This cylindrical coil mode of operation does not contribute to the deep central heating desired and has been suppressed in the more recent coil designs of the US Control Division of Radiological Health (CDRH). It has been reported for the CDRH coil that the heating pattern is tipped off axis in symmetric tissue phantoms. This is caused by a nonuniform E-field being radiated from the coil. This coil is operated with an air coupling region between the tissue surface and the inside region of the coil. This enhances patient comfort and facilitates patient insertion. However, not using a high dielectric coupling medium potentially increases stray fields, and this can lead to increased heating of tissues overlying bony protrusions. The major flaw in this type of design is the lack of heat pattern steering and adjustable focusing to steer the heating more selectively into a tumor region. Future deep heating devices will be required to provide more selective tumor heating. It is not expected that helical coil devices will be utilized widely in deep hyperthermia systems of the future, although benefit may be obtained in central tumors in the human torso or in poorly perfused tumor regions.

Phased Array Deep Heating

The phased array technology has emerged as the most promising method for achievement of deep focal hyperthermia. The EM phased array techniques have been utilized with both interstitial and non-invasive methods since 1979 with the BSD-1000 hyperthermia system. Flat or planar phase arrays have been reported to yield very limited improvement in heating at depth compared to other flat surface applicators (Turner and Kumar 1982; Johnson et al. 1985; Andersen 1985). Therefore, the use of crossfire or overlapping radiating energy beams which are also phase-coherent in the target focal zone has become the most recognized EM method of obtaining deep selective tumor heating. A special publication summarizing current work in this was edited by Lin (1986). The physics of localizing the power field focus is limited to a minimum heating zone diameter of about one-third of the average tissue wavelength in the focal area. Since these deep regions are normally dominated by tissues with a high water content, such as tumor, muscle, and critical organs, the minimum focal diameters are about 20 cm at 50 MHz, 10 cm at 100 MHz, and 1.2 cm at 915 MHz.

   As operating frequency increases, the focal zone size decreases and the penetration depth also decreases. This results in trade-off decisions involving larger heating field zones when heating deep and also requires larger applicators to heat deeply (Turner and Kumar 1982). It is expected that substantial improvement in the deep selective tumor heating at depth will soon be realized clinically with the use of frequencies moderately higher than previously employed. This results in a trade-off of depth versus focusing. To choose the best frequency, the tumor size, location, and tissue cross-section size must be considered. The most widely utilized phased array device is the Annular Phased Array

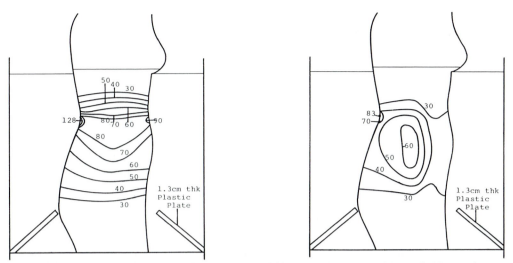

Fig. 1 *(left).* APAS relative power density to an arbitrary level, measured by E-field scanning at 60 MHz: 0.3% saline-filled thin latex mannequin surrounded by a water bolus

Fig. 2 *(right).* APAS relative power density to an arbitrary level, measured by E-field scanning at 100 MHz: 0.3% saline-filled thin latex mannequin surrounded by a water bolus

System (APAS), previously described in detail (Turner 1984b). This cylindrical array of radiators creates much the same E-field distribution, at similar operating frequencies, as the CDRH helical coil. However, to minimize leakage field and to reduce excessive heating of tissues overlying prominent bony regions, a water bolus is normally used to fill the space between the patient's body and the apparatus. At the normal operating frequencies, 60–70 MHz, the heating is very broadly spread across the whole heated region, as shown in Fig. 1. This is the range of operation frequency utilized during the patient treatments to date. This results in a significant amount of heating of adjacent normal tissues. Figure 2 shows the effect of raising the frequency to 100 MHz to increase the central focusing. The diameter of the central focal zone is seen to reduce substantially at the higher frequency. The clinical studies in the phase I evaluation of the APAS has shown that the primary clinical difficulties have been the slow treatment setup, the discomfort to the patient caused by the confining, non-stretching PVC water bolus bags, the systemic heating, and patient pain. Utilization of a more selectively placed focal heating zone would reduce systemic heating by less heating of normal tissues. It would also reduce patient heating pain in non-target tissues.

Another prototype design, the SIGMA 60 (Fig. 3), is being tested and introduced (Turner 1986a). This is a transparent cylindrical plastic tube along which are placed eight EM dipole antennas each aligned with the body axis so as to create an axially aligned E-field just like the APAS (the device in current clinical use). When these two array designs are operated in a balanced mode with equal amplitude and phase, the heating patterns are almost identical. The SIGMA 60 is primarily intended to improve heat pattern steering control as well as patient comfort.

The SIGMA 60 contains a single annular flexible bolus similar to that in the mini annular phased array (Turner 1986b). This decreases patient restraint and pressure and thus improves comfort over the APAS, and is expected to substantially improve patient

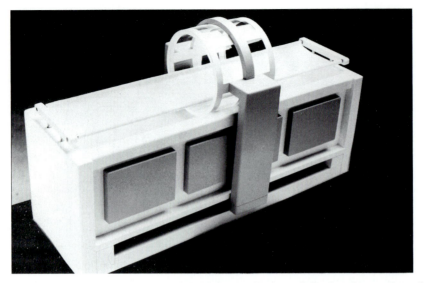

Fig. 3. SIGMA 60 (prototype model): eight equally spaced dipoles along a clear plastic cylinder

tolerance to treatments. The transparent plastic cylinder and bolus of the SIGMA 60 enable the clinician to completely view the surfaces of the patients body inside the device before and during treatment. The thin shell applicator appearance is also expected to be much less frightening to the patient, thereby reducing anxiety and stress during treatment. The annular bolus design fills and drains rapidly from the bottom of the cylinder. The excessive air in the liner is automatically removed through a hole in the top of the cylinder. When the air is rapidly drained this hole is automatically closed to force the emptied bolus liner to conform tightly to the plastic cylinder wall, providing a large unobstructed opening for patient insertion and removal. This applicator shell is light in weight and can be manually positioned by rolling mechanisms under the cylinder to slide out of the way for patient placement on the treatment sling. These mechanical modifications are expected to greatly improve the speed, ease, and acceptance of the SIGMA 60 treatments.

One of the most powerful developments in phased array technology is the multiple channel synchronous amplifier. This enables electronically controlled phase and amplitude steering of the heating pattern of these phased arrays. One such four-channel amplifier has been developed which is capable of phase and amplitude steering from 60 to 220 MHz of each of four 500-W channels. This capability has been found to provide remote electronic targeting and steering of the heating field by optimizing the phase and amplitude levels of the APAS devices. The steering functions are expected to enable even computer-controlled optimization methods with this class of phased arrays.

The next challenge in these annular or cross-firing phased arrays is to determine the best frequency, amplitude, and phase balance of these various EM radiators. One pair of researchers have suggested that the principal of reciprocity could be utilized to optimize the phase and amplitude settings (Andersen 1985). This technique involves placing a small radiating antenna inside the target tumor zone and receiving the relative amplitude and phase arriving at each of the phased array antennae. After the removal of the inserted targeting antenna the amplitude balance received is set for each antenna and the conjugate phase is applied to each antenna. Most recent reports have claimed that this technique is accurate in obtaining the highest power gain or transfer to the target point, but that it may do so without regard to the possible excessive heating in other non-target tissues in the body. Numerical studies have been reported by others which are intended to predict the resultant SAR or power fields with various amplitude, phase, and frequency selections. Most of these have been two-dimensional models of the EM fields and antennae arrays (Iskander et al. 1982; Strohbehn 1984a). The authors have also included modeling of the effects of thermodynamics with these numerical methods. Of particular interest, Strohbehn (1984a) has actually incorporated an optimization method to iteratively optimize the phase and amplitude excitation of an annular phased array, attempting to improve the predicted thermal uniformity within the tumor volume by numerical means. This is still a two-dimensional solution, but three-dimensional modeling of this technique is reported to be under development. Other researchers reporting the development of three-dimensional models predicting the EM fields and tissue heating have also indicated that such solutions are becoming feasible with the computational speeds available in present day computers. Although it will not be possible to know all the thermodynamic properties of the patients tissues, these numerical models will enable a useful method by which to predict initial settings for frequency, amplitude and phase for such arrays. Their usefulness in the clinical situation is yet to be determined. Blood flow has a major effect in altering the temperature patterns during treatment; this effect, however, is not even constant during the same or consecutive treatments. Therefore, bioheat or thermal models are not going to be correct unless real-time blood flow or thermal washout information is obtained during treatment to provide some blood flow input to these calculations. It has been reported that even discrete temperature measurements can be used to estimate other not measured temperature zones. This may become useful for numerical model inputs for the blood flow in the treatment zone and also provide a real-time display of thermal patterns from a finite number of measurements. At the present such calculations require several minutes or hours of processing time, which could cause major delays in the display of such patterns. However, this may provide useful treatment guidance at some time in the future for the deep heating techniques.

Another method which has been reported to be useful in optimizing the frequency, phase, and amplitude of APA type devices is that of internal E-field sensors (Turner 1987). Figure 4 shows an example where an internal sensor was placed at the location indicated by the asterisk and the phase and amplitude balance was adjusted to peak the detected field at that point. After this was done the power density fields were scanned to determine how well the sensor guided the heat pattern steering. This figure plots the relative power density or SAR as a thermal equivalent SAR. This means that the maximum measured SAR value was normalized and plotted as an 8° C rise above the normal body temperature of 37° C. The other data points were plotted, with zero SAR being 37° C and the maximum being shown as 45° C. The equation used for these plots is the following:

thermal equivalent $SAR = (SAR/SAR \ max) \times 8 + 37° C$

This provides clinicians with a clearer understanding of the expected initial heating rates, but should not be expected to relate directly to the expected temperature patterns in dynamic living tissue. The thermodynamic effects are not included. The numbers shown around the cylindrical phantom indicate the external E-field sensor readings which are proportional to the power density. Figure 5 shows an example of an offset targeting position similar to that in Fig. 4. This involves placing an electric field sensing probe interstitially into the tumor site or in a natural orifice passing through or near the

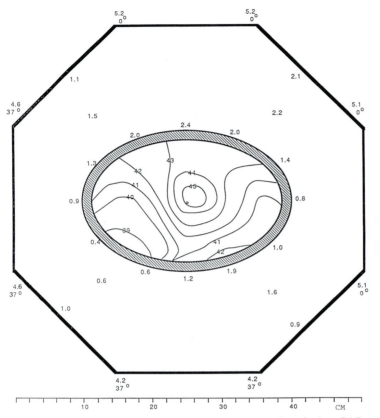

Fig. 4. Dipole array (SIGMA 60) at 105 MHz, thermal equivalent SAR pattern in ° C (initial phantom temperature rise), measured by E-field scanning in saline, diel = 78.5, cond. = 0.486 mho/m. Power and phase lag are balanced as shown on octagon bolus corners

tumor site. This single internal sensor can also be used to adjust frequency, phase and amplitude balance to optimize the power gain at the target point of the sensor. Small E-field sensor designs have been published by Batchman and Gimpelson (1983). This alone may not create the best heating within the tumor, but would achieve phase and amplitude settings similar to those in the reciprocity method. The advantage of this method is that the optimization could be performed during the early phases of the heat induction period or at any time during the treatment. The small inserted antenna required with the reciprocity method should probably be removed during active treatment to prevent the metallic coaxial feed cable affecting the heating pattern. The use of external surface E-field levels has provided additional information about the balance of the heating field which can further provide feedback to enable better setting of the phase and amplitude balance. With all these measurements, it will still become necessary at times to make some modifications of these phase and amplitude settings to compensate for the unpredictable thermodynamic tissue conditions. Therefore, an interactive computer modification of the initial phase and amplitude balance should be considered a desirable part of future deep hyperthermia systems. This type of temperature feedback would require that temperature resolution be better than the normal 0.1° C so that smaller changes in tissue temperature can be resolved when these control parameters are be-

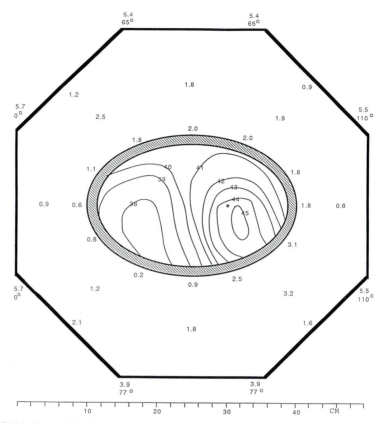

Fig. 5. Dipole array (SIGMA 60) at 105 MHz, thermal equivalent SAR pattern in ° C (initial phantom temperature rise), measured by E-field scanning in saline, diel = 78.5, cond. = 0.486 mho/m. Power and phase lag imbalance shown on octagon bolus corners

ing adjusted to optimize the heating by the observed changes in the temperature pattern. The Bowman probe is the only minimally perturbing temperature sensor which has been reported as being capable of thermal resolution as low as 0.01° C. Other optical techniques have such high system noise and drift that higher accuracy is not currently possible. This type of resolution improvement in the software control algorithm used to optimize the thermal pattern would speed up the thermal optimization process. Continual optimization would therefore be feasible if adequate thermometry were used and if adequate limits were placed on the range over which software could alter these parameters from the initial predetermined phase and amplitude settings.

Ultrasound

Future trends will include the use of US applicators. The fundamental increase in penetration depth and focusing will make US the treatment of choice for some specific treatment sites. The number of sites available for US will be quite limited due to the dominant interaction with bone and air regions. This eliminates the lung and many abdominal masses. Some recent developments have been reported by Cain and Umemura (1986) regarding a concentric-ring and sector-vortex US phased array which appears to offer some promise for heating of tumors of limited size (about 5 cm diameter) at depths of up to 9 cm. In another interesting method being evaluated by Hynynen et al. (1986), four heating transducers have been incorporated with a US imaging system. Corporate developments include the Varian Helios, the Labthermics planar array, the Clinitherm US transducers, and the BSD US transducers. These systems are being found to have limited application in the presence of air and bone, but are able to heat well in some deep tissue regions. It is unlikely that a clinical setting will be able to rely solely upon US for most deep heating, but as the techniques evolve there will be a role for US in certain deep-seated tumors. Efforts will continue in this area to utilize the precision and depth advantage of US.

Three-dimensional treatment planning will also play an important role in effective use of these US devices to characterize the limitations of the devices in planned treatment sites, thereby quantifying the effects of tissue variation to optimize the choice of treatment setup and modality. The highly focused US arrays may also be used to heat portions of the tumor to higher temperatures ($> 50°$ C) for shorter exposure times (< 60 s) to provide more specific and quicker tumor treatment, as proposed by DW Pounds (personal communication). This method would require very careful imaging to track such a highly focused heating zone.

Conclusion

The future developments in hyperthermia are technically challenging and the current trends quite exciting. It appears that the microwave interstitial technique will be the method of choice for many treatment sites which can be adequately implanted, but this substantial invasion is not practical for many deep tumors. Therefore, more refined and focused external EM phased arrays will be most generally used. Deep focused US will evolve to become the method of choice for some of the deep tumor sites, but with severe limitations on the allowable beam portal, which must avoid bone and air regions such as lung and bowel. Improvements in thermometry are expected to provide many more invasive measurement locations. Non-invasive thermometry will become feasible for some treatment methods, but the temperature accuracy and spatial resolution associated with invasive sensors will not be commercially available for at least 5 years at reasonable costs. Therefore, early non-invasive thermometry will probably serve only to complement invasive measurements for this period of time. It is clear, with the improved control, flexibility, and thermometry of future systems, that the role of and demands on computers will greatly increase. Pretreatment planning will also require the newest minicomputer innovations to be utilized.

Summary

Hyperthermia techniques and equipment have developed into a level of sophistication requiring special training and operator skills. Recent advances electromagnetic and ultrasound heating techniques are multi-applicator arrays, both external and invasive. These arrays are most effective when individual channel power control can be utilized to improve the heating pattern control during treatment. A preview of future capabilities shows that more selective tumor heating is possible with the advanced equipment. The role of computer pretreatment planning is a valuable aid in operator training as well as optimization of parameters. The current development of regional heating is in the deep focal heating by external focused arrays. The SIGMA 60 cylindrical dipole array deep steerable focal zone is described and test data presented showing the selective focusing obtained. A small invasive targeting E-field sensor is used to optimize phase and amplitude of the array to improve the selective heating in the target zone. These patented techniques and devices are expected to reduce systemic toxicity and increase tumor temperatures with lower power input. Future equipment development is predicted to include many types of hyperthermia devices and techniques. This trend suggests that systems must be designed to simplify the operator's tasks and improve the heating pattern controls. The computer power and control will become much more important in these future systems.

References

Andersen JB (1985) Theoretical limitations on radiation into muscle tissue. Int J Hyperthermia 1 (1): 45–55

Batchman TE, Gimpelson G (1983) An implantable electric field probe of submillimeter dimensions. IEEE Trans Microwave Theory Tech 31 (9): 745–751

Bolomey JC, Jofre L, Peronnet G (1983) On the possible use of microwave – active imaging for remote thermal sensing. IEEE Trans Microwave Theory Tech 31 (9): 777–781

Bowman RR (1976) A probe for measuring temperature in radiofrequency heated material. IEEE Trans Microwave Theory Tech 25 (1): 43–45

Cain CA, Umemura SI (1986) Concentric-ring and sector-vortex phased-array applicators for ultrasound hyperthermia. IEEE Trans Microwave Theory Tech 34 (5): 542–552

Cetas TC, Roemer RB (1984) Status and future developments in the physical aspects of hyperthermia. Cancer Res [Suppl] 44: 4894 s–4901 s

Christensen DA (1982) Current techniques for non-invasive thermometry. Physical aspects of hyperthermia. Am Assoc Phys Med Monogr 8: 266–279

Chute FS, Vermeulen FE (1981) A visual demonstration of the electric field of a coil carrying a time-varying current. IEEE Trans Educ 24 (4): 278–283

Clegg ST, Roemer RB, Cetas TC (1985) Estimation of complete temperature fields from measured transient temperatures. Int J Hyperthermia 1 (3): 265–286

Dunscombe PB, McLellan J (1986) Heat production in microwave irradiated thermocouples. Med Phys 13 (4): 457–461

El-Sayed EM, Abdel-Hamid TK (1981) Use of sheath helix slow-wave structure as an applicator in microwave heating systems. J Microwave Power 16 (3): 283–288

Geyser AC (1916) The physics of the high frequency current. NY Med J 4 Nov 1916, pp 891–892

Gibbs FA (1983) Thermal mapping in experimental cancer treatments with hyperthermia: description of semiautomatic system. Int J Radiat Oncol Biol Phys 44 (10): 1057–1063

Gibbs FA (1984) Regional hyperthermia: a clinical appraisal of noninvasive deep-heating methods. Cancer Res [Suppl] 44: 4765 s–4770 s

Hagmann M, Levin RL (1985) Coupling efficiency of helical coil hyperthermia applications. IEEE Trans BME 32 (7): 539–540

Hagmann MJ, Levin RL, Turner PF (1985) A comparison of the annular phased array to helical coil applicators for limb and torso hyperthermia. IEEE Trans Biomed Eng 32 (11): 916–927

Hill JC, Goldner RB (1985) The thermal and spatial resolution of a broad-band correlation radiometer with applicator to medical microwave thermography. IEEE Trans Microwave Theory Tech 33 (8): 718–722

Hiraoka M, Jo S, Akuta K, Takahahi M, Abe M (1985) Effectiveness of RF capacitive hyperthermia in the heating of human deep-seated tumors. Proceedings of the 1st Meeting of the Japanese Society of Hyperthic Oncology. Mag, Tokyo, pp 98–99

Hynynen K, Roemer R, Moros E, Johnson C, Anhalt D (1986) The effect of scanning speed on temperature and equivalent thermal exposure distributions during ultrasound hyperthermia in vivo. IEEE Trans Microwave Theory Tech 34 (5): 552–560

Iskander MF, Turner PF, Dubow JB, Kao J (1982) Twodimensional technique to calculate the EM power deposition pattern in the human body. J Microwave Power 17 (3): 175–185

Johnson RJ, Andrasic G, Smith DL, James JR (1985) Field penetration of arrays of compact applicators in localized hyperthermia. Int J Hyperthermia 1 (4): 321–336

Kanda M, Driver LD (1987) An isotropic electric-field probe with tapered resistive dipoles for broad-band use, 100 KHz to 18 GHz. IEEE Trans Microwave Theory Tech 35 (2): 124–130

Kuhn PK, Christensen DA (1986) Influence of temperature probe sheathing materials during ultrasonic heating. IEEE Trans Biomed Eng 33 (5): 536–538

Lin JC (1984) Special issue on phased arrays for hyperthermia treatment of cancer. IEEE Trans Microwave Theory Tech 34 (5): 481–644

Mechling JA, Strohbehn JW (1986) A theoretical comparison of the temperature distributions produced by three interstitial hyperthermia systems, Int J Radiat Oncol Biol Phys 12: 2137–2149

Morand A, Bolomey JC (1987) A model for impedance determinations and power deposition characterization in three-electrode configurations for capacitive radio frequency hyperthermia, parts A and B. IEEE Trans Biomed Eng 34 (3): 217–233

N'Guyen DD, Mamouni A, Leroy Y, Constant E (1979) Simultaneous microwave local heating and microwave thermography possible clinical applications. J Microwave Power 14 (2): 135–137

Oleson JR (1984) Regional power deposition for hyperthermia theoretical approaches and considerations. Cancer Res [Suppl] 44: 4761s–4764s

Oleson JR, Sim DA, Jackie C, Fletcher AM, Gross EJ (1986) Results of a phase I regional hyperthermia device evaluation: microwave annular array versus radio frequency induction coil. Int J Hyperthermia 2 (4): 327–336

Overgaard J (1984) Rationale and problems in the design of clinical studies. In: Overgaard J (ed) Hyperthermia oncology, vol 2. Taylor and Francis, London, pp 325–338

Paliwal BR, Su SL, Wang GB, Steeves R, Partington B, Adams CA (1987) The effect of perfusion, spatial arrangement and Curie point on the thermal distribution from ferromagnetic needles. Poster, NAHG group meeting, Feb 1987

Ruggera PS, Kantor G (1985) Helical coil for diathermy apparatus. US patent no 4527550

Strohbehn JW (1984a) Calculation of absorbed power in tissue for various hyperthermia devices. Cancer Res [Suppl] 44: 4781s–4787s

Strohbehn JW (1984b) Summary of physical and technical studies. In: Overgaard J (ed) Hyperthermia oncology, vol 2. Taylor and Francis, London, pp 353–369

Trembly BS, Wilson AH, Sullivan MJ, Stein AD, Wong TZ, Strohbehn JW (1986) Control of the SAR pattern within an interstitial microwave array through variation of antenna driving phase. IEEE Trans Microwave Theory Tech 34 (5): 568–572

Turner PF (1984a) Invasive hyperthermia apparatus and method. US Patent no 4448198 (filed 1979)

Turner PF (1984b) Regional hyperthermia with an annular phased array. IEEE Trans Biomed Eng 31 (1): 106–114

Turner PF (1986a) Apparatus for creating hyperthermia in tissue. US Patent no 4589423

Turner PF (1986b) Mini-annular phased array for limb hyperthermia. IEEE Trans Microwave Theory Tech 34 (5): 508–514

Turner PF (1987) Electric field probe. US Patent no 4638813

Turner PF, Kumar L (1982) Computer solution for applicator heating patterns. Natl Cancer Inst Monogr 61: 521–525

Percutaneous Placement of Catheters for Temperature Measurement During Hyperthermia

H. Berger[1], G. Markl[1], R. D. Issels[2], and J. Lissner[1]

[1] Radiologische Klinik und Poliklinik, Klinikum Großhadern, Ludwig-Maximilians-Universität, Marchioninistraße 15, 8000 München 70, FRG
[2] Medizinische Klinik III, Klinikum Großhadern, Ludwig-Maximilians-Universität, Marchioninistraße 15, 8000 München 70, FRG

Introduction

Thermometry in local hyperthermia treatment still presents problems (Chan et al. 1984). Thermal dosimetry should be developed such that the temperature distribution within the heated volume can be predicted (Cosset et al. 1985). At present the method of choice for temperature monitoring is direct measurement of the tissue temperature by invasive techniques (Nilsson et al. 1982). In superficial tumors, temperature probes are easily introduced into the tumor at different positions or are placed on the skin (Marmor et al. 1979; Lindholm et al. 1982; Nilsson et al. 1982). In the case of deep-seated tumors, temperature is checked by means of catheters inserted into the vagina, the rectum and the bladder (Marchal et al. 1985).

   The purpose of this paper is to demonstrate the effectiveness and safety of percutaneously placed catheters for thermal measurement within deep-seated pelvic tumors. Temperature probes are applied through these catheters during hyperthermia treatment. Surgery is not necessary for placing the catheters. Two types of catheters proved to be suitable and safe for guided percutaneous placement.

Material and Methods

Diagnostic Procedures. CT is performed before placement of catheters for precise localization of tumors and adjacent vital structures (bowel, vessels, ureter, bladder). The access route is determined on the basis of the information obtained by continuous CT scans in the tumor area.

Patient Preparation. Epidural anesthesia is required for insertion of catheters, as well as determination of blood clotting. For puncture the patient is placed on a fluoroscopic table, bowel and bladder are visualized by water soluble contrast medium.

Puncture. A C-arm fluoroscopy system is used to guide the catheter placement, providing a biplanar fluoroscopic control. Location of the tumor relative to the body surface determines the entry site of the puncture.

Recent Results in Cancer Research, Vol. 107
© Springer-Verlag Berlin · Heidelberg 1988

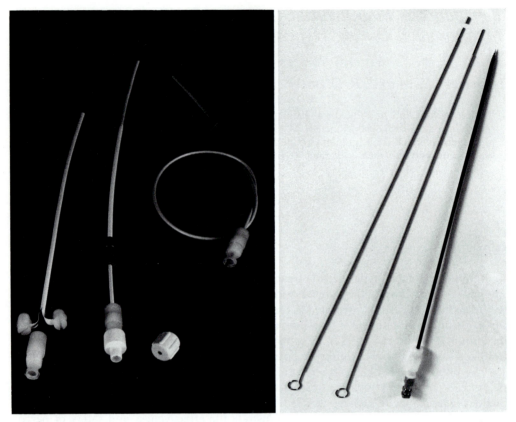

Fig. 1 *(left).* Catheters without endhole are introduced using a peel-away sheath or a vessel introducer set in Seldinger's technique with a trocar needle

Fig. 2 *(right).* Catheters with an endhole are conically tapered at the distal end. The endhole is sealed after insertion by a fitting teflon cap, which is delivered by a metal stylet

Catheters. Catheters with and without endhole, especially designed for this purpose, are used. They are now commercially available. The diameter of the catheters is at least 2 mm in order to provide an easy insertion of the temperature probes. Trocar catheters with endhole can be sealed by a special device.

Results and Discussion

28 catheters for temperature monitoring have been inserted percutaneously in seven patients for 17 hyperthermia treatments. The tumors were all located in the pelvis: two recurrences of rectal carcinoma, one prostate sarcoma, one rhabdomyosarcoma, 1 schwannoma, 1 histiocytoma, 1 leiomyosarcoma.

Epidural anesthesia was required in 16 of 17 procedures. In local skin anesthesia catheter insertion was not tolerated, because the catheters had to be positioned centrally within the tumor in a longitudinal axis, providing thermal mapping along the axis of the catheter. In deep-seated tumors it is not possible to place numerous 6-F catheters as

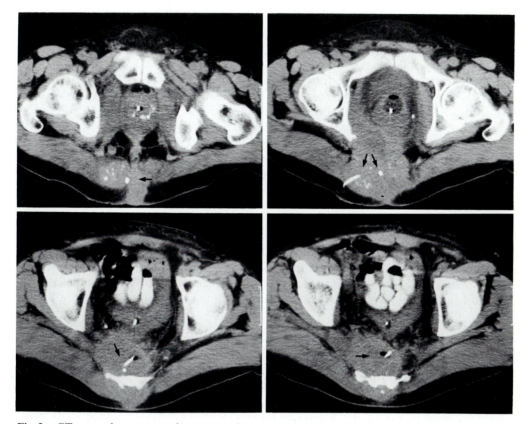

Fig. 3. CT scans demonstrate the correct placement of the catheters *(arrows)* within the tumor in the posterior region of the pelvis. One catheter is placed in a longitudinal axis within the tumor, the second in a perpendicular direction

usually applied in superficial tumors. To provide representative temperature monitoring of the heated tissue, up to three catheters were inserted, one in a longitudinal axis within the tumor and one or two in a perpendicular direction. Additionally, non-traumatic temperature measurement inside the vagina, the bladder and the rectum is possible, as suggested by some authors (e.g., Marchal et al. 1985).

With the availability of CT, precise localization of tumors and adjacent vital structures has become common, enabling highly accurate planning and execution of access routes for puncture or drainage procedures (van Sonnenberg et al. 1982; Mueller et al. 1984).

Especially in cases with pelvic neoplasms, surgical catheter placement for thermometry is avoidable. It is an effective alternative to operation and especially valuable in critically ill patients and in those who would be at high risk under general anesthesia. Although CT can be used for guidance, we prefer the advantages of direct fluoroscopic visualization when manipulating the catheters and guide wires. C-arm fluoroscopy provides an overall view and constant control of the puncture needle. Especially when Seldinger's technique is applied, fluoroscopic control is essential (Berger et al. 1986; van Sonnenberg et al. 1982).

After catheter placement CT was performed within 24 h to check catheter position and to rule out complications, such as bleeding, perforation and infection. One catheter had to be repositioned because of inadequate placement; local inflammation around the puncture site was observed in three cases.

The catheters were in place for between 2 and 10 days (average 6.2 days). Bacteriologic studies of the catheters were negative in all cases except one. It has to be stressed that catheters with an endhole should be sealed to prevent infection, because they are in position for several days. Trocar catheters used in this study had conically tapered ends which were sealed by a small Teflon cap after correct placement. Catheters without endhole are positioned using a peel-away sheath introduced percutaneously by means of Seldinger's technique over a puncture needle.

Conclusions

Surgery can be avoided in placement of catheters for temperature monitoring of deep-seated tumors during local hyperthermia. Suitable catheters are available and can be placed percutaneously with fluoroscopy or CT guidance without high risk of complications. CT is essential for determination of tumor location and puncture site.

References

Berger H, Gebauer A, Hertlein H, Lissner J (1986) Results of CT, US and fluoroscopy guided percutaneous drainage of abscess and fluid-collections in the abdomen (Abstr) 4th European Congress of Interventional Radiology, May 24–27, Athens

Chan KW, Miller W, Roemer RB, Williamson J, Cetas TC (1984) Thermal dosimetry of RF interstitial hyperthermia (Abstr). 4th International Symposium on Hyperthermic Oncology, July 2–6, Aarhus

Cosset JM, Dutreix J, Haie C, Gerbaulet A, Janoray P, Dewar JA (1985) Interstitial thermoradiotherapy: a technical and clinical study of 29 implantations performed at the Institut Gustave-Roussy. Int J Hyperthermia 1 (1): 3–13

Lindholm CE, Kjellen E, Landberg T, Nilsson P, Persson B (1982) Microwave-induced hyperthermia and ionizing radiation. Preliminary clinical results. Acta Radiol Oncol 21: 241–254

Marchal C, Bey P, Jacomino JM, Hofstetter S, Gaulard ML, Robert J (1985) Preliminary technical, experimental and clinical results of the use of the HPLR 27 system for the treatment of deep-seated tumors by hyperthermia. Int J Hyperthermia 1 (2): 105–116

Marmor JB, Pounds D, Postic TB, Hahn GM (1979) Treatment of superficial human neoplasms by local hyperthermia induced by ultrasound. Cancer 43: 188–197

Mueller PR, van Sonnenberg E, Ferruci JT (1984) Percutaneous drainage of 250 abdominal abscesses and fluid collections. Radiology 151: 343–347

Nilsson P, Persson B, Kjellen E, Lindholm CE, Landberg T (1982) Technique for microwave-induced hyperthermia in superficial human tumors. Acta Radiol Oncol 21: 235–238

Van Sonnenberg E, Ferruci JT, Mueller PR, Wittenberg J, Simeone JF (1982) Percutaneous drainage of abscesses and fluid collections: technique, results, and applications. Radiology 142: 1–10

CT-Guided Placement of Temperature Probes in Pulmonary Cancer

H.-D. Piroth and G. Brinkmann

Abteilung für Onkologie und Radiologie, Marien-Hospital, Rochusstraße 2, 4000 Düsseldorf 40, FRG

Introduction

The clinical application of hyperthermia is based on experimental results of regional tumor heating. A temperature of 41.5°–42° C is assumed to be effective for therapeutic procedures in oncology. It has been shown that hyperthermia improves tumor regression, especially in combination with radiotherapy or chemotherapy.

Clinical studies, started in the last years, should be based on clear knowledge of the temperature which can be reached in a human tumor by the application of an external high frequency. It is especially of great interest to know if a tumor temperature of 41.5° C can be reached in vivo in *all* patients by using a standard technique of hyperthermia, and if hyperthermia can be used even in more deeply localized tumors without temperature measurement.

On the other hand, we need a method to compare noninvasive methods of temperature measurement with invasive direct measurements that can be done without operation even on an outpatient basis. Noninvasive techniques to measure tumor temperature were described by Beuter (1985), Edrich (1985), and Hermeking (1983). Their methods have to be compared with direct temperature measurement in deep localized tumors, especially in thoracic malignancies.

Methods

Based on our experience with CT-guided tumor biopsy, especially in cases of pulmonary carcinoma, we employed CT to place a Teflon tube 1.5 mm in diameter transthoracally into the bronchial carcinoma. Figure 1 depicts the procedure.

In the first step we looked for the slice representing the shortest distance between tumor and chest wall. If there was atelectasis between tumor and pleura, we chose this slice as the basis for the tube's placement (see Fig. 1). Figure 1b demonstrates the direction and length of the puncture. The tube then was marked by a clip to avoid deeper penetration and thereafter the puncture was done under local anesthesia. If the position of the needle was correct, the temperature probe was directed through the tube to the tip of the cannula, which then was replaced immediately. The last position of the temperature probe, guaranteed to have an accuracy of 0.1° C and having a diameter of 0.8 mm, was documented by CT (Fig. 1d).

Recent Results in Cancer Research, Vol. 107
© Springer-Verlag Berlin · Heidelberg 1988

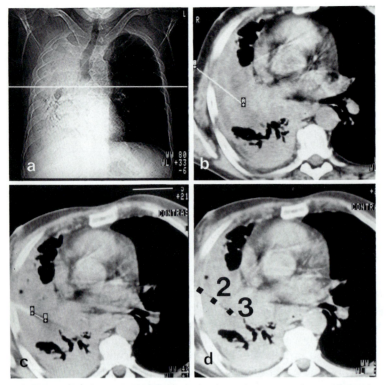

Fig. 1 a–d. Placement of a wire with three probes (see points and number of probe). **b** Measurement of the distance and direction from surface to tumor. **d** Final position of the wire with three probes *(solid squares). 3,* deepest probe.

Hyperthermia was done with a high frequency machine (HMS 200, Strassner) with two generators (13.57 and 27.12 MHz).

We used two pairs of water-cooled capacitive applicators. We tried to reach 150 W for 30 min and then we applied hyperthermia for another 30 min. Heating time was 60 s with a pause of 10 s. A measuring time of 2 s was followed by heating time of 10 s.

Results

Temperature probes were inserted in 13 patients. In seven patients it was impossible to measure temperature as the probes were either defect or dislocated even before the first examination could be performed. We developed a plastic clip to attach the measuring wire of the probe to the skin by suture. With this clip all probes remained within the tumor, but another three wires were broken after several days of placement. Although we obtained sufficient information in only six patients, we were still able to find out some interesting results.

An insufficient rise in temperature was found in four patients in spite of more than 150 W high frequency over 30 min, in spite of different techniques of application (opponent and/or crossover techniques), and in spite of raising the high frequence energy to the pain threshold.

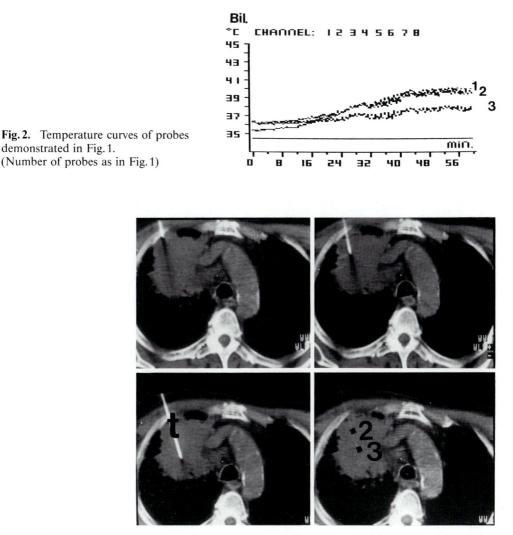

Fig. 2. Temperature curves of probes demonstrated in Fig. 1. (Number of probes as in Fig. 1)

Fig. 3. Placement of temperature probes in an adenocarcinoma of the right upper lobe. *t*, teflon tube

Figure 2 shows the temperature curves of the patient from Fig. 1. It can be clearly seen that the temperature range measured in the region of the tumor only rises from 36.7° C to 38° C, whereas probes at points 1 and 2 at the surface of the chest wall reach higher temperature values.

Figures 3 and 4 refer to another patient with an adenocarcinoma of the right upper lobe. We found that the temperature in the tumor rose from 36° to 39° C, while probes 1 and 2 near the chest wall showed higher temperature values too.

Figure 5 shows the positioning of a probe in a tumor infiltrating the chest wall. Figure 6 shows two curves, Fig. 6a after dislocation, showing a high increase in temperature, and Fig. 6b, showing a low temperature within the tumor immediately after the positioning of the probe within the tumor.

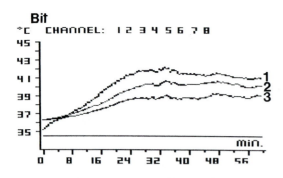

Fig. 4. Curves of probes in Fig. 3

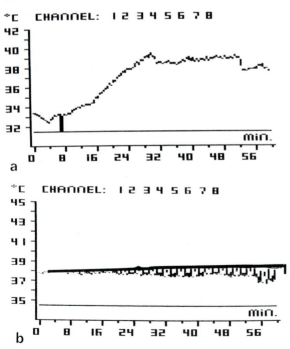

Fig. 5. Squamous cell carcinoma, infiltrating the chest wall. *p*, probe at the end of the wire; *t*, teflon tube

B M

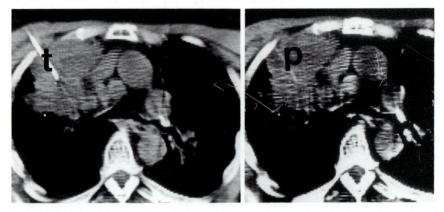

Fig. 6a, b. Curves of probe *p* (see Fig. 5). **a** After dislocation; **b** before dislocation

Discussion

Even though we observed only a small number of patients, we have to postulate that

1. It is absolutely necessary to measure the temperature in the tumor.
2. It is possible to reach temperatures of more than 40° C, even if we consider several nonsufficient temperature increases in some patients by use of high-frequency (13.57 and 27.12 MHz) and capacitive applicators.
3. If this technique of hyperthermia is used, a direct measurement of temperature is necessary not only in deep-situated tumors but also in superficial tumors, because temperature enhancement cannot be guaranteed in either superficial or deep positions and therefore temperature has to be measured.

Our technique of insertion of temperature probes in pulmonary cancer was not been described elsewhere. The method of CT-guided puncture in pulmonary cancer is well documented. Poe and Sinner (1980), Sinner (1979), and Perez et al. (1984) have published accounts of their experience with transpulmonary puncture or the need for direct measurement of temperature.

We have performed 77 diagnostic biopsies in lung cancer using of CT and saw no pneumothorax or evidence of bleeding if the tumor or an associated atelectasis was in contact with the pleural space in which the puncture was made.

In total we found four cases of pneumothorax in diagnostic pulmonary biopsy; two of them had to be treated, but complications occurred if normal lung tissue had to be penetrated.

We must conclude that the insertation of temperature probes – controlled by CT – even in pulmonary cancer can be executed without risk for the patient, if certain conditions of patient selection are made.

References

Beuter K (1986) Möglichkeiten der noninvasiven Temperaturmessung mit Ultraschall. In: Streffer C, Herbst, Schwabe (eds) Lokale Hyperthermie. Deutscher Ärzte-Verlag, Cologne, pp 111–120

Bowen T et al. (1979) Measurement of the temperature dependance of the velocity of ultrasound in soft tissues. National Bureau of Standards, Washington (Special publication no 525)

Cetas TC, Connor WG (1978) Thermometry considerations in localized hyperthermia. Med Phys 5 (2)

Edrich J, Ließ HD (1986) Microwave thermography for control of microwave induced hyperthermia: Potentials, limitations and preliminary results. In: Streffer C, Herbst, Schwabe (eds) Lokale Hyperthermie. Deutscher Ärzte-Verlag, Cologne, pp 121–125

Kanzenbach J, Köhler J, Lüdecke KM (1982) Mikrowellen Thermographie für die medizinische Diagnostik. Bundesministerium für Forschung und Technik, Bonn (Final report) (FKZ 01Z5059-ZK/NT/IMT 294)

Lüdecke KM, Röschmann P (1986) Drahtlose Temperaturmessung im Körpergewebe mittels implantierter FMR-Sonden. In: Streffer C, Herbst, Schwabe (eds) Lokale Hyperthermie. Deutscher Ärzte-Verlag, Cologne, pp 126–138

Perez CA, Emami B, Nussbaum GH (1984) Clinical experience with external local hyperthermia in treatment of superficial malignant tumours. Front Radiat Ther Oncol 18: 83–102

Poe RH, Tobin RE (1980) Sensitivity and specifity of needle biopsy in lung malignancy. Am Rev Respir Dis 112: 725–729

Sinner WN (1979) Pulmonary neoplasms diagnosed with transthoracic biopsy. Cancer 43: 1533–1536

Subject Index